2019 International Symposium on VLSI Technology, Systems and Application (VLSI-TSA 2019)

AA001031

Hsinchu, Taiwan
22 – 25 April 2019

IEEE Catalog Number: CFP19846-POD
ISBN: 978-1-7281-0943-5

**Copyright © 2019 by the Institute of Electrical and Electronics Engineers, Inc.
All Rights Reserved**

Copyright and Reprint Permissions: Abstracting is permitted with credit to the source. Libraries are permitted to photocopy beyond the limit of U.S. copyright law for private use of patrons those articles in this volume that carry a code at the bottom of the first page, provided the per-copy fee indicated in the code is paid through Copyright Clearance Center, 222 Rosewood Drive, Danvers, MA 01923.

For other copying, reprint or republication permission, write to IEEE Copyrights Manager, IEEE Service Center, 445 Hoes Lane, Piscataway, NJ 08854. All rights reserved.

*** This is a print representation of what appears in the IEEE Digital Library. Some format issues inherent in the e-media version may also appear in this print version.

IEEE Catalog Number: CFP19846-POD
ISBN (Print-On-Demand):   978-1-7281-0943-5
ISBN (Online):   978-1-7281-0942-8
ISSN: 1930-8868

Additional Copies of This Publication Are Available From:

Curran Associates, Inc
57 Morehouse Lane
Red Hook, NY 12571 USA
Phone: (845) 758-0400
Fax: (845) 758-2633
E-mail: curran@proceedings.com
Web: www.proceedings.com

TABLE OF CONTENTS

SLOW-DOWN IN POWER SCALING AND THE END OF MOORE'S LAW? 1
Ghavam Shahidi

JOURNEY TO 5G 2
Li Fung Chang

SEMICONDUCTOR FOR 5G 3
Yujun Li

SILICON TECHNOLOGIES FOR NEXT GENERATION 5G ARCHITECTURES AND
APPLICATIONS 4
Shankaran Janardhanan

SILICON PROCESS IMPACT ON 5G NR MMWAVE FRONT END DESIGN AND
PERFORMANCE 5
*Chuan-Cheng Cheng ; Jeremy Dunworth ; Sriram Kalpat ; Haitao Cheng ; Gang Liu ; Ming-Ta Yang ; Wing Sy ;
Joseph Wang ; Kamal Sahota ; Pr. Chidi Chidambaram*

R&D ACTIVITIES FOR 5G RADIO ACCESS TECHNOLOGIES USING SHF BANDS AND CO-
CREATION OF NEW SERVICES USING 5G 7
Yukihiko Okumura

R&D ACTIVITIES FOR CAPACITY ENHANCEMENT USING 5G ULTRA HIGH-DENSITY
DISTRIBUTED ANTENNA SYSTEMS 8
Takashi Dateki

DEVELOPMENT AND EVALUATION OF LOW-SHF-BAND C-RAN MASSIVE MIMO SYSTEM
FOR 5G 9
Yasushi Maruta

EUVL OPTICS: STATUS AND FUTURE PERSPECTIVES 10
Winfried Kaiser

AUTONOMOUS DRIVING TECHNOLOGIES AND COMPUTING PLATFORM 11
Peter Hsieh

CRISP: CENTER FOR RESEARCH ON INTELLIGENT STORAGE AND PROCESSING-IN-
MEMORY 12
Yuan Xie

A PERSPECTIVE ON NVRAM TECHNOLOGY FOR FUTURE COMPUTING SYSTEM 13
Katsuhiko Hoya ; Kosuke Hatsuda ; Kenji Tsuchida ; Yohji Watanabe ; Yusuke Shirota ; Tatsunori Kanai

NEW MEMORY TECHNOLOGY, DESIGN AND ARCHITECTURE CO-OPTIMIZATION TO
ENABLE FUTURE SYSTEM NEEDS 15
Arnaud Furnemont

EMBEDDED MEMORIES FOR SILICON-LN-PACKAGE: OPTIMIZATION OF MEMORY
SUBSYSTEM FROM LOT TO MACHINE LEARNING 16
Fatih Hamzaoglu

EVOLUTION AND ADVANCES OF THE NONVOLATILE MEMORIES AND APPLICATIONS 17
Yan Li

EMBEDDED MEMORY: THE FUTURE OF EMERGING MEMORIES 18
Feng-Min Lee

FUTURE OF COMPUTING AND SENSING SYSTEMS FOR EMBEDDED APPLICATIONS 19
Thomas Ernst

PROGRESS IN NEUROMORPHIC COMPUTING : DRAWING INSPIRATION FROM NATURE
FOR GAINS IN AI AND COMPUTING 20
Mike Davies

ADVANCES IN QUANTUM DOT LASERS FOR SILICON PHOTONICS 21
Yasuhiko Arakawa

SILICON PHOTONICS AND INTEGRATED PHOTONIC SWITCHES 22
Ming C. Wu

SILICON PHOTONIC BASED HIGH-SPEED TRANSCEIVERS FOR FIBER-OPTIC SHORT-
REACH APPLICATIONS 23
Peter Ossieur

LAYOUT AUTOMATION FOR INTEGRATED PHOTONICS 24
Alan Sherman

TWIN-BIT RESISTIVE RANDOM ACCESS MEMORY IN FINFET CMOS LOGIC TECHNOLOGIES............25
Chieh Lee ; Yu-Ting Hung ; Cheng-Jun Lin ; Ya-Chin King ; Chrong Jung Lin

HOT-CARRIER INJECTION-INDUCED DISTURB AND IMPROVEMENT METHODS IN 3D NAND FLASH MEMORY............27
Wei-Liang Lin ; Wen-Jer Tsai ; C. C. Cheng ; Chun-Chang Lu ; S. H. Ku ; Y. W. Chang ; Guan-Wei Wu ; Lenvis Liu ; S. W. Hwang ; Tao-Cheng Lu ; Kuang-Chao Chen ; Tseung-Yuen Tseng ; Chih-Yuan Lu

HIGH DENSITY NV-SRAM USING MEMRISTOR AND SELECTOR AS TECHNOLOGY ASSIST............29
S. S. Teja Nibhanupudi ; Jaydeep P. Kulkarni

THE IMPACT OF FORMING TEMPERATURE AND VOLTAGE ON THE RELIABILITY OF FILAMENTARY RRAM............31
G. Y. Chen ; F. M. Lee ; Y. Y. Lin ; P. H. Tseng ; K. C. Hsu ; D. Y. Lee ; M. H. Lee ; H. L. Lung ; K. Y. Hsieh ; K. C. Wang ; C. Y. Lu ; M. C Wu

STOCHASTIC FILAMENT FORMATION ON THE CYCLING ENDURANCE OF BACKFILLED CONTACT RESISTIVE RANDOM ACCESS MEMORY CELLS............33
Yun-Feng Kao ; Chrong Jung Lin ; Ya-Chin King

A VARIABILITY SOURCE FOR NANOSHEET GAA TRANSISTORS FOR SUB-7NM NODES............35
P. Harsha Vardhan ; Swaroop Ganguly ; Udayan Ganguly

IMPACT OF MULTI-DOMAIN INTERACTION ON ON-STATE CHARACTERISTICS OF MFIS-TYPE 2D NEGATIVE-CAPACITANCE FETS............37
Po-Sheng Lu ; Chia-Chen Lin ; Pin Su

ACCURATE MEASUREMENT OF SNEAK CURRENT IN RERAM CROSSBAR ARRAY WITH DATA STORAGE PATTERN DEPENDENCIES............39
Yaqi Shang ; Takashi Ohsawa

SELECTIVE ETCHING OF NATIVE SILICON OXIDE IN PREFERENCE TO SILICON OXIDE AND SILICON............41
Christopher Ahles ; Jong Choi ; Raymond Hung ; Namsung Kim ; Srinivas Nemani ; Andrew Kummel

ENHANCING IGZO THIN FILM TRANSISTOR SCALABILITY THROUGH TUNNELING CONTACT............43
Zichao Ma ; Xintong Zhang ; Clarissa Prawoto ; Lining Zhang ; Longyan Wang ; Mansun Chan

CELL ARRAY DESIGN WITH ROW-DRIVEN SOURCE LINE IN BLOCK SHUNT ARCHITECTURE APPLICABLE TO FUTURE 6F2 1T1MTJ MEMORY............45
Tongshuang Huang ; Takashi Ohsawa

EVALUATION OF 2D NEGATIVE-CAPACITANCE FETS FOR LOW-VOLTAGE SRAM APPLICATIONS............47
Kuei-Yang Tseng ; Wei-Xiang You ; Pin Su

ANALYSIS OF TRANSIENT EFFECT ON SUPER-STEEP SS PN-BODY TIED SOI-FET............49
Takayuki Mori ; Jiro Ida ; Hiroki Endo

A LARGE DYNAMIC RANGE CURRENT SENSOR USING MAGNETIC TUNNEL JUNCTION ON 8" SI PROCESS LINE............51
D. Y. Wang ; I. J. Wang ; C. S. Lin ; J. W. Su ; H. H. Lee ; Y. C. Hsin ; S. Y. Yang ; Y. J. Chang ; Y. C. Kuo ; Y. H. Su ; S. Z. Rahaman ; G. L. Chen ; S. H. Li ; J. H. Wei ; K. C. Huang ; C. I. Wu

ARCHITECTURE EVALUATION FOR STANDALONE AND EMBEDDED 1T-DRAM............53
Md. Hasan Raza Ansari ; Nupur Navlakha ; Jyi-Tsong Lin ; Abhinav Kranti

TIPS-PENTACENE:PS BLEND ORGANIC FIELD-EFFECT TRANSISTORS WITH HYBRID GATE DIELECTRIC ON PAPER SUBSTRATE............55
Vivek Raghuwanshi ; Deepak Bharti ; Ajay Kumar Mahato ; Ishan Varun ; Shree Prakash Tiwari

ORIGIN OF FIXED CHARGES AND DIPOLE IN GEO_X/AL_2O_3 GATE STACK BASED ON GE............57
Lixing Zhou ; Xiaolei Wang ; Xueli Ma ; Jinjuan Xiang ; Chao Zhao ; Tianchun Ye ; Wenwu Wang

CONTRADICTION BEHAVIORS BETWEEN I-V AND C-V CURVES AFTER SELF-HEATING STRESS IN A-IGZO TFT WITH TRIPLE-STACKED CHANNEL LAYERS............59
Yu-Ching Tsao ; Mao Chou Tai ; Ting-Chang Chang

DEFECT LOCALIZATION AND ELECTRICAL FAULT ISOLATION FOR METAL CONNECTION USING HELIUM ION MICROSCOPE............61
Deying Xia ; Shawn McVey ; Wilhelm Kuehn

THERMAL ATOMIC LAYER ETCHING OF AMORPHOUS AND CRYSTALLINE HAFNIUM OXIDE, ZIRCONIUM OXIDE, AND HAFNIUM ZIRCONIUM OXIDE............63
Jessica A. Murdzek ; Steven M. George

THERMAL STABILITY OF SHALLOW GE N^+-P JUNCTION WITH THIN GESN TOP LAYER............65
Hsiu-Hsien Liao ; Yi-Ju Chen ; Bing-Yue Tsui

SELECTIVE ATOMIC LAYER DEPOSITION OF TIO_2............67
Christopher Ahles ; Jong Choi ; Keith Wong ; Srinivas Nemani ; Andrew Kummel

SWCNT AND SWCNT-BASED HETEROSTRUCTURES FOR DEVICES..69
Rong Xiang ; Shigeo Maruyama

TOWARD HIGH-MOBILITY AND LOW-POWER 2D MOS_2 FIELD-EFFECT TRANSISTORS........................71
Zhihao Yu ; Ying Zhu ; Weisheng Li ; Yi Shi ; Gang Zhang ; Yang Chai ; Xinran Wang

TWO-DIMENSIONAL MATERIALS ELECTRON DEVICES: CONTACT AND DOPING..................73
Yang Chai

SYNTHESIS AND ELECTRONIC DEVICES OF ATOM-THIN TRANSITION METAL DICHALCOGENIDES..75
Jiadong Zhou ; R. Govindan Kutty ; Lixing Kang ; Xiaowei Wang ; Zheng Liu

INTERFACE ENGINEERING FOR 2D LAYERED SEMICONDUCTORS.................................77
Kosuke Nagashio

2D LAYERED SEMICONDUCTORS BEYOND MOS_2...79
Wen-Hao Chang

ELECTRONIC DEVICES WITHIN SINGLE ATOMIC LAYER - DEVELOPMENT OF 2D LATERAL JUNCTIONS..80
Jr-Hau He

CAN MAGNETIC MEMORY (MRAM) DISPLACE DRAM?...81
Denny Tang

INTERFACIAL ENGINEERING OF SOT-MRAM TO MODULATE ATOMIC DIFFUSION AND ENABLE PMA STABILITY >400 ^{o}C...82
Chong Bi ; Shy-Jay Lin ; Xiang Li ; Telem Simsek ; M. Song ; Wilman Tsai ; Shan X. Wang

COMPUTATIONAL RANDOM ACCESS MEMORY (CRAM) AND APPLICATIONS.................84
Jian-Ping Wang

COMPREHENSIVE RELIABILITY STUDY OF STT-MRAM DEVICES AND CHIPS FOR LAST LEVEL CACHE APPLICATIONS (LLC) AT 0X NODES..85
Jian Zhu ; Yuan-Jen Lee ; Huanlong Liu ; Son Le ; Jodi Iwata-Harms ; Sahil Patel ; Ru-Ying Tong ; Vignesh Sundar ; Santiago Serrano-Guisan ; Dongna Shen ; Renren He ; Jesmin Haq ; Jeffrey Teng ; Vinh Lam ; Yi Yang ; Yu-Jen Wang ; Tom Zhong ; Luc Thomas ; Hideaki Fukuzawa ; Guenole Jan ; Po-Kang Wang

SPIN-ORBIT TORQUE DRIVEN ONE-BIT MAGNETIC RACETRACK DEVICES - MEMORY AND NEUROMORPHIC APPLICATIONS...87
See-Hun Yang ; Chirag Garg ; Timothy Phung ; Charles Rettner ; Brian Hughes

DEVICE STRUCTURAL EFFECTS, SPICE MODELING AND CIRCUIT EVALUATION FOR NEGATIVE-CAPACITANCE FETS...89
Pin Su ; Wei-Xiang You

FERROELECTRIC SI-DOPED HFO_2 CAPACITORS FOR NEXT-GENERATION MEMORIES............91
Ava J. Tan ; Zhongwei Zhu ; Hwan Sung Choe ; Chenming Hu ; Sayeef Salahuddin ; Alex Yoon

THE GUIDELINE ON DESIGNING A HIGH PERFORMANCE NC MOSFET BY MATCHING THE GATE CAPACITANCE AND MOBILITY ENHANCEMENT...93
Y. C. Luo ; F. L. Li ; E. R. Hsieh ; C. H. Liu ; Steve S. Chung ; T. P. Chen ; S. A. Huang ; T. J. Chen ; Osbert Chenz

IMPACT OF GATE STACK DESIGN ON IMPROVING SUBTHRESHOLD SWING BEHAVIORS IN FERROELECTRIC-GATE FIELD-EFFECT TRANSISTORS...95
Shinji Migita ; Hiroyuki Ota ; Akira Toriumi ; Takashi Matsukawa

FABRICATION OF Ω-GATED NEGATIVE CAPACITANCE FINFETS AND SRAM....................97
P.-J. Sung ; C.-J. Su ; D. D. Lu ; S.-X. Luo ; K.-H. Kao ; J.-Y. Ciou ; C.-Y. Jao ; H.-S. Hsu ; C.-J. Wang ; T.-C. Hong ; T.-H. Liao ; C.-C. Fang ; Y.-S. Wang ; H.-F. Huang ; J.-H. Li ; Y.-C. Huang ; F.-K. Hsueh ; C.-T. Wu ; Y.-C. Huang ; W. C.-Y. Ma ; K.-P. Huang ; Y.-J. Lee ; T.-S. Chao ; J.-Y. Li ; W.-F. Wu ; W.-K. Yeh ; Y.-H. Wang

ELECTRON ENHANCED ATOMIC LAYER DEPOSITION (EE-ALD).......................................99
Steven M. George

ANALYTICAL ESTIMATION OF LER-LIKE VARIABILITY IN GAA NANO-SHEET TRANSISTORS...100
Amita ; Ajinkya Gorad ; Udayan Ganguly

LOW TEMPERATURE JUNCTIONLESS DEVICE STACKING ENABLED BY LEADING EDGE SEQUENTIAL 3D INTEGRATION...102
Guillaume Besnard ; Gweltaz Gaudin ; Walter Schwarzenbach ; Ludovic Ecarnot ; Ionut Radu ; Bich-Yen Nguyen ; Anne Vandooren ; Nadine Collaert

NEW OBSERVATION AND ANALYSIS OF LAYOUT DEPENDENT EFFECTS IN SUB-40NM MULTI-RING AND MULTI-FINGER NMOSFETS FOR HIGH FREQUENCY APPLICATIONS.................104
Zu-Cheng Li ; Jyh-Chyurn Guo ; Jinq-Min Lin

BACKSIDE SI PASSIVATION: LEADING TO HIGH PERFORMANCE UTB GEOI STRUCTURES FOR MONOLITHIC 3D INTEGRATIONS...106
Wen Hsin Chang ; Toshifumi Irisawa ; Hiroyuki Ishii ; Noriyuki Uchida ; Tatsuro Maeda

HYPER-SELECTIVE CO METAL ALD ON METALS VS. SIO$_2$ WITHOUT PASSIVATION 108
Steven Wolf ; Mike Breeden ; Scott Ueda ; Andrew Kummel

NOVEL FINE-GRAIN BACK-BIAS ASSIST TECHNIQUES FOR 14NM FDSOI TOP-TIER SRAMS INTEGRATED IN 3D-MONOLITHIC 110
D. Bosch ; F. Andrieu ; L. Ciampolini ; A. Makosiej ; O. Weber ; X. Garros ; J. Lacord ; J. Cluzel ; E. Esmanhotto ; M. Rios ; S. Lang ; B. Giraud ; R. Berthelon ; G. Cibrario ; L. Brunet ; P. Batude ; C. Fenouillet-Béranger ; D. Lattard ; J. P. Colinge ; F. Balestra ; M. Vinet

NOVEL VERTICALLY-STACKED TENSILY-STRAINED GE$_{0.85}$SI$_{0.15}$ GAA N-CHANNELS ON A SI CHANNEL WITH SS=76MV/DEC,DIBL=36MV/V, AND I$_{ON}$/I$_{OFF}$=1.2E7 112
Yu-Shiang Huang ; Fang-Liang Lu ; Hung-Yu Ye ; Ya-Jui Tsou ; Yi-Chun Liu ; Chien-Te Tu ; C. W. Liu

THE FIRST GESN GATE-ALL-AROUND NANOWIRE P-FET ON THE GESNOI SUBSTRATE WITH CHANNEL LENGTH OF 20 NM AND SUBTHRESHOLD SWING OF 74 MV/DECADE 114
Yuye Kang ; Kaizhen Han ; Eugene Y.-J. Kong ; Dian Lei ; Shengqiang Xu ; Ying Wu ; Yi-Chiau Huang ; Xiao Gong

COMPARISON OF VERTICALLY DOUBLE STACKED POLY-SI NANOSHEET JUNCTIONLESS FIELD EFFECT TRANSISTORS WITH GATE-ALL-AROUND AND MULTI-GATE STRUCTURE 116
Meng-Ju Tsai ; Kang-Hui Peng ; Yu-Ru Lin ; Yung-Chun Wu

VIRTUAL SOURCE BASED I-V MODEL FOR CRYOGENIC CMOS DEVICES 118
Hazem Elgabra ; Brandon Buonacorsi ; Christopher Chen ; Jeff Watt ; Jonathan Baugh ; Lan Wei

CORE-SHELL TFET DEVELOPMENTS AND TFET LIMITATIONS 120
M. Passlack ; P. Ramvall ; T. Vasen ; A. Afzalian ; C. Thelander ; K. A. Dick ; L.-E. Wernersson ; G. Doornbos ; M. Holland

3D INTEGRATION 122
Piyush Gupta

ADVANCED STACKING TECHNOLOGIES FOR HETEROGENEOUS DEVICE INTEGRATION 123
Ionut Radu

HIGH DENSITY W2W AND D2W DBI® HYBRID BONDING FOR STACKED APPLICATIONS 124
Sitaram Arkalgud

SILICON PHOTONICS AS A POST MOORE TECHNOLOGY 126
Koji Yamada

VLSI RESEARCHES FOR MACHINE LEARNING AND NEUROMORPHIC COMPUTING 128
Atsuya Okazaki

MACHINE LEARNING SOLUTIONS FOR PROCESS CONTROL IN SEMICONDUCTOR MANUFACTURING 129
Eugen Foca

DEPLOYING NEW NODES FASTER WITH MACHINE LEARNING FOR IC DESIGN AND MANUFACTURING 130
Chris Schuermyer

DESIGNING AND MODELING ANALOG NEURAL NETWORK TRAINING ACCELERATORS 132
Sapan Agarwal ; Robin B. Jacobs-Gedrim ; Christopher Bennett ; Alex Hsia ; Michael S. Van Heukelom ; David Hughart ; Elliot Fuller ; Yiyang Li ; A. Alec Talin ; Matthew J. Marinella

FULL MEMORY ENCRYPTION WITH MAGNETOELECTRIC IN-MEMORY COMPUTING 134
Albert Lee ; Kang -L. Wang

INTEGRATED PHOTONICS OF TRANSISTOR LASER, DETECTOR AND ACTIVE LOAD FOR ALL OPTICAL NOR GATE 136
Ardy Winoto ; Junyi Qiu ; Dufei Wu ; Yu-Ting Peng ; Milton Feng

A NOVEL RRAM BASED WATERMARK TECHNIQUE UTILIZING THE IMPACT OF FORMING CONDITIONS ON RESET DISTRIBUTION 139
Yachuan Pang ; Huaqiang Wu ; Bin Gao ; Bohan Lin ; He Qian

Author Index

Slow-Down in Power Scaling and the End of Moore's Law?

Ghavam Shahidi

IBM

This talk tracks the evolution of drop in chip power in recent CMOS technology nodes (relative to the previous node), with focus on high performance microprocessors. It argues that in the more recent nodes, the total chip power (energy-per-operation) has scaled much less than that of earlier CMOS nodes. If the present trends continue, it will spell challenge in continuing Moore's Law. Improving the chip power scaling in the future nodes is critical, especially as it relates to high performance microprocessors.

Journey to 5G

Li Fung Chang
Ministry of Economic Affairs

The cellular industry has evolved to 5th generation (5G) since its first generation launched in 1978, Roughly, a generation of cellular system is created every 10 years. Since ITU officially defined service category and key performance indicators in 2015 (2-3 years after the concept was 1st brought up), the progress of the standard specifications and industrial development have moved fairly fast. 3GPP has released R15 5GNR technical specification for NSA (Non-Stand Alone) and SA (Stand Alone) by now and is expected to release R16 specification by the end of 2019. The industry is geared for 5G field trials, and launch of commercial services in 2019 and 2020. In the meantime, field trials/testbeds for non-traditional vertical applications are also being deployed worldwide to explore new services and to create new business models associated with these services. Though commercial launches of 5G services are expected in 2020-2021 time frame, it is a general consensus that eMBB service will be offered initially. It is also believed that 4G and 5G networks will likely co-exist and remain complementary for many years from operator point of view. Once 5G commercial system is launched, what will be next beyond 2020-21?In this talk, an overview of world-wide 5G activities including standardization, trials, spectrum policy will be presented first followed by summary of on-going/planned field trials/testbeds in Taiwan. We will then discuss "what next" from short term and long term perspective.

Semiconductor for 5G

Yujun Li

TSMC

In this speech, I will first cover the 5G market trends and technology requirement; then talk about the anticipated semiconductor innovation to meet the 5G requirement; and finally I will spend most of my time talking about the silicon content increase in 5G that drives accelerated technology migration in leading edge silicon technology, increasing use of specialty technology to address power integrity and connectivity needs, and the ever more important integration with advanced packaging.

Silicon Technologies for Next Generation 5G Architectures and Applications

Shankaran Janardhanan

GLOBALFOUNDRIES

RF is seen as THE key enabler for connected intelligence applications. Products and applications made possible by RF connectivity include mobile devices, the internet of things, automotive radar, connected cars, the network infrastructure and optical modules. This talk examines the many RF architecture approaches being pursued to keep pace with today's new levels of connectivity. We will also explore the semiconductor offerings, IP and design enablement that GLOBALFOUNDRIES brings to the marketplace to serve these next-generation applications, and how they can be used to address 5G architectural challenges.

Silicon Process Impact on 5G NR mmWave Front End Design and Performance

Chuan-cheng Cheng, Jeremy Dunworth, Sriram Kalpat, Haitao Cheng, Gang Liu, Ming-Ta Yang, Wing Sy, Joseph Wang, Kamal Sahota, PR. Chidi Chidambaram

Qualcomm Technologies, Inc., San Diego, USA
E-mail: chuanche@qti.qualcomm.com

ABSTRACT

5G NR (New Radio) standard allows utilization of higher frequency spectrum into millimeter wave (mmWave) bands (24GHz to 71GHz and above) to support wider modulated signal bandwidths from 100MHz to more than 2GHz, higher data rate communication, and lower latency. RF chip design in mmWave frequency to support 5G NR has been realized to overcome many challenges in system integration, circuit design, process technology, and reliability. In this paper, we review the impact of Silicon process performance on mmWave front end transceiver solutions. A dual band (28GHz, 39GHz), multi-element transceiver SOC has been successfully integrated with antenna module and is ready for commercialization in 2019. [1]

SYSTEM REQUIREMENT

Next generation mobile communication (5G NR) standard has been released (3GPP Rel 15 2017). Carriers and mobile phone manufacturers are enabling infrastructure and user equipment to support faster communication speed and lower latency. Primary 5G frequency bands have moved up from crowded and fractured LTE bands to higher 3.5GHz bands and into mmWave spectrum to allow wider bandwidth support. Due to high path loss in mmWave frequency, multiple antenna elements design and beamforming technique are widely deployed to transmit directional power and boost receiver sensitivity. Overall system supports larger fractional bandwidth, lower latency, and fine beamforming resolution. As free space wavelength for standardized 3GPP Rel 15 FR2 bands is shorter (~4-12.5mm range), it is possible to integrate multiple antenna elements with RFIC into single module design. Different mmWave spectrums are currently planned and released by different countries and regions. To globally support various spectrums with single RFIC, design challenges have significantly increased not only to support multiple frequency bands but also to cover wider fractional bandwidths. Many system topologies have been proposed and reviewed. The antenna module design poses another challenge to meet wide bandwidth, small form factor, thermal control, and cost target. To consider integrated RFIC with antenna module, it's obvious that single SOC approach will be preferred to reduce the antenna module design complexity in module routing, provide accurate beamforming control, and scaling form factor design to enable mobility.[2][3][4] However, SOC architecture greatly increases chip design complexity to include power amplifier (PA), low noise amplifier (LNA), phase shifter (PS), local oscillator (LO), mixer, and digital control blocks inside one chip. Each block has its own key performance requirement. In this paper, we will review process technology impact to the performance of key components to successfully implement front-end SOC to meet 5G NR requirement.

PROCESS REQUIREMENT

SOC chip contains many RF components like PA, LNA, PS, LO and each block has their special device requirement. To support mmWave frequencies, device Ft and Fmax above 200GHz are the basic requirement. Advanced CMOS or high speed SiGe BiCMOS processes are capable to reach higher device Ft and Fmax. However, smart digital control is usually applied to achieve fast beamforming switch control and various multi-element calibration. Based on the long-term trend in 2G, 3G, and 4G transceiver development, we expect CMOS to outperform over other process choices because finer lithography node enables lower power and highly compact digital control blocks for smaller area resulting in simplification of system integration and improvement of overall form factor design.

Advanced CMOS node devices tend to have lower breakdown voltage due to scaling which is detrimental to power amplifier design. Though individual PA output power for 5G mmWave is not as demanding as GSM or LTE PA due to multiple element antenna gain, PA linearity to support both QPSK and 64QAM modulation with required EVM is very important. Differential stack PA design or power combining are typical design approaches to achieve Pout target but suffer lower PAE (power added efficiency) with larger area and require careful trade-off. [5][6] Power combining techniques add extra transistors, inter-stage matching and output transformer area which is not preferred especially for multi-element design. Reliability, performance, and efficiency are the most critical factors for PA design. Devices suitable for PA usually meet both high Fmax to delivery power efficiently at mmWave frequency and reliability such as ruggedness over high VSWR and long-term device lifetime caused by HCI and TDDB degradation, which should be carefully examined to ensure product quality. Device SPICE model and layout parasitic extraction (LPE) to match Si performance at mmWave frequency are essential for RF design. Mature models for large signal simulation and reliability projection allow PA designers to examine circuit design at early stage to trade off performance and reliability. We have successfully correlated PA device performance and aging behavior thru reliability modeling by close collaboration between RF design team, process technology team, modeling team, reliability team, and foundry support.

In advanced CMOS process nodes, HKMG (High K gate dielectric and metal gate) is applied to reduce gate leakage for device scaling. Though finer gate length usually improves transistor gm, device Fmax didn't improve accordingly due to increasing gate resistance (Rg) with narrow gate length and higher parasitic with tighter gate pitch. Special layout optimization is required to minimize Fmax impact. Advanced CMOS device has gm and gain improvement which help lower noise figure (NF), suitable for good low noise amplifier design. However, TR switch between LNA and PA requires careful trade-off to minimize PA transmit loss and provide LNA matching network over wide bandwidth at the same time. Ron*Coff (or Insertion Loss and Isolation) is typical switch device KPI. Device parasitic capacitances creates extra lossy paths for traditional series and shunt switches at mmWave frequency while it may not be too critical at lower frequency. Optimized pCell layout and special isolation scheme can help reduce parasitic capacitance effect to improve switch performance.

978-1-7281-0943-5/19 $31.00 © 2019 IEEE

Phase shifter is an essential block for mmWave beamforming implementation. There are several architectures to implement phase shifter to trade-off resolution and power such as active or passive vector modulator or passive switched-LC. [2][3] VCO (voltage control oscillator) and mixer operating at mmWave frequency consume higher power, to minimize device parasitic will help reduce switching loss. In order to cover wider bandwidth, high Q, high tuning ratio varactor or low loss switch capacitor can be used to widen VCO tuning range. Devices with low flicker noise and thermal noise will enable low phase noise VCO design. Noise characterization and modeling at mmWave frequency remain challenge.

Low loss passives (inductor, capacitor) are as important as active devices. At mmWave frequency, transmission line and transformer are essential components. High Q, low loss passives can be implemented with carefully chosen metal stack. Thicker metal and larger distance to substrate will help improve passive Q and reduce loss. Device isolation need to be carefully evaluated especially for the combination of multi-elements simultaneous usage. Overall, superior process technology for mmWave RFIC is capable to support versatile device choices with better gain, high Ft/Fmax, better reliability, low noise, low-loss passives, and good switching characteristics under excellent manufacturing control. We have successfully designed and implemented dual band (28GHz, 39GHz), multiple PA/LNA element mmWave SOC using 28nm bulk CMOS and ready for 5G commercial application in 2019.

CONCLUSION

5G NR raises many new challenges in system integration, chip design, process technology, and reliability. However, it also opens up the new era for traditional RF chip designers. We have demonstrated suitable process technology to implement mmWave SOC to cover 5G NR spectrum and performance requirement and ready for 5G mobile commercialization.

ACKNOWLEDGEMENT

Authors would like to thank team members from Qualcomm Technologies Inc. in RF system, RF IC design, Corporate R&D, process technology, device reliability, and characterization lab for their valuable contribution.

REFERENCES

[1] Qualcomm 5G QTM052 antenna module announcement, Oct. 2018.

[2] J. D. Dunworth, et al, "A 28GHz Bulk-CMOS Dual-Polarization Phased-Array Transceiver with 24 Channels for 5G User and Basestation Equipment," *ISSCC*, pp. 70-71, Feb. 2018.

[3] K. Sahota, "5G Radio Design for Mobile Products," *IMS2017 5G Summit*, Jun. 2017.

[4] J. D. Dunworth, et al, "28GHz Phased Array Transceiver in 28nm Bulk CMOS for 5G Prototype User Equipment and Base Stations," *IMS, pp. 1330-1333,* Jun. 2018.

[5] K. Entesari, "Millimeter-wave CMOS Power Amplifiers for 5G Applications," *RFIC2018 WMD-6,* Jun. 2018

[6] H. Wang, "Broadband, Linear, and High-Efficiency Mm-Wave Power Amplifiers and Co-Designs with Antennas", *RFIC2018 WSC-4,* Jun. 2018

Reference to Poly/SiON	Poly/SiON	HKMG	Thin SOI	Thick SOI	SiGe BiCMOS
Gate Stack	Poly/SiON	HKMG-GL	HKMG-GF	Poly/SiON	Poly/SiON
Min Lg (nm)	27	27	18	40	130
Substrate	Bulk	Bulk	FDSOI	PDSOI	SiGe-BiCMOS
Digital Area	ref	↗	↗↗	↘↘	↘↘↘
Analog gain gm/gds	ref	↗	↗↗	→	
Peak Ft (GHz) (NMOS/NPN)	ref	↗	↗↗	→	↘↘
Peak Fmax (GHz) (NMOS/NPN)	ref	↘↘	↘↘	↘↘	↘↘↘
Ron*Coff (fs) @ 30GHz	ref	↗	↗↗↗	↗↗↗	↗
Insertion Loss (dB) @ 30GHz	ref	→	↗↗	↗↗↗	
Nfmin (dB) @ 30GHz	ref	↗	↘		→
BVDSS (V)	ref	→	↘	↘↘	↘↘↘
MOM (fF/um²)	ref	→	↗	↘↘↘	↘↘

↗ Better ↘ Worse → Similar

Table 1. Device RF KPI benchmark table

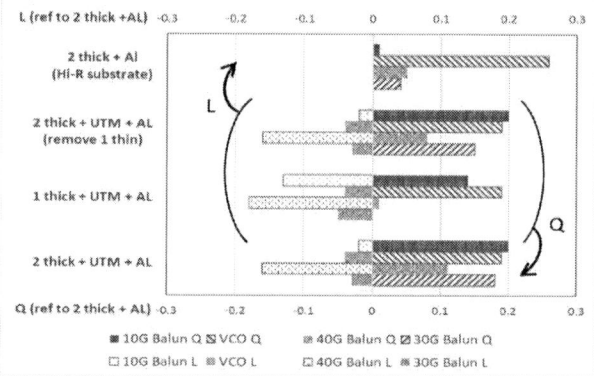

Figure 2. Top metal stack options comparison table

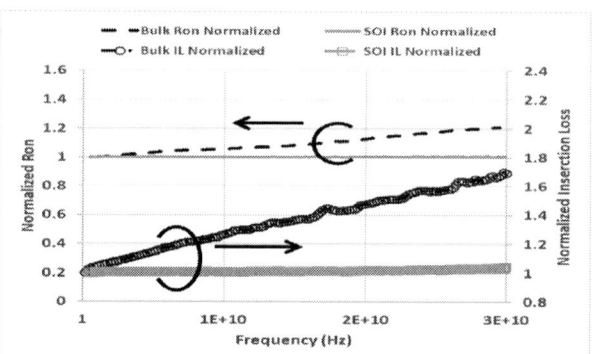

Figure 3: Bulk and SOI switch Ron and insertion loss trend

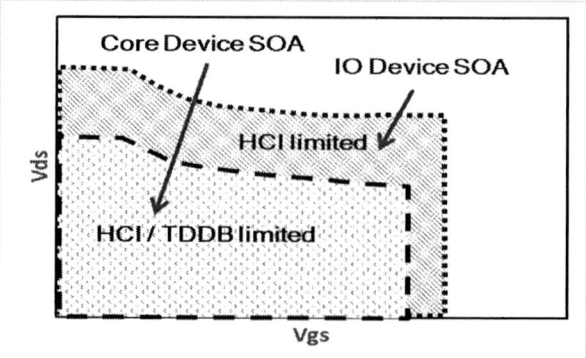

Figure 4: Device reliability SOA

R&D Activities for 5G Radio Access Technologies Using SHF Bands and Co-Creation of New Services Using 5G

Yukihiko Okumura

NTT DOCOMO, Inc.

Toward the introduction of various new mobile services and applications since 2020, the 5th generation mobile communication system (5G) aims to further improve the capability from the 4th generation mobile communication system, and R&D activities of 5G technologies have been promoted to achieve higher capacity, higher data speed and lower power consumption. In this talk, we introduce overview and major results of R&D on 5G radio access technologies using SHF bands jointly carried out with domestic companies, and we will also introduce our latest activities on co-creation of new service using 5G and its field trials with partners in a wide range of industries.

R&D Activities for Capacity Enhancement Using 5G Ultra High-Density Distributed Antenna Systems

Takashi Dateki

Fujitsu Limited

Mobile data traffic is rapidly increasing along with the popularization of smartphone. In order to accommodate the mobile data traffic, which is anticipated to reach 30 ExaBytes per month in 2020, it is necessary to significantly increase system capacity compared with 4G in fifth generation mobile communication system (5G). In this talk, we introduce our R&D results for ultra-high-density distributed antenna systems to support high capacity in 5G era.

Development and Evaluation of Low-SHF-Band C-RAN Massive MIMO System for 5G

Yasushi Maruta

NEC Corp.

To achieve capacity enhancement and low power consumption for 5G, we focus on utilizing Massive MIMO with digital beamforming in the low super-high-frequency (SHF) band, which is expected to be used for 5G commercial bands relatively soon. Also we focus on inter access point coordination based on Centralized-RAN (C-RAN) to improve cell edge performance. We introduce a prototype low-SHF-band Centralized-RAN Massive MIMO System and evaluation results.

EUVL Optics: Status and Future Perspectives

Winfried Kaiser

Carl Zeiss SMT GmbH

For more than 50 years Moore's Law has been driving the steady progress in semiconductor technology. This was enabled by strong resolution improvements of lithography optics. Currently the next big technology step is happening: The introduction of EUV Lithography into chip making. High performance EUVL optics is now produced in volume. Concurrently the next generation, EUV High NA, is in development.

Autonomous Driving Technologies and Computing Platform

Peter Hsieh

ARM Taiwan

Autonomous Driving is one of the most complex problems automotive and high-tech industries are facing today. This presentation first reviews the autonomous driving business landscape. Then it discusses current technologies used in HD mapping, localization, perception, prediction, motion planning and decision. Next, the presentation discusses future technology breakthroughs needed in order for autonomous vehicles to operate efficiently and safely in complex metropolitan areas. Finally the presentation discusses computing platform to support autonomous driving subsystems, AI training, safety and security requirements.

CRISP: Center for Research on Intelligent Storage and Processing-in-memory

Yuan Xie

University of California at Santa Barbara

In this talk, I will give an overview of the CRISP: center for research on Intelligent Storage and Processing-in-memory, a center of JUMP program, sponsored by SRC and DARPA. As technology scales, data movement between the processing units (PUs) and the memory is becoming one of the most critical performance and energy bottlenecks in various computer systems, ranging from cloud servers to end-user devices. As we enter the era of big data, many emerging data-intensive workloads become pervasive and mandate very high bandwidth and heavy data movement between the computing units and the memory. To close the gap between the computing and the memory, the CRISP center has done extensive work to explore possible solutions, with a holistic approach from system software to device innovations.

A perspective on NVRAM technology for future computing system

Katsuhiko Hoya, Kosuke Hatsuda, Kenji Tsuchida, Yohji Watanabe,
Yusuke Shirota*, Tatsunori Kanai*

Toshiba Memory Corporation, Japan
*Toshiba Corporation, Japan
E-mail: katsuhiko.hoya@toshiba.co.jp

INTRODUCTION

Memory hierarchy (Fig. 1) is one of the most essential component of any computing system. Non-volatile random access memory (NVRAM) technologies, such as PCRAM, STT-MRAM and ReRAM, have been massively produced in recent years and expected to bridge the bandwidth gap between DRAM (main memory) and NAND flash (storage). NVRAM also has been expected to provide the revolution of existing computing system in terms of density, performance and persistence management in the near future.

In this paper, we provide the overview of the impact of the NVRAM technology on the future memory systems and devices. We examine the suitable device structure and the implications of NVRAM in future memory system, such as in-memory database (IMDB), including the possibility of applying NVRAM to the conventional operating memory system.

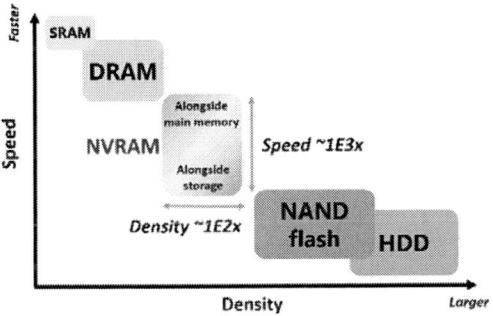

FIG. 1. MEMORY HIERARCHY

SELECTOR REQUIREMENT FOR HIGH DENSITY NVRAM

Many researchers have already proposed several approaches to the integration of high density NVRAM [1-3]. Figure 2 shows an example of NVRAM core structure which has two-deck crossbar memory array with CMOS circuit placed under array (CUA). Multi-level metal layers can be inserted between the memory array and the CUA. This chip structure has potential to realize lower chip cost than DRAM, since this structure enables to stack multi-crossbar memory arrays of a cell size as small as $4F^2$ (where F is the minimum feature size). The core control circuits, such as row/column decoders, sense amplifiers, and the data writing circuits, would be also placed under the memory array, therefore, the chip area can be minimized. The key to realize high-density crossbar memory array is to develop a vertical switch like a selector. Currently, two-terminal selector devices have been developed intensively, because of their high on current, high on/off current ratio and the suitability for the implementation to large crossbar memory array. There are several candidates for selector devices, including Chalcogenide Glass based selector (CG) [4-10], Volatile Conducting Bridge based selector (VCB) [3, 11-13] and Mott effect based selector (Mott) [14-16].

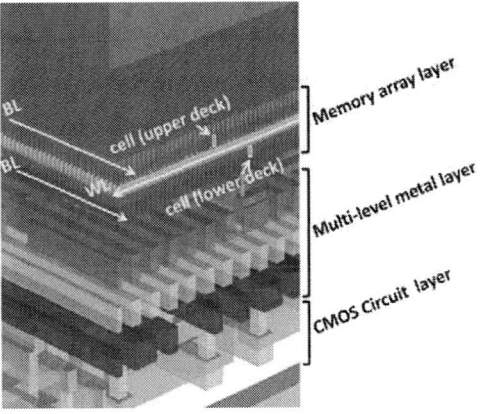

FIG. 2. SCHEMATICS OF NVRAM CORE STRUCTURE
WITH TWO-DECK MEMORY ARRAY,
MULTI-LEVEL METAL LAYERS AND CMOS CIRCUIT

The selector device especially featuring low leakage current becomes an essential solution to achieve stable read/write operation with large memory array due to inherent sneak paths. When the writing voltage is applied to the selected cell during writing operation, around a half of writing voltage is applied to the half-selected cell as shown in Fig. 3(a). To reduce the leakage current in the memory array due to the half-selected cell, the leakage current from the selector (Ihalf) should be suppressed. Compared with several selectors reported previously as shown in Fig. 3(b), The CG exhibits the tendency of declining leakage current, as the threshold voltage of the selector increases. The VCB represents the low leakage characteristics albeit lower threshold voltage of the selector compared to the CG. The suitable selector is different among each NVRAM technology. In addition to these requirements, other requirements of each NVRAM, such as dynamic behavior or reliability, have to be also met according to each target application.

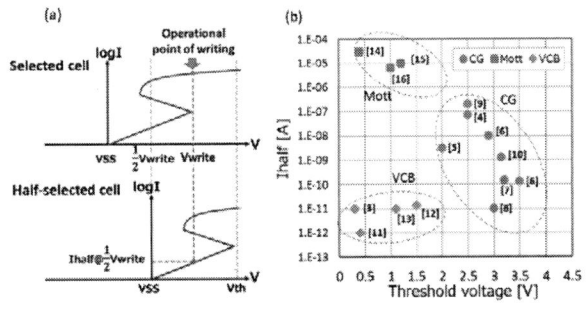

FIG. 3. (A) OPERATIONAL WRITING POINT DEFERENCE BETWEEN A
SELECTED CELL AND A HALF-SELECTED CELL
(B) COMPARISON OF IHARF AND THRESHOLD VOLTAGE

978-1-7281-0943-5/19 $31.00 © 2019 IEEE

EXISTING PROCESSOR MEMORY BUS ALONGSIDE DRAM

High density NVRAM technology has the possibility to be implemented into the main memory area alongside DRAM. The NVRAM is capable of a byte addressable and read latency of 100 ns to 1000 ns. Even though NVRAM has slower read speed in one order of magnitude than DRAM, connecting the processor memory bus alongside DRAM can improve the access latency of non-volatile devices. It is because, if the NVRAM is implemented into the storage area alongside NAND flash, the several microseconds of hardware / software overheads will remain. The processor memory bus also takes advantage of a small random byte access rather than storage access with a minimum granularity of more than 512 byte.

In order to understand how the existing memory system access the DRAM located on the processor memory bus, we used commercial IMDB with TPC-B workloads which is suitable for stress test in the IMDB applications, such as online transaction processing. Figures 4 (a) and (b) show the amount of read and write access per DRAM page on processor memory bus, respectively. There are many sharp peaks, which indicate both temporal and spatial locality for memory access. Covering a part of DRAM access with NVRAM could improve the total system read/write performance especially at higher read/write frequency.

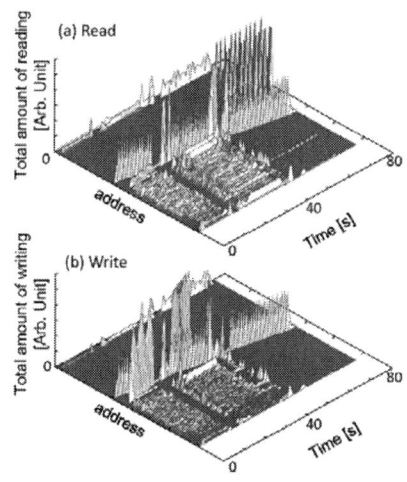

FIG. 4. THE TOTAL AMOUNT OF DRAM ACCESS USING COMMERCIAL IN-MEMORY DATABASE SERVER WITH TPC-B WORKLOADS (A) READ ACCESS, (B) WRITE ACCESS

IMPACT OF NONVOLATILE MAIN MEMORY ON SYSTEM PERFORMANCE

We further evaluated the impact of NVRAM on system performance using a commercial IMDB with TPC-B workloads from another aspect. Figure 5 (a) shows a block diagram of IMDB system. In IMDB, logging to secure persistence impacts on the response time of transactions, since each transaction must wait for at least write operation until the committing process is completed. The log data getting involved in committing is stored at the log storage through the log buffer in the background processing.

Figure 5 (b) shows the number of transactions using different hierarchical devices as log storage. The IMDB using NVMe-SSD can improve system performance 11.3 times better compared with the method using HDD, because of the shorter process time of committing.

When the log buffer in the main memory layer can be act as the log storage using NVRAM, system performance can be improved 13.7 times better compared with the method using NVMe-SSD. Furthermore, we implemented our newly optimized database library for NVRAM. System performance with new database library can be improved 178.7 times better compared with the method using NVMe-SSD.

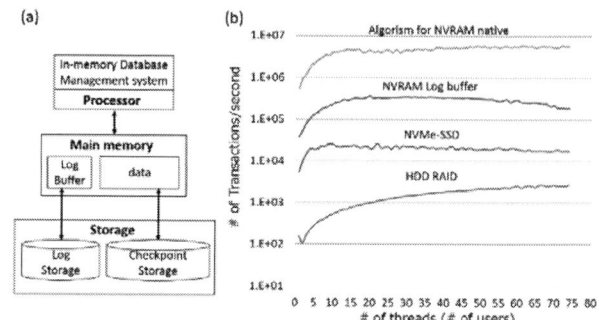

FIG. 5. (A) BLOCK DIAGRAM OF IN-MEMORY DATABASE (B) SYSTEM PERFORMANCE USING DIFFERENT HIERARCHICAL DEVICES AS LOG STORAGE

CONCLUSION

In this paper, the core structure of high density NVRAM, including crossbar memory array, is discussed. The key is the development of suitable selector devices to realize high density NVRAM. The advantages of NVRAM for the total system performance are also studied, assuming the implementation of NVRAM on the processor memory bus alongside DRAM. Memory access for DRAM have both temporal and special locality in the existing memory system, such as IMDB. Therefore, NVRAM, as a non-volatile main memory, should be a promising solution for realizing high density and high speed persistence in any memory systems. In near future, NVRAM should be expected to revolutionize the computing system performance.

ACKNOWLEDGEMENT

The authors would like to thank M. Momodomi, K. Ishimaru, T. Maruyama, Y. Oowaki, K. Hashimoto, S. Inaba and all the members of NVRAM development team in Toshiba Memory Corp. for helpful comments and suggestions.

REFERENCES

[1] T. Kim et al., *IEDM Tech. Dig.,* pp.851-854, Dec. 2018
[2] S. G. Kim et al., *IEDM Tech. Dig.,* pp.24-27, Dec. 2017
[3] H. Yang, et al., *IEDM Tech. Dig.,* pp.836-839, 2017
[4] M. J. Lee et al., *IEDM Tech. Dig.,* pp.33-36, Dec. 2012
[5] A. Verdy, et al., *IEDM Tech. Dig.,* pp.863-866, Dec. 2018
[6] H. Y. Cheng, et al., *IEDM Tech. Dig.,* pp.859-862, Dec. 2018
[7] S. Yasuda, et al., *Symp. on VLSI Tech. Dig.,* pp.30-33, 2017
[8] A. Verdy, et al., *IEEE IMW,* pp.978-981, 2017
[9] G. Navarro, et al., *Symp. on VLSI Tech. Dig.,* pp.94-97, 2017
[10] H. Y. Cheng, et al., *IEDM Tech. Dig.,* pp.28-31, 2017
[11] Q. Luo, et al., *IEDM Tech. Dig.,* pp.253-256, 2015
[12] N. Shukla, et al., *IEDM Tech. Dig.,* pp.866-869, 2016
[13] H. Yang, et al., *Symp. on VLSI Tech. Dig.,* pp.130-133, 2015
[14] M. Son, et al., *IEEE ELECTRON DEVICE LETTERS,* VOL. 32, NO. 11, pp.1579-1581, Nov. 2011
[15] E. Cha, et al., *IEDM Tech. Dig.,* pp.268-271, 2013
[16] W. G. Kim, et al., *Symp. on VLSI Tech. Dig.,* pp.138-141, 2014

978-1-7281-0943-5/19 $31.00 © 2019 IEEE

New Memory Technology, Design and Architecture

Co-optimization to Enable Future System Needs

Arnaud Furnemont

Imec

While conventional scaling at device level becomes increasingly difficult, the system still requires lower power and lower cost for a given functionality and performance. It is time to bridge the gaps between the technology, device and system teams. Emerging concepts can only make sense by studying their impact on power and performance at system level. This needs to happen early enough in the development to be able to give feedback on the device specifications, hence the need of system simulators. In particular, STT-MRAM has a lot of potential to reduce energy of cache level, and we'll dive deeper into its impact on HPC and mobile applications. DRAM roadmap and even storage needs also seem to accelerate faster than the device roadmap, we'll also open the discussion around these concepts.

Embedded Memories for Silicon-In-Package: Optimization of Memory Subsystem from IoT to Machine Learning

Fatih Hamzaoglu

Intel

Traditional memory subsystem consisting of SRAM, DRAM and SSD/HDD has served the needs of electronics industry for decades. With the rapid increase in Graphics, IoT and Machine Learning applications, several new memories have been innovated to optimize the memory hierarchy. Optane memory is deployed to close the performance/area gap between DRAM and SSD. Similarly, embedded DRAM inserted in products as L4 Cache to close the gap between SRAM and DRAM. Lately, MRAM and ReRAM are also brought to reality to target wide range of applications, covering embedded Non-Volatile Memory and Flash buffer at platform level. This talk will go through these innovate memories and how one needs to optimize the memory subsystem for the best application. There's no memory that fits all, but design and architecture opportunities exist for targeted applications.

Evolution and Advances of the Nonvolatile Memories and Applications

Yan Li

Western Digital

With smart phone using more and more flash memory for storage, the nonvolatile memory enjoyed the golden years of rapid advance from 2D NAND to 3D NAND. The technology is still evolving towards forever more dense and large capacity. This talk will reflect on the evolution and future directions of the NAND memory and introduce new non-voltile memories and its applications.

As data increase dramatically, the cost of access data dominates the energy and throughput. Many new applications emerged to use memory for computing or move computation close to data. This talk will also touch on the new paradigm of applications of the memories.

Embedded Memory: The Future of Emerging Memories

Feng-Min Lee

Macronix

NAND is the dominant memory technologies currently in use. For NAND flash, with the migration from 2D to 3D, is projected to go from today's announced 96 layers to over 500 layers over the next few years. Decreasing the price of conventional memory and increasing the memory capacity, make it hard for emerging memories to replace them. In this situation, where will emerging memories gain a foothold?Among emerging memories, MRAM is very fast, has write and read speeds that are very close to DRAM and has very high endurance. All the major foundries (including TSMC, UMC, Samsung and Global Foundries) have teamed up with MRAM companies to make embedded products using MRAM. There are many other storage and other systems that are using a bit of MRAM to improve overall system performance. MRAM will be particularly useful where power consumption is an issue in IoT devices.For PCM, Intel's Optane memory is to be available in DIMM form for populating computer and server memory buses and will be available in 2019. This is supposed to provide improved performance for Intel servers. Besides MRAM and PCM, RRAM could also compete against current memories, such as: split gate, used in embedded applications. The standalone market for memory chips will provide some of the projected volume of emerging memories, but embedded applications, particularly for mobile and AI application, will likely be the big drivers for embedded emerging non-volatile memory. In conclusion, in this talk, we will give an overview of current status and future vision of emerging memory, especially for MRAM, RRAM and PCM.

978-1-7281-0943-5/19 $31.00 © 2019 IEEE

Future of Computing and Sensing Systems for Embedded Applications

Thomas Ernst

CEA-LETI

Today, the omnipresence of "big data" and worldwide social interactions requires technologies capable of "intelligent" features in order to analyzing complex objects (such as sounds, images or videos), in real time, and interact with humans in a cognitive way. The advent of the Internet-of-Things has also introduced a new paradigm that supports a decentralized and hierarchical communication architecture in which a great deal of analytics processing should be done at the edge and at the end-device instead of in the cloud. In this talk, we will discuss a novel research strategy targeting specialized low-power computing architectures, based on innovative components, such as emerging resistive memories, advanced CMOS and 3D technologies. Moreover, novel applications in the field of embedded systems, will be discussed thanks to the coupling of brain-inspired computing technologies with new sensing systems.

Progress in Neuromorphic Computing : Drawing Inspiration from Nature for Gains in AI and Computing

Mike Davies

Intel

In 2018, Intel published its Loihi neuromorphic research chip and launched the Intel Neuromorphic Research Community (INRC). As an exploratory AI research chip inspired by the principles of neural computation in nature, Loihi explores a very different design regime compared to the Von Neumann and matrix-acceleration architectures in wide use today. Loihi supports sparse and irregular communication between its units, packetized in the form of event-driven "spikes". Programmable plasticity processes distributed throughout the chip allow Loihi to rapidly adapt and learn in response to stimuli. A novel asynchronous design methodology provides maximal energy efficiency by exploiting the architecture's pervasive sparse activity.Members of the INRC are now collaboratively advancing the state-of-the-art in neuromorphic algorithms, software, and applications. Exciting recent results to date suggest that Loihi's combination of fine-grain parallel architecture, its spike-based computational model, and its asynchronous design implementation can provide compelling gains in computing efficiency, scalable to orders of magnitude, for the right kinds of adaptive dynamic problems. These include many exciting applications at the forefront of AI that call for highly efficient real-time interaction with real-world data.

Advances in Quantum Dot Lasers for Silicon Photonics

Yasuhiko Arakawa

The University of Tokyo

Since the first proposal of the concept of the quantum dot in 1982 by Arakawa *et al.*, growth and physics of the quantum dots have been intensively investigated. Development of high-quality InAs/GaAs quantum dots has been already led to commercialization of the quantum dot lasers for telecom, silicon photonics, and other applications. In particular, high temperature operation and high feedback-noise tolerance of the quantum dot lasers are advantageous characteristics for light sources in the silicon photonics.

Quantum dot lasers integrated into silicon circuit systems by the flip-chip bonding enabled high bandwidth-density over 15Tbps/cm^2 at 125 °C in 2014, which was lead to realization of I/O Core chips of 5x5mm^2 transceiver/receiver with 5mW/Gbps, 25Gbps/ch and 12ch parallel transmission. By using the wafer bonding method, a hybrid silicon quantum dot laser, evanescently coupled to a silicon waveguide, was demonstrated above 100°C. Moreover, we realized direct epitaxial growth of quantum dot lasers on ox-axis (100) silicon substrate only using MBE growth.

In this presentation, we discuss the current states of the art for the quantum dot lasers integrated onto silicon photonic circuit systems, focusing our achievements.

Silicon Photonics and Integrated Photonic Switches

Ming C. Wu

University of California, Berkeley

Silicon photonics has emerged as a promising solution to address the interconnect bottleneck in high performance computing (HPC) systems and data centers. Silicon photonics provides unprecedented I/O bandwidth, enabling ultrahigh aggregated bandwidth (~ 10 Tbps), high bandwidth density (~ Tbps/mm), and high energy efficiency (~ pJ/bit). In addition, silicon photonics also enable optical switching with large radix and short switching time. This talk will provide an overview of the state of the art of silicon photonic switches, with emphasis on new micro-electro-mechanical-system (MEMS)-actuated switching mechanism. Large scale switches with 64x64 and 128x128 port counts have been demonstrated, as well as wavelength-selective switches with 8x8 ports and 8 wavelengths. Future scaling to even larger port count will be discussed.

Silicon Photonic Based High-speed Transceivers for Fiber-optic Short-reach Applications

Peter Ossieur

Ghent University - imec IDLab

Next generation transceivers intended for short-reach fiber-optic links inside data centers will need to transport 800Gb/s to 1.6Tb/s capacities, through the use of space or wavelength division multiplexing of >56Gbaud lanes. Such capacities will need significant advances in energy efficiency (targeting 1pJ/bit), integration density and cost (targeting <<1$/Gb/s). In this paper, we present transmitters and receivers with record performance (either bitrate, energy efficiency or sensitivity), realized through close co-development of the drive/receive electronics on one hand and Silicon Photonic Integrated Circuits (PICs) on the other hand. Examples using SiGe BiCMOS or FDSOI CMOS drivers and receiver chips are provided, demonstrating a path to 100Gb/s per-lane bandwidths.

Layout Automation for Integrated Photonics

Alan Sherman

Mentor Graphics, A Siemens Business

As integrated photonic designs are increasing in size, automation is next logical step for productivity improvements. The integrated photonic design community is maturing so larger foundries are starting to participate in PDK offerings. The discussion will provide details on Mentor Graphics approach and solution to both automation and foundry enablement.

Twin-Bit Resistive Random Access Memory in FinFET CMOS Logic Technologies

Chieh Lee, Yu-Ting Hung, Cheng-Jun Lin, Ya-Chin King, and Chrong Jung Lin,
Institute of Electronics Engineering, National Tsing Hua University, Hsinchu, Taiwan
Phone/Fax: +886-3-5162182, E-mail: cjlin@ee.nthu.edu.tw

Abstract

In this paper, a Twin Bit FinFET RRAM fully compatible to advanced FinFET CMOS logic process technology is proposed. Through the new array arrangement and an operation scheme, the set and reset voltage can be effectively separated. In addition, the number of pulses needed to complete a set and reset operation in cycling tests is significantly less than that found in single-bit RRAM cell. Two bits in one cell can be independently set/reset/read without disturbing the other side. Cycling endurance over 10K and data stability in addition to its excellent read/set/reset disturb immunity are demonstrated.

I. INTRODUCTION

As consumer electronic products become more prevalent, the demands for portable data storage devices raises rapidly. As the CMOS technology node keep scaling down, charge-storage based non-volatile memory (NVM), faces limitations of minimum feature size and the reliability challenges [1-2]. RRAMs have attracted many interests for the next-generation non-volatile memory development. In our previous work [3], FinFET Dielectric RRAM (FIND RRAM) which is fully compatible to advanced CMOS FinFET technologies has been proposed and demonstrated high feasibility. The circuit schematic of this cell is shown in figure1 (a), where the HfO_2-based TMO film between drain and gate of the cell transistor is used as the storage node. To further improve its operation window, preventing the occurrences of overset and over-reset, a new Twin-Bit FinFET Dielectric RRAM (TB FIND RRAM) is proposed here, as illustrated in figure 1(b). TB FIND RRAM is also fully compatible to FinFET CMOS logic process, and can be developed without any extra mask or processing step. Using this new cell/array architecture, TB FIND RRAM can not only obtained double data density, but also allows for enhance operation window and reliability performance.

II. CELL STRUCTURE AND OPERATION

Figure 2 shows the transmission electron microscopy (TEM) picture of TB FIND RRAM. The transition metal oxide is the high-k gate dielectric stack used in standard FinFETs. Metal gate (MG) served as top electrode and epi-SiP act as bottom electrode for the RRAM storage node. With one shared select transistor, both the source and drain side of a cell transistor can serve as storage nodes, see the 3D illustration in figure 3(a). TB FIND RRAM cells can be arranged in an array as shown in figure 3(b), where source line on alternate bits can shared. For forming and reset operations, the array can be operated without selection under blanket mode. The array can be operated in a page mode. The operation conditions for parallel set of bits on WL_m are summarized in Table 1. Four steps are needed to complete of writing all bits on the same WL, while reading of data can be done by two stages. Both TB FIND RRAM and SB FIND RRAM are both unipolar RRAM, as shown in the figure 4. The difference in the switching characteristics between SB FIND RRAM and TB FIND RRAM can be explained by the band diagrams illustrated in Figure 5. When high positive voltage is applied on the metal gate during forming, the tunneling barrier width become narrowed. In contrast, the barrier when apply positive bias on the epi side remain the same. Therefore, it is easier for electrons passing through the dielectric stacks with positive bias applied on the metal gate [4-5]. Accordingly, the operation voltage for SB FINDRRAM is found to be lower than that for TB FIND RRAM. However, it may cause the issues of overset and over reset, as reported in figure 6. In addition to much tighten set voltage distribution, the difference between set voltage and reset voltage on the SL is much higher for TB FIND than that found in SB FIND RRAM. Hence, it is expected that TB FIND RRAM will less probable to experience overset and reset when placed in an array operation. Furthermore, the forming operation of TB FIND RRAM is stronger, which requires higher voltage on the SL. Figure 7 shows both SB FIND RRAM and TBFIND RRAM are set up to stable read windows for 104 cycles. These cycling test are down using ISPP [6] algorithm to ensure a target read current window can be obtained. The number of set/reset pulses required to reach the target cell current level is much less than that for a SB FIND RRAM Device, as shown in figure 8.

III. DEVICE CHARACTERISTICS

As figure 9 shows, there is no significant difference the read current distributions between left and right bits. TB FIND RRAM can be quickly set to LRS within 40ns and reset process takes about 40μs . The twin bit in a single cell can be operate independently without disturb the operation of the other bit, as shown in figure 10. Figure 11 shows that a stable LRS and HRS states are found after continuous DC stress, proving that TB FIND RRAM is immune to read disturb. To test its data stability, TB FIND RRAM cells were baked at high temperature of 200 °C for 500 hours. As shown in figure 12, the read window of TB FIND RRAM maintains stable. It can indicate that the TB FIND RRAM has an excellent data retention capability. Data in figure 13 reveals that when the left bit is selected to experience 10K cycles, the read current of the right bit remain stables at either state.

IV. CONCLUSIONS

In this paper work, a Twin-bit FIND RRAM is proposed and demonstrated. With the new array arrangement and operation scheme, overset and over-reset problem in conventional FIND RRAM can be alleviated. In addition, number of iterations needed for ISPP operation during AC cycling test is found to be much less in TB FIND RRAM. With excellent reliability, both bits sharing the same cell transistors are found to be immune to array disturbance from the operations of the other bit.

REFERENCES

[1] S. Lee, IEEE International Memory Workshop, 2012, pp. 1-4.
[2] S. Hong, IEDM, 2010, pp. 12.4.1-12.4.4.
[3] H. W. Pan, et al, IEDM, 2015, pp. 10.5.1-10.5.4.
[4] N. Alimardani, et al, Applied Physics Letters, 2013.
[5] P. Maraghechi, et al, Appl. Phys. Lett. 100, 2012.
[6] K. D. Suh, et al, JSSC, 1995,vol. 30, no. 11, pp. 1149-1156.

978-1-7281-0943-5/19 $31.00 © 2019 IEEE

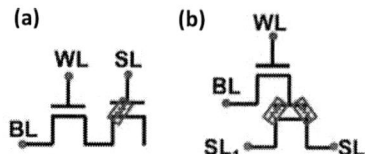

Figure1 Circuit Schematics illustrating (a) a single-bit FIND RRAM (SB FIND RRAM) cell and that of (b) a twin-bit FIND RRAM(TB FIND RRAM) cell.

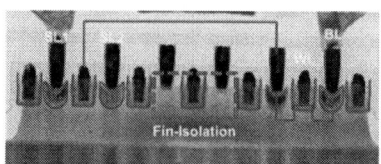

Figure 2 Cell structure and TEM picture of Twin-bit FIND RRAM cell.

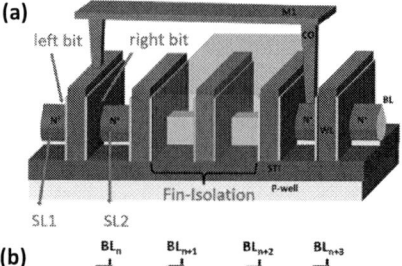

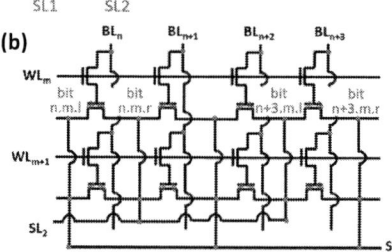

Figure 3 (a) 3D Illustration of Twin-bit FIND RRAM (b) Array Architecture of TB FIND cell 4×4 array

	Set				Read	
	Bits (n,r) (n+2,r) (n+4,r) ...	Bits (n+1,r) (n+3,r) (n+5,r) ...	Bits (n+1,r) (n+3,r) (n+5,r) ...	Bits (n,r) (n+2,r) (n+4,r) ...	Bits (n,r) (n+1,r) (n+2,r) (n+3,r) ...	Bits (n+1,r) (n+2,r) (n+3,r) (n+4,r) ...
WL$_m$	1V				0.8V	
BL$_n$, BL$_{n+2}$...	0V	F	F	0V	0V	0V
BL$_{n+1}$, BL$_{n+3}$...	F	0V	0V	F	0V	0V
SL$_1$	F	F	4V	4V	1V	0V
SL$_2$	4V	4V	F	F	0V	1V

TABLE 1 UNITS FOR MAGNETIC PROPERTIES

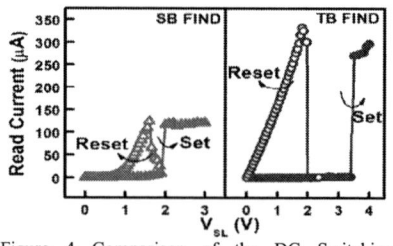

Figure 4 Comparison of the DC Switching Characteristics between FIND and TB FIND RRAM cells

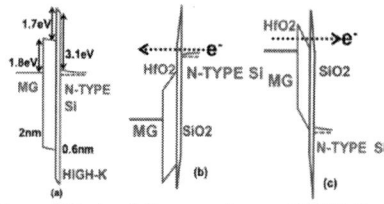

Figure 5 The band diagrams of across RRAM film gate illustrating when there is (a) no bias, with (b) 4V at metal gate and (c) 5.5V at N+ epi

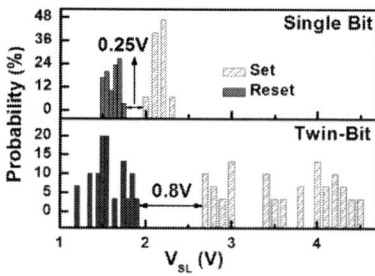

Figure 6 Set and reset voltage distributions of FIND and TB FIND cells are compared. Difference between Set and Reset V$_{SL}$ is much larger in TB FIND cells.

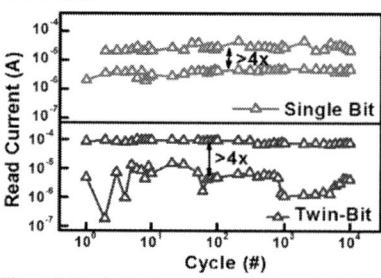

Figure 7 Read window on cycling characteristics of single bit FIND RRAM and Twin-bit FIND RRAM cells are compared.

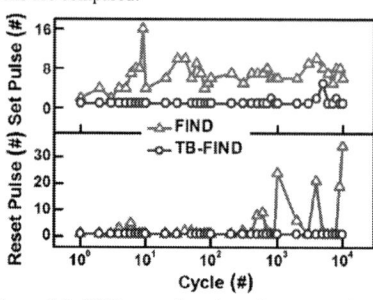

Figure 8 In ISPP operation, the pulse count of each SET/RESET is monitored. It is found that Twin-bit cell operation can effective reduce number of pulses required to successfully set/reset the RRAM cells.

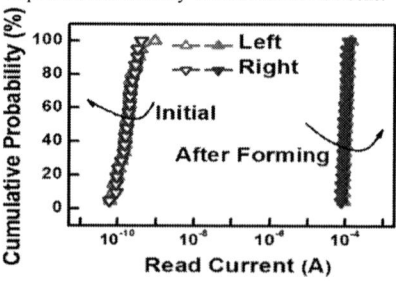

Figure 9 Read current distribution of the left and right bits in its initial state and after forming. No observable difference is found between left and right bits.

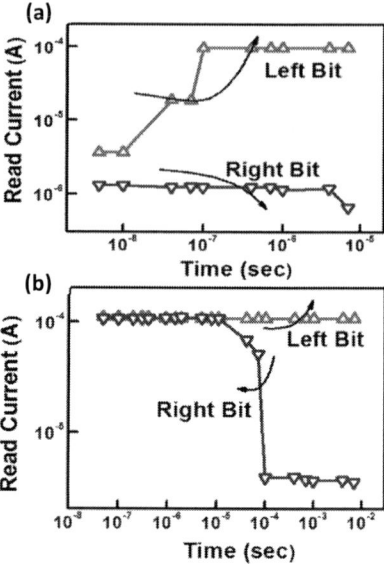

Figure 10 (a) Set operation can be achieved within 40ns while another bit keeps undisturbed. (b) Reset operations can be achieved within 40μs while another bit keeps undisturbed.

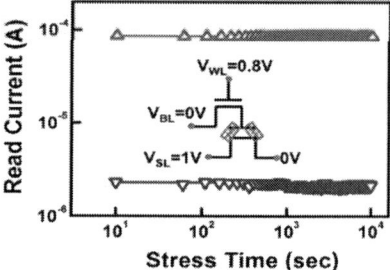

Figure 11 Read stress of twin bit FIND RRAM. Both states are undisturbed after 104 seconds of continuous read stress.

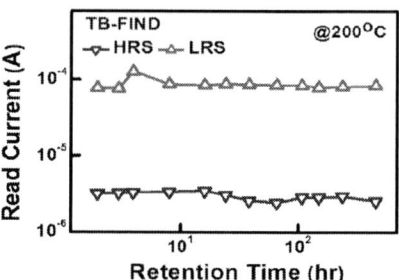

Figure 12 Data Retention characteristics show stable read window after 500 hours of 200℃ bake.

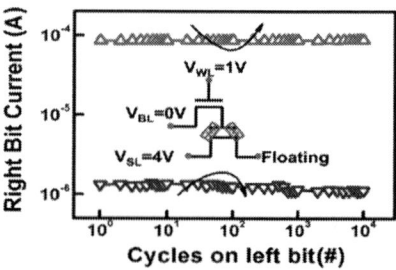

Figure 13 Set disturb test for unselected bit. Read current remain unchanged for the right bit when left bit experienced 10k cycles.

978-1-7281-0943-5/19 $31.00 © 2019 IEEE

Hot-Carrier Injection-Induced Disturb and Improvement Methods in 3D NAND Flash Memory

Wei-Liang Lin[*,†], Wen-Jer Tsai[*], C.C. Cheng[*], Chun-Chang Lu[*], S.H. Ku[*], Y.W. Chang[*], Guan-Wei Wu[*], Lenvis Liu[*], S.W. Hwang[*], Tao-Cheng Lu[*], Kuang-Chao Chen[*], Tseung-Yuen Tseng[†], and Chih-Yuan Lu[*]

[*]Macronix International Company Ltd., No. 16, Li-Hsin Road, Science Park, Hsin-Chu, Taiwan, R. O. C
[†]Department of Electronics Engineering & Institute of Electronics, National Chiao-Tung University, Hsinchu, Taiwan
Phone: +886-3-5786688-78039; E-mail: suzukilin@mxic.com.tw

ABSTRACT

We investigate a hot-carrier injection-induced program disturb in a 3D NAND flash memory. As there exist specific coding patterns, a "down-coupling" region and a "pre-charge" regions are formed during program-verify and the following program phases, respectively, in the inhibit cell strings. A high heating field is built nearby the PGM wordline. Hot carriers may inject into the inhibit cells as Vpgm is applied. Soft ramp-down and pre-turn-on schemes are proposed to mitigate this disturb.

INTRODUCTION

In addition to increasing the stacking layers, TLC/QLC technologies have been adopted in 3D NAND flash technologies to boost the bit density and greatly reduce the bit cost [1, 2]. To realize TLC/QLC technologies, it is important to well control the Vt distribution of each level. It also means that any disturb should be minimized. For example, a pre-charge scheme is utilized to alleviate the Y-/XY-mode disturb in BiCS [3]. Recently, "down coupling" is reported to be unique phenomena in the 3D NAND string [4]. It may cause read disturb [5] and degrade PGM inhibit [4] such that the memory window is detrimentally shrunk. In this paper, we would reveal another hot-carrier induced program disturb mode. Coding patterns and bias effects are comprehensively studied, and TCAD analysis is utilized to clarify the mechanism in detail. We then propose novel bias schemes to greatly mitigate this disturb.

EXPERIMENTAL DETAILS

32-WL 3D NAND cell strings with $Al_2O_3/SiO_2/Si_3N_4/SiO_x$ gate stack and poly-Si channel was adopted in this study [6] (Fig.1(a)). These cell strings are arranged in an array (Fig.1(b)). Fig.2(a) depicts the ISPP scheme with program-verify (PV) phase inserted between two successive program (PGM) phases. The detailed waveform of each ISPP step, which are implemented by 6-channel arbitrary linear waveform generator in Aligent B1500, is shown in Fig.2(b). The 2-phase waveform is taken as a "shot" and applied to the cell string. Vt shift of the victim cell is monitored after a predefined number of program (NOP) shots is applied.

RESULTS AND DISCUSSION

As shown in Fig.1(b), these cell strings construct a "block", in which cells of the same stacking layer share the same WL signal. Each SSL controls the cell strings of the same X-position (they consist of a sub-block), whereas each BL connects to cell strings of the same Y-position but of different sub-block. During PGM operation, Vpgm is applied to all the cells of the same layer in the block. Except for the PGM cells (SSL=Sel, BL=PGM), other cells may suffer from X-mode (SSL=Sel, BL=Inhibit), Y-mode (SSL=Un-Sel, BL=PGM), or XY-mode (SSL=Un-sel, BL=Inhibit) disturb. The typical Y-mode disturb is shown in Fig.3(a) (square), which is caused by soft-FN injection as NOP increases. However, we found an abnormal Vt jump (Fig.3(a) (circle)) as there is a specific coding pattern in the string (Fig.3(b)). Usually, the coding sequence proceeds sub-block-by-sub-block for the same WL layer, and from bottom to top layers (i.e. firstly WL0 from SSL0 to SSLn, and then WL1 from SSL0 to SSLn, and so forth). In this case, WL0, WLn-2,

and WLn-1 have been programmed to high-Vt state before WLn. We further clarify that the abnormal disturb would occur only when both PV and PGM phases are applied for each PGM shot (Fig.4).

We utilize TCAD simulation [7] to analyze the channel potential evolution during the PGM operation (Fig.5). As reported by Kim et al. [4], a "down coupling" region between the high-Vt cells (WLn-k-to-WLn-1) is formed at the end of PV phase, since the 3D string inherits a floating body structure. At the beginning of the next PGM phase, the pre-charge operation would raise the channel potential below WLn to SSL. A high heating field is built in the channel by the "down coupling" and "pre-charge" potential. As Vpgm is applied, electrons are heated and inject to WLn, which causes the observed disturb. As the coding cells (e.g. WLn-1, WLn-2 etc) are at lower Vt state, disturb is smaller because the down coupling level is shallower (Fig.6). Higher Vpgm would worsen the disturb because the vertical field is enhanced (Fig.7). As the pre-charge bias (VBL) is higher, it shows a smaller and a larger disturbs at NOP=10 and 10K, respectively (Fig.8). It's because that, as referring to Fig.3(a), at NOP=10, the cells suffer the typical soft-FN disturb. Higher pre-charge level would raise the channel potential and alleviate the vertical field. While at NOP=10K, the disturb is caused by hot carrier effect, which is enhanced as pre-charge level is higher.

Two methods are proposed to mitigate this disturb (Fig.9(a)). For soft-ramp-down approach, the Vpass recovery level at the end of PV phase is raised, such that the down-coupling level is shallower. Besides, the softly turn-on WLn-1/n-2 cells bridge the "down coupling" region to the BL end. For pre-turn-on approach, we turn on WLn-1/n-2 cells such that the channel potential could be "equalized" during pre-charge period. Both method can reduce the heating field and raise the conduction barrier between down-coupling and pre-charge regions (Fig.10), thus effectively mitigate the disturb (Fig.9(b)). It should be noted that as there are complex coding patterns in practice, the proposed pre-turn-on method may introduce disturb at other cells (Fig.11). It's because the large channel potential difference tends to exist at the intersection of high-/low-Vt cells in the string. If the pre-turn-on WL happens to be at low-Vt state, they could be disturbed as the pre-turn-on level is too high (e.g. 8V). By controlling the pre-turn-on level (e.g. 5V, which can still "turn-on" the high-Vt state), this disturb can be well managed.

CONCLUSION

The hot-carrier injection-induced disturb in the 3D NAND flash memory is study extensively. Experimental evidence and TCAD analysis clearly clarify that the coding-pattern-dependent "down coupling" effect play an important role, and the "pre-charge" scheme further worsen this disturb. We also suggest that "soft-ramp-down" and "pre-turn-on" methods can effective mitigate this disturb.

REFERENCES

[1] C. Kim, et al, ISSCC, pp.202-203, 2017.
[2] S. Lee, et al, ISSCC, pp.340-341, 2018.
[3] R. Yamashita, et al, ISSCC, pp.196-197, 2017.
[4] Y. Kim et al, IEEE-EDL, p.1566, 2016.
[5] Y. Zhang et al, IEEE-EDL, p.1669, 2017.
[6] W.J. Tsai, et al., IEDM Tech. Dig., pp. 288–291, 2016.
[7] Sentaurus, Synopsys, Version 2017.09.

978-1-7281-0943-5/19 $31.00 © 2019 IEEE

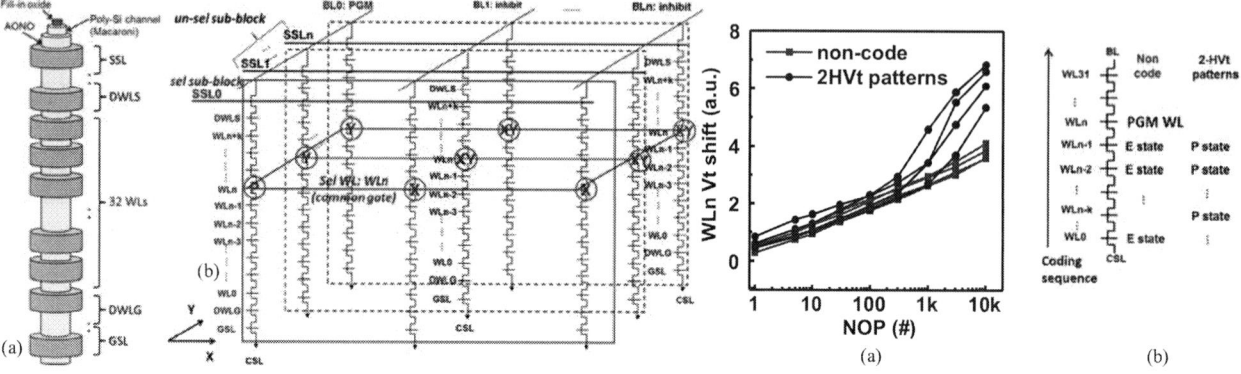

Fig.1 (a) Device structure of the 3D NAND cell string. (b) Array structure of the 3D NAND flash. The cell at WLn of SSL0/BL0 is selected to program, and X/Y/XY-mode victim cells suffer from program disturb.

Fig.3(a) Vt shift v.s. NOP of the disturbed cell strings of different coding patterns. (b) The schematics of coding pattern in the strings.

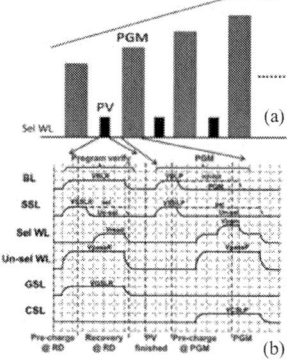

Fig.2 (a) The ISPP scheme with PV operation. (b) Detailed applied waveforms to the cell string.

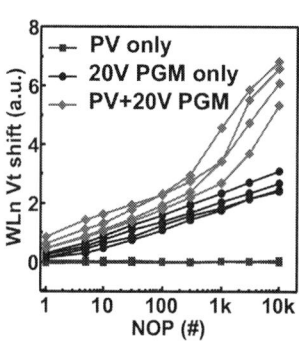

Fig.4 Vt shift v.s. NOP at various PGM- and PV-phase combinations.

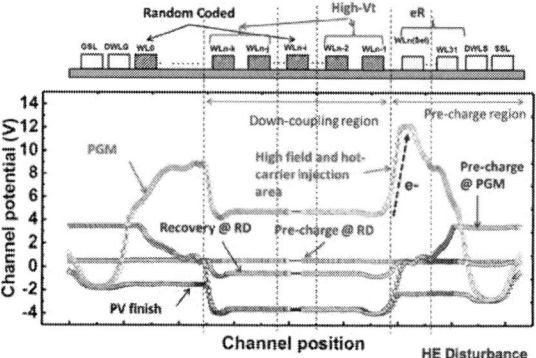

Fig.5 Simulated channel potential evolution during PV and PGM phases. Lateral field was built nearby the "down coupling" region and the "pre-charge" region.

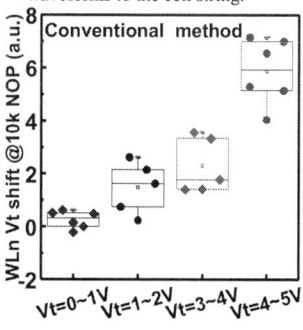

Fig.6 Disturb Vt shift v.s. NOP as coding cells are at different Vt levels.

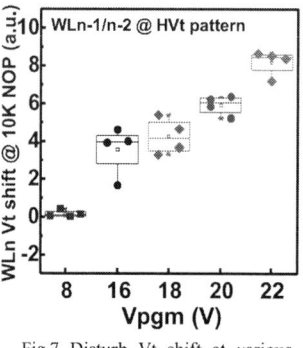

Fig.7 Disturb Vt shift at various Vpgm level.

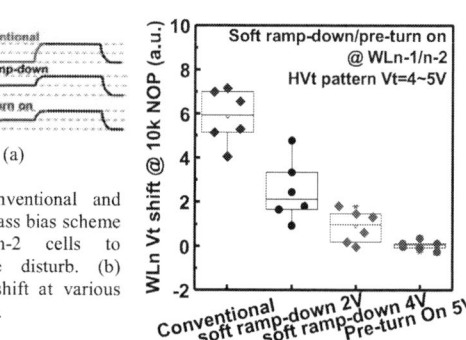

Fig.9(a) Conventional and proposed Vpass bias scheme at WLn-1/n-2 cells to mitigate the disturb. (b) Disturb Vt shift at various bias schemes.

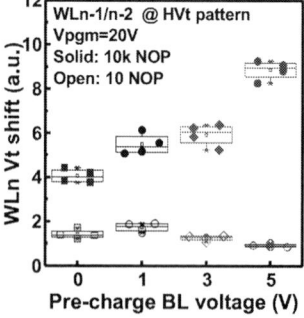

Fig.8 Disturb Vt shift at various pre-charge VBL levels.

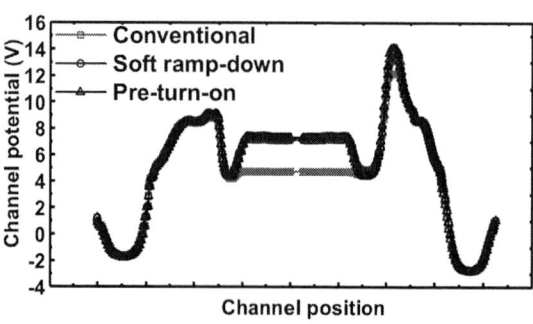

Fig.10 Channel potential as Vpgm is applied (refer to Fig.2(b), at time "PGM") for various bias schemes.

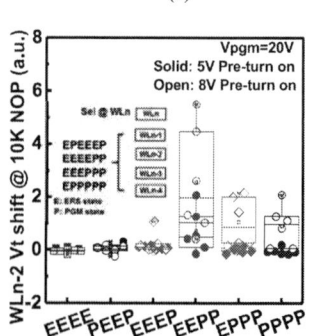

Fig.11 Vt shift of WLn-2 at various coding patterns and pre-turn-on levels.

978-1-7281-0943-5/19 $31.00 © 2019 IEEE

High density NV-SRAM using memristor and selector as technology assist

S.S. Teja Nibhanupudi and Jaydeep P. Kulkarni

Department of Electrical and Computer Engineering

The University of Texas at Austin, USA

E-mail: subrahmanya_teja@utexas.edu, jaydeep@austin.utexas.edu

Abstract − This work proposes 6T-2R-2S Non-Volatile (NV)-Static Random Access Memory (SRAM) bitcell for state retention applications with minimal sneak path current without incurring active cell area overhead. Various operating modes are described and Vmin/power comparisons with the baseline 6T SRAM are presented.

I. INTRODUCTION

Total power consumption is of paramount importance in energy constrained battery operated sensor nodes and IoT devices. Leakage power dominates the total power consumption in such devices due to very low activity factor. Large capacity embedded SRAMs consume significant leakage power while in retention mode, during the standby mode. For improved energy efficiency 2-level macro with SRAM for active mode storage and an off-chip non-volatile memory (e.g. FLASH) for standby mode storage has been used. However, this approach incurs significant energy and latency overhead due to data movement between SRAM-FLASH arrays during the state store/restore operation. This limits the overall power savings and frequency of activating such state retention modes. To mitigate these issues, multiple Non-Volatile SRAM (NV-SRAM) topologies such as 8T-2R, 7T-1R, 6T-2R (T= Transistor, R= memristor), have been proposed for in-situ state retention using different memristors [1-3]. However, additional transistors in 8T-2R, 7T-1R result in active bitcell area growth while 6T-2R bitcell results in sneak path current across the bitcells. In this work, we propose a novel 6T-2R-2S NV-SRAM bitcell for in-situ state retention featuring (1) zero standby leakage (2) zero active area growth (3) minimal sneak path current.

II. PROPOSED 6T-2R-2S BITCELL

A. Bitcell configuration

The proposed bitcell configuration utilizes CMOS 6T-SRAM, 2 memristors (2R) and 2 selector devices (2S) as shown in Fig.1(a). Fig.1(b) and Fig.1(c) show the I-V characteristics of a typical memristor (RRAM in this case) and the selector device (Phase Transition Material (PTM) in this case) respectively. The selector device is designed to exhibit large off-state resistance which reduces the sneak path current across the bitcells. Also, the selector switch can be engineered such that its insulator to metallic phase transition voltage (V_{IMT}) is higher than $V_{CC}/2$. The 2R-2S device stack is accessed through a Control Line (CL) which is shared across all bitcells routed along the bitcell row, parallel to the wordline direction. The 2R-2S stack can be formed over the bitcell storage nodes and can mitigate the SRAM bitcell active area growth.

B. State store operation

State retention is accomplished through a specific sequence of voltage signals applied to the control line (CL) (Fig.2.a, 2.d, Table.2). In the first step, the CL node shared along the bitcell row is raised to V_{CC}. This results in a current flow from CL to the node storing '0' (VL) whereas no current flows from CL to the node storing '1' (VR). The selector device transitions into metallic phase when voltage across the PTM exceeds the transition threshold and connects the memristor to the CL node. This initiates current flow in the bitcell from CL→ PTML (metallic phase) → RL → VL → NL → Vss. The voltage developed across RL memristor ($V_{RL} > V_{RESET}$) is large enough to program it to a High Resistance State (HRS). In the next step, the CL node voltage is lowered to Vss. This would result in a current flow from the node storing '1' (VR) to CL. Again, selector device (PTMR) transitions to metallic phase and large current flows from PR→ VR → RR → PTMR (metallic phase) → CL. The voltage developed across RR memristor ($V_{RR} > V_{SET}$) is large enough to program it into a Low Resistance State (LRS).

C. State restore operation

Restore operation is performed in three steps. First, all CLs are initialized to $V_{CC}/2$ to make sure PTM devices are in insulating phase to avoid any false programing of the memristors corrupting the stored resistance state. Next, as the bitcell-V_{CC} is ramped up, the SRAM storage nodes (VL and VR) are initialized with random values. In the second step, the WL is triggered with both bitlines at Vss. This would result in forcing '0' on both storage nodes. Note that CL is held at $V_{CC}/2$ to maintain the PTM selector in the insulating phase and isolate the 2R-2S stack (especially the LRS memristor) which would otherwise increase the crowbar current significantly during bit-cell V_{CC} ramp up. In the third step, as soon as WL is transitioned low, the CL is also transitioned to Vss. With bitcell-V_{CC} enabled, both bitcell PMOS transistors start charging VL and VR nodes towards V_{CC}. However, one of the sides is connected to Vss (through CL) with low resistance state of RRAM and other side is connected to Vss (through CL) with high resistance state of RRAM. This difference in the RRAM resistances results in the differential voltage development across the storage nodes (VL and VR) which then resolve to full-rail voltage due to positive feedback effect of the cross coupled inverters. Once the storage nodes restore the RRAM states, the CL is maintained at $V_{CC}/2$ to isolate 2R-2S stack from the subsequent active SRAM mode operations. Thus, row wise sequential restore mechanism with unselected CLs held at $V_{CC}/2$ can mitigate the peak current and IR drop along the bitcell-V_{CC} grid at the expense of increased restore time.

D. Active SRAM – mode

In this mode, the state retention component (2R-2S) is kept inactive to minimize the sneak path among bitcells and to prevent any disturbance to the subsequent active mode SRAM operations (read/write). All CLs are biased at $V_{CC}/2$ (Fig. 2.e) to maintain both selector devices (PTML, PTMR) in the insulating phase ($V_{IMT} > V_{CC}/2$).

III. SIMULATION RESULTS

Baseline 1-1-1 fin 6T SRAM bitcell and the proposed 6T-2R-2S bitcell are compared using 7nm predictive FinFET, and memristor /selector Verilog-A models (Table 1). Zero standby mode leakage attribute in the 6T-2R-2S bitcell can be utilized to lower FinFET Vt to achieve lower active-Vmin (minimum operating voltage). The 6T-2R-2S bitcell with 50mV reduced Vt shows 100(120) mV lower read(write) Vmin (@1e-6 P_{FAIL}) compared the baseline 6T SRAM requiring high Vt transistors for leakage reduction (Fig.4.a, 4.b). Lower active Vmin in the 6T-2R-2S bitcell results in lower total power for the activity factor (α) below 18%, making it very attractive for memory leakage power sensitive IoT designs (Fig. 5.a). Power savings increase for large arrays for small activity factors (Fig. 5.b). Table 3 compares 6T-2R-2S bitcell with prior approaches.

ACKNOWLEDGMENTS

This research is funded by Semiconductor Research Corporation # 2824.001 and National Science Foundation award # 1815616.

REFERENCES

[1] P.F. Chiu et al., IEEE VLSIC, pp229-230, June 2010 [2] A. Lee et.al., IEEE VLSIC, pp C76-C77, 2015 [3] W. Wang et.al., IEEE IEDM, pp 1-4, 2006 [4] http://ptm.asu.edu/ [5] Jiang,Z., Wong H.P, (2014). Stanford University RRAM Verilog-A model. nanoHUB doi: 10.4231/D37H1DN48 [6] W.Y. Tsai et.al., IEEE MSCS, pp30-48, 2016 [7] S.S. Sheu, et.al., IEEE ASSCC, pp.245-248, Nov 2013 [8] W.Wei et.al., IEEE T.NANO, pp.905-916, Sep 2014

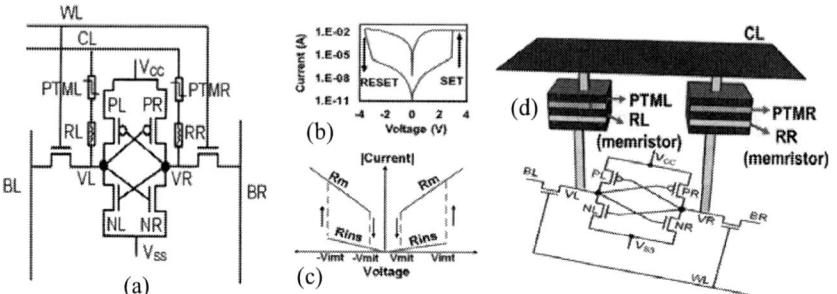

Fig. 1 (a) Proposed 6T-2R-2S SRAM bitcell with memristor and selector devices (b) Memristor (Resistive-RAM) (c) selector (Phase Transition Material) I-V characteristics (d) 3D conceptual view of the proposed 6T-2R-2S bitcell. The non-volatile memory stack can be fabricated in BEOL process.

Table 1: FinFET, memristor and selector models

Simulation technology node	7nm Predictive FinFET models [4]
Simulation tool	HSPICE
SRAM design	All 1 fin transistor SRAM
Memristor specifications	SET/RESET – 0.6/-0.6 ; ratio HRS/LRS - 100; Verilog-A model [5]
Selector specifications	V_{IMT}/V_{MIT} – 0.5/0.04 ; ratio met/ins - 10^4; Verilog-A model [6]

Table 2: Bitcell operating conditions

Mode	CL	WL	BL/BR
SRAM mode	$V_{CC}/2$	Normal operation	Normal operation
Store-0	V_{CC}	V_{SS}	Floating
Store-1	V_{SS}	V_{SS}	Floating
Restore – step 1	$V_{CC}/2$	V_{SS}	Floating
Restore – step 2	$V_{CC}/2$	> V_{CC}	V_{SS}
Restore – step 3	V_{SS}	V_{SS}	Floating

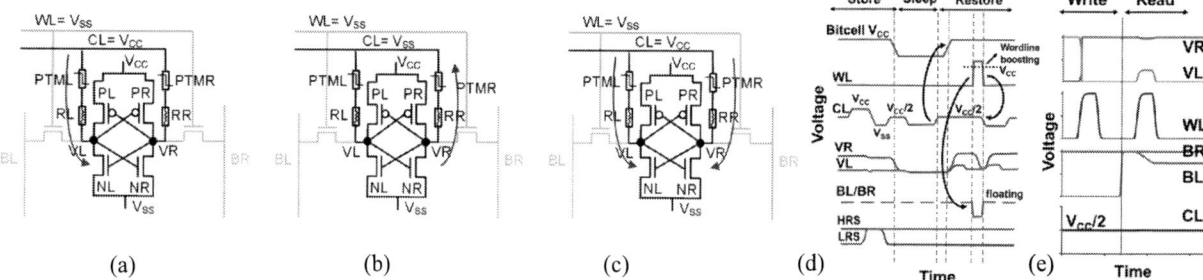

Fig. 2 Schematic showing the current flow direction in (a) & (b) State store operation (c) State restore operation (d) Timing diagram for store/restore operation (e) Timing diagram for SRAM mode. CL is held at $V_{CC}/2$ for minimal sneak path current.

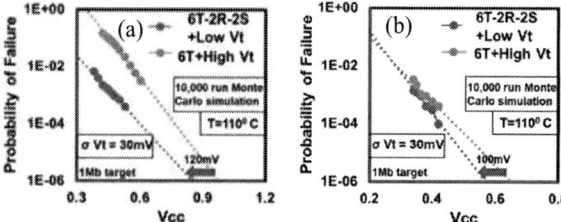

Fig.3: 6T-2R-2S SRAM array floorplan: CLs can be routed along the wordline and can be activated sequentially to limit the peak current and IR drop during store/restore operations.

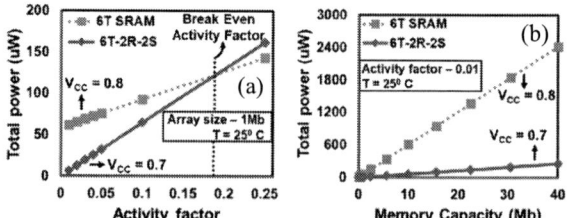

Topology	6T2R[3]	7T2R[7]	8T2R[1]	7T1R[8]	7T1R[2]	6T-2R-2S (This Work)
Cell Schematic						
Restore Method	Differential R Sensing	Differential R Sensing	Differential R Sensing	Single ended sensing	Initialization-and-overwrite	Differential R Sensing
Restore Yield	High	High	High	Low	High	High
SRAM mode short circuit current	Yes	Yes	No	No	No	No

Table 3. 6T-2R-2S bitcell comparison with prior NV-SRAM topologies

Fig.4 (a) Write Vmin improvement for the proposed 6T-2R-2S bitcell with 50mV reduced Vt compared to the baseline 6T SRAM bitcell which would require higher Vt to minimize leakage during retention mode. Write failure is quantified as the number of bit flips for a constant wordline pulse-width (500ps) (b) Read Vmin improvement for the proposed 6T-2R-2S bitcell with 50mV Vt reduction. Read failure is quantified as the number of bit flips during a read operation for a constant wordline pulse width of 500ps employing wordline underdrive (WLUD) of 0.8Vcc (110⁰C used for worst case analysis).

Fig.5 (a) Total power consumption variation with activity factor for the baseline 6T SRAM* using high Vt FinFETs and the proposed 6T-2R-2S SRAM$ using 50mV lower Vt FinFETs operating at 100mV lower V_{CC}. Breakeven activity factor is 0.18 beyond which 6T-2R-2S bitcell array consumes higher power than 6T bitcell array (b) Total power consumption as a function of memory capacity.

* 6T SRAM $P_{Total} = \alpha*(P_{active\_dyn\_0.8V} + P_{active\_ret\_0.8V}) + (1- \alpha)*P_{ret\_0.32V}$

\$ 6T-2R-2S $P_{Total} = \alpha*(P_{active\_dyn\_0.7V} + P_{active\_ret\_0.7V}) + \alpha*(P_{store\_0.7V}+ P_{restore\_0.7V})$

The Impact of Forming Temperature and Voltage on the Reliability of Filamentary RRAM

G.Y. Chen[1,2], F.M. Lee[1], Y.Y. Lin[1], P.H. Tseng[1], K.C. Hsu[1], D.Y. Lee[1], M.H. Lee[1], H.L. Lung[1], K.Y. Hsieh[1], K.C. Wang[1], C.Y. Lu[1] and M.C Wu[2]

[1] Macronix Emerging Central Lab., Macronix International Co., Ltd.
16 Li-Hsin Rd. Hsinchu Science Park, Hsinchu, Taiwan, ROC
[2] Institute of Electronics Engineering, National Tsing-Hua University (NTHU), Hsinchu 300, Taiwan
Phone : +886-3-5715131 Ext : 34028 E-mail : s106063528@m106.nthu.edu.tw

Abstract

The forming voltage and temperature for creating the first filament in the RRAM device are found having impacts on RRAM reliability. The correlation between the forming temperature and the required voltage was evaluated on the WOx/TiOx RRAM devices, and then the reliability tests including cycling endurance and data retention were performed. A model was proposed to explain the results.

Introduction

Transition metal oxide resistive random-access memory (TMO RRAM) is one of the promising next generation nonvolatile memories in light of its high speed, good scalability, and low process cost. In our previous works [1], the 1 Mb WOx/TiOx RRAM chip (Fig.1) was developed based on a 180nm standard CMOS logic platform with an RRAM process module inserted between the contact and the first metal layer. The TEM image in Fig.2 shows the 1T1R cell with the memory cell connected to the drain side of an NMOS. Nice array yield has been achieved on this vehicle [1].

Nevertheless there are still issues that need to be handled before delivering high quality products, one of which is the reliability performance. Although RRAM devices typically are good on reliability, the stochastic issues inherited from the filamentary conduction mechanism may still result in tail-bits on both the resistance distribution as well as reliability performances. Thus the quality of the conduction filament (CF), which is the assembly of oxygen vacancies created in the TMO layer by the electrical 'forming' procedure, are ultimately critical for RRAM. This paper studies the effects of high temperature forming operation with the 1Mb WOx/TiOx RRAM chip and demonstrates that executing the forming procedure at a high temperature not only reduces the forming voltage but also affects the tail-bit distribution in reliability tests. A model is proposed to explain the results, which may as well suggest directions to improve product performances.

Forming Voltage

Fig.3 shows the forming procedure. The forming process is conducted under an assigned temperature via a fixed BL voltage with a pulse/verify loop consists of a current-mode forming pulse and the set/read steps. If the read resistance of RRAM cell is higher than the verify level (50kΩ), the current pulse compliance will be increased for the next loop. The forming procedure stops when the read resistance is lower than 50kΩ or the compliance current level attains 280μA.

Fig.4(a) and (b) show the read R distribution after performing the forming process with different applied BL voltage at room temperature and at 110°C, respectively. The probability for fail-to-form (R>5Mohm) increases as the applied voltage decreases. High forming temperature effectively reduces the voltage required to form the devices. For example, at 4.1V applied voltage, all the cells in 1Mb RRAM chip can be formed at 110°C, whereas the fail rate remains at around 100ppm if the forming procedure is conducted at room temperature (Fig.5).

The result can be explained by the Poole-Frenkel effect [2]. At high temp. although the current of the transistor reduces slightly due to lattice oscillation, the 1T1R cell current still increases from the reduction of the TMO resistance. Electrons can achieve higher kinetic energy under high temperature, and pass the barriers in the oxide layer at a lower bias condition. As a result the first conduction path can be setup and the following up higher conducting current from self-heating further enhances the filament structure and "forms" the device.

Endurance

Three conditions (Table.1) are selected to evaluate the effect of the forming condition (including temperature and applied voltage) on endurance and retention performances. Fig.6 shows the cumulative resistance distributions from a 1Mb WOx/TiOx RRAM chip. The high resistance state (HRS) and low resistance state (LRS) read window can be maintained with raw bit error rate (RBER) as low as 1E-5 within 40 RESET/SET cycles. The forming temperature and applied voltage has almost no effect on RRAM endurance characteristics.

Retention

The RRAM chips formed at different temperatures/voltages were first cycled for 10 times at room temperature before the retention tests. Instead of looking at the median (at 0.5 RBER) of the distribution, the behavior of tail bit (at 1E-4 RBER) is more critical for RRAM products due to their stochastic nature. Fig.7 shows the cumulative resistance distribution from the retention tests, and Fig.8 further shows the time-dependent resistance level change at the point of 1E-4 RBER during 150°C and 180°C retention tests for chips from different forming conditions. Results show that the applied voltage has no effect on RRAM retention characteristics, while the RRAM chip formed at a high temperature has better retention in LRS but worse retention in HRS.

In order to explain the effect of forming temperature, we investigate the resistance distributions from three forming conditions (Fig.9) and a related model (Fig.10) is proposed. The RRAM resistance distribution formed at high temperature not only has higher baseline R level than that formed at room temperature but also extends its tail toward high R side. While the baseline R level difference is from the different effective verify level (since the Poole-Frenkel conduction mechanism resulting in lower cell R at high temp. and pass the verify step), the extended tail is due to the fast relaxation of the conduction filament at the high temp. environment during chip forming operation: The oxygen vacancies with high kinetic energy tends to diffuse-out and resulting in a larger but less dense filament [3]. The consecutive SET operations fill the wide-but-loose filament with more vacancies and resulting in a wide-and-dense filament with stable retention behavior. However, for the high resistance state after RESET, the remaining oxygen vacancies in or close to the wide filament have higher chance to move to the sites that create new conduction paths and reduce the cell R.

Conclusion

In conclusion, although high temperature forming operation can reduce the required forming voltage, reliability characteristics especially data retention will be affected. High temperature forming operation tends to create wider filaments, which may be good for LRS retention but not for HRS.

References

[1] C.H. Wang *et al.*, "Reliable and low forming voltage RRAM enabled by contact shrinking and pre-soldering baking", SSDM, 2018
[2] Ch. Walczyk *et al.*, "Impact of temperature on the resistive switching behavior of embedded HfO2-based RRAM devices", IEEE Transaction on El. Devices, vol. 58, no. 9, 2011.
[3] T. Cabout *et al.*, "Effect of SET temperature on data retention performances of HfO2-based RRAM cells", IMW, 2014

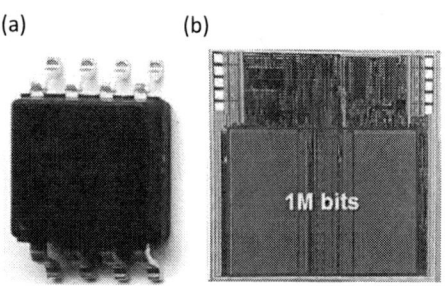

Fig.1. (a) 1Mb ReRAM chip with standard SPI interface and (b) die image.

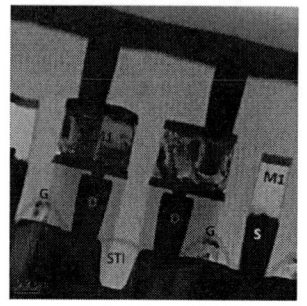

Fig.2.TEM image of 1T1R structure.

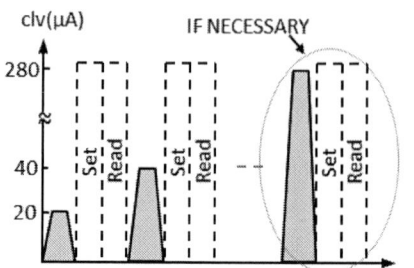

Fig.3.Forming procedure with a fixed BL voltage and followed with the set/verify-read step. If device resistance is still higher than 50kΩ, the current compliance level will increase for the next forming loop.

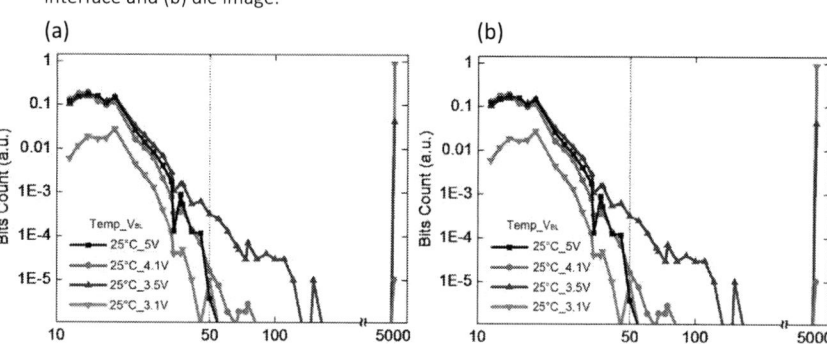

Fig.4.Read resistance distribution after forming with different applied voltage at (a) room temperature and (b) 110°C. RRAM cells which are fail to form are indicated at 5 Mohm (sensing limit).

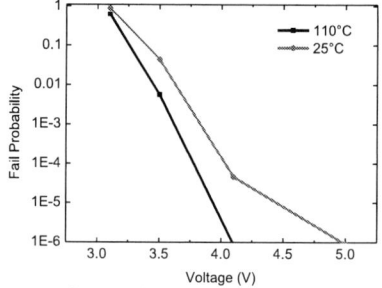

Fig.5. The probability of fail-to-form under different forming BL voltage. High temperature forming helps to reduce required forming voltage.

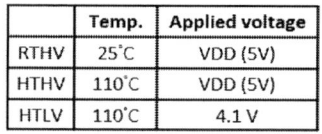

	Temp.	Applied voltage
RTHV	25°C	VDD (5V)
HTHV	110°C	VDD (5V)
HTLV	110°C	4.1 V

Table.1. Three forming conditions chosen for the reliability tests.

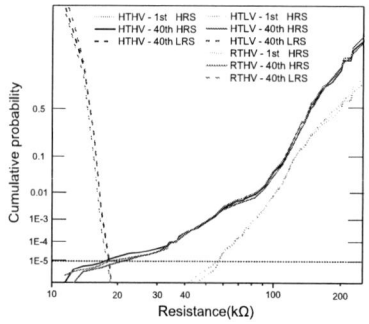

Fig.6. Cumulative read R distributions after cycling endurance tests (1st and 40th cycle) for chips having different electrical forming conditions.

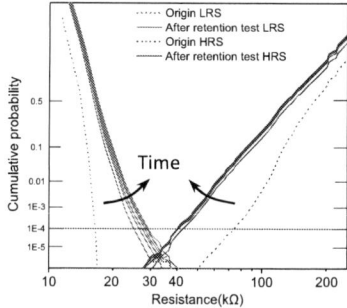

Fig.7. Cumulative R distribution evolves with time in the retention tests, especially at the 1E-4 (100ppm) tail.

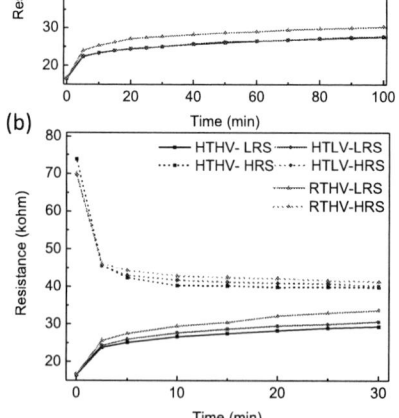

Fig.8. Retention performances at (a) 150°C and (b) 180°C for the 100ppm tail. The RRAM chip formed at high temp. showed better LRS retention but worse HRS retention comparing with low temp. ones.

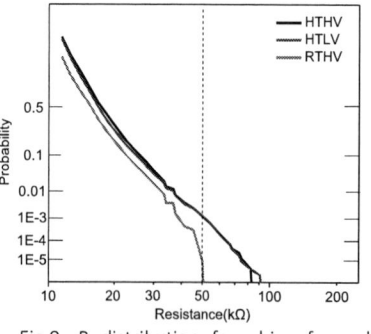

Fig.9. R distribution for chips formed with different conditions, after being kept at room temperature for 1 hour.

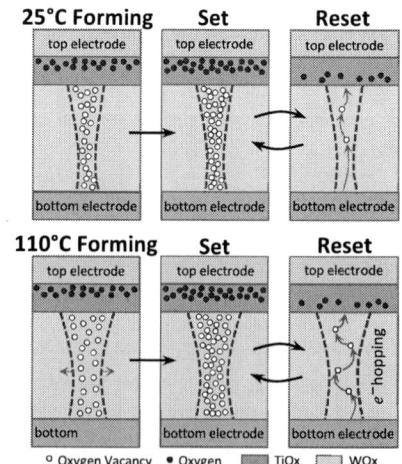

Fig.10. High temperature forming procedure increases the kinetic energy of the oxygen vacancies and resulting in wide-but-loose filament, which significantly affects the filament structures, and thus the reliability behavior, in the following SET/RESET operations.

Stochastic Filament Formation on the Cycling Endurance of Backfilled Contact Resistive Random Access Memory Cells

Yun-Feng Kao, Chrong Jung Lin, Ya-Chin King*

Microelectronics Laboratory, Institute of Electronics Engineering, National Tsing Hua University, Hsinchu 300, Taiwan
*Phone/Fax: +886-3-5162219/+886-3-5721804, E-mail: ycking@ee.nthu.edu.tw

ABSTRACT

Repeated and fast switching between on/off state without read window degradation is key for the development of resistive random access memory (RRAM) to meet reliable nonvolatile storage demands. In cycling study of the cell, we first established an incremental step pulse programing (ISPP) algorithm for 1T-1R backfilled contact resistive random access memory (BCRRAM) array. Experimental data reveals a strong correlation between reset time and the types of conductive filament (CF) formed. A training methodology is proposed to reduce the chance of the stochastic CF growth inside BCRRAM to maintain efficient reset operations and extend the endurance capability.

1. Introduction

In recent year, explosive growth on smart phone and wearable devices drive the demands of various data storage medium. The mainstream data storage products, i.e., Flash memory maintained the lead as the 3D NAND technologies matures [1]. On the other hand, as the need of embedded memories grows, many new emerging memory technologies, such as, RRAM, MRAM, have been investigated extensively [2]. As a result of the high compatibility of RRAM to CMOS logic process, it has become the first solution which provides viable memory IPs for integrated systems [3]. While implementing RRAM technologies into sizeable memory array, one faces challenges in endurance and variability issues, which must be thoroughly analyzed and resolved [4]. In this study, a new ISPP cycling algorithm is built for stable switching between states in 1T-1R BCRRAM cells. The influence of stochastic CF formation during cycling stress has been comprehensively discussed. Finally, a training reset scheme is first proposed and implemented to extend the cycling capability of the cell under high speed operations.

2. BCRRAM Structure

The BCRRAM cell array investigated in this study is fabricated by 0.18μm CMOS logic process, on which a series of cycling procedures are tested and characterizations of the stressed cells are performed [5]. As can be seen in the TEM image in Figure 1, storage node of BCRRAM is serially connected with a n-channel select transistor to prevent irreversible switching and sneak current path in a memory array. The TMO layer analyzed by EDS in Figure 1, is composed of $TiN/TiON/SiO_2$ sandwiched between top W-plug and Si bottom electrodes defined by the n+ diffusion.

3. ISPP Algorithm

An optimized ISPP algorithm for controlling large resistance variation has to be established to ensure good cyclability [6] between the read window of 10μA and 1μA when read at $V_{WL}=1.1V$ and $V_{BL}=0.5V$. First the BCRRAM cells forced with same first shot pulse, are then subject to 2 different set pulses subsequently, as listed in Figure 2(a). As shown in Figure 2(b), increasing WL voltage during the 2nd set pulse ensures that electric field can be fully exerted on dielectric rather than limited by the transistor. This prevents the cells to stuck in middle states and pushes the cells to lower resistance states. On the other hand, the lower HRS current states shown in Figure 3 can be obtained by applying a higher SL voltage. Based on these two principles, an ISPP algorithm, illustrated in Figure 4, is determined. As shown in Figure 5(a), BCRRAM can be successfully cycled for 1K times by employing proposed ISPP algorithm with a long reset pulse of 0.1msec. To enable high speed switching, reducing reset time is detrimental to the cycling capability, as shown in Figure 5(b) and (c).

4. Impact of CF Topography and Training Methodology

The types and shapes of conductive filaments in RRAM are known strongly influence resistive switch characteristics of RRAM [7]. In order to extend the reliability of the BCRRAM cells, in-depth investigating the properties of CF is critical to understand why and how the cells fail during endurance tests. To understand what types of CFs dominate the RRAM, random telegraph noise (RTN) found on the read current is used [8]. The types of RTN observed after forming operations are compared in Figure 6. If one dominant CF governs the conductive path, bi-level RTN is more likely to occur. On the other hand, if multiple conductive paths exist, the chances of having multi-level RTN raises [9]. In unipolar RRAM, the reset process is determined by temperature inside CF. Considering the joule-heating model, the simulated temperature profile is obtained to compare the heating in the CFs of the two types of cells during reset [10]. As expected and shown in Figure 7, higher temperature is generated for cells with single CF, leading to more effective reset operation. Therefore, unnecessary CFs produced by stochastic forming process need to be minimized to ensure efficient reset. As shown in Figure 8, the fresh cell after forming can be shifted to higher resistance states with 5 intense training reset pulses. The aim is to minimize the redundant CFs from the beginning. The types of CFs inside TMO layer are further examined based on statistical data obtained from a cell array. To better identify the types of RTN in a cell, lag plots used to monitor the shift of current levels are shown in Figure 9 [11]. Data in Figure 10 suggests that the probability of having multi-level RTN decreases dramatically for cells experienced the training reset process. The cycling characteristics of trained/untrained cells are compared in Figure 11(a), 11(b). For the trained cells, a good read window during cycling tests can be maintained under much shorter reset time. Comparing to untrained group, trained cells shows much lower failure rate during 1000 cycling tests, see Figure 11(c).

5. Conclusions

In this study, an optimized ISPP algorithm for operating 1T-1R BCRRAM memory array is demonstrated. To obtain high speed and good endurance, the problem of long reset pulses caused by multiple CFs generation can be relieved by introducing a training procedure.

ACKNOWLEDGEMENT

The authors would like to thank the support from the Ministry of Science and Technology (MOST), Taiwan.

REFERENCE

[1] H. T. Lue et al., in VLSI Symp. Tech. Dig., pp.1, 2010
[2] L. Baldi et al., in ESSDERC, pp.30, 2013
[3] H. S. P. Wong et al., in IEEE Proc., vol. 100, no. 6, pp.1951, 2012
[4] S. Balatti et al., in IRPS, pp. 5B.3.1, 2015
[5] H. Y. Chen et al., in VLSI-TSA, pp.1, 2016
[6] S. Y. Kim et al., in JJAP, vol. 53, no. 04ED15, pp.1, 2014
[7] C. Y. Liu et al., in IEEE EDL, vol. 35, no. 8, pp.829, 2014
[8] Z. Chai et al., in VLSI Symp. Tech. Dig. pp.1, 2016
[9] Y. H. Tseng et al., IEDM Tech. Dig., pp. 28.5.1, 2010
[10] U. Russo et al., in IEEE TED, vol. 56, no. 2 pp. 193, 2009
[11] M. B. Gonzalez et al., IEEE TED, vol. 63, no. 8, pp. 3116, 2016

978-1-7281-0943-5/19 $31.00 © 2019 IEEE

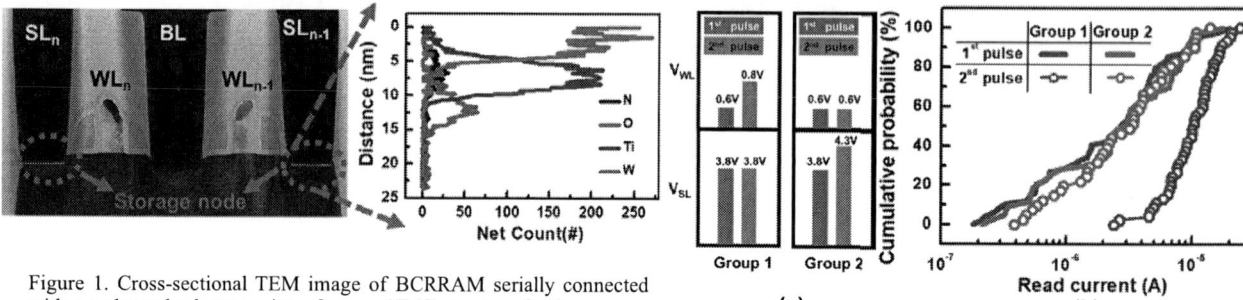

Figure 1. Cross-sectional TEM image of BCRRAM serially connected with a n-channel select transistor forms a 1T-1R structure for the purpose of selection in a memory array. The transition metal oxide layer of BCRRAM is composed of TiN/TiON/SiO₂ as revealed in the EDS results, sandwiched between top tungsten plug and bottom Si electrode defined by the diffusion region of the transistor.

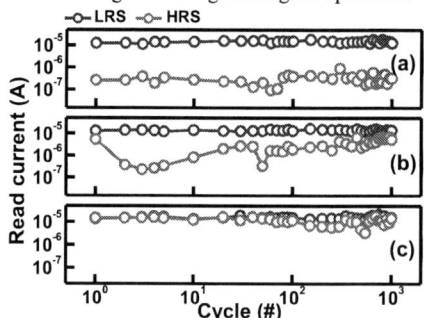

Figure 2. (a) Operation conditions of the set pulses. (b) Read current distributions of 90 BCRRAM devices after 1st 5μsec set pulse at V_{WL}=0.6V, V_{SL}=3.8V, and different 2nd 5μsec set pulses, showing the effect of raising WL voltage during set operation.

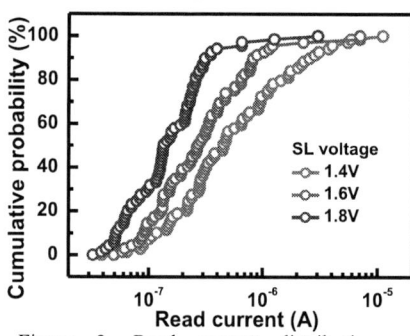

Figure 3. Read current distributions of BCRRAM cells experiencing different reset SL voltage while V_{WL}=1.2V, pulse width=100μsec. As expected, low current states can be better controlled by increasing SL voltage.

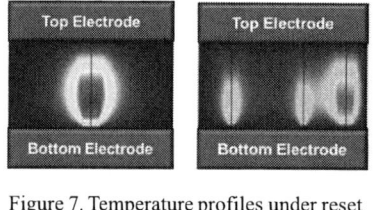

Figure 4. ISPP algorithm for the following endurance tests. V_{SL} at 3.8V and pulse width of 5μs are fixed during ramping WL voltage in set operation. V_{WL}=1.2V when ramping SL voltage in reset operation.

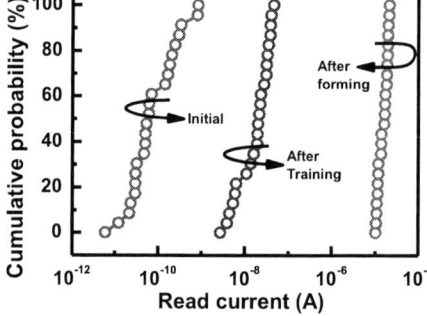

Figure 5. Cells under ISPP cycling tests with reset pulse width (a) 0.1ms (b) 10μs (c) 1μs. Dramatic read window degradation is found when reset pulse shortens, preventing cells to be applied on high speed operations.

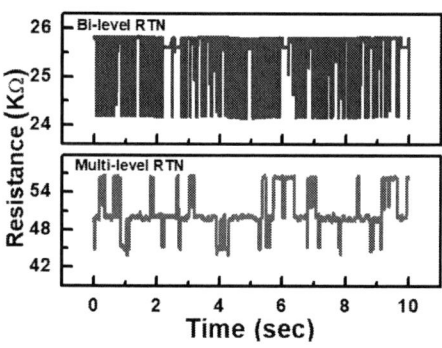

Figure 6. Prominent RTN signals of bi-level and multiple level current fluctuations are found in the BCRRAM after forming, suggesting high variability on the types of conductive paths generated in the TMO layers of BCRRAM.

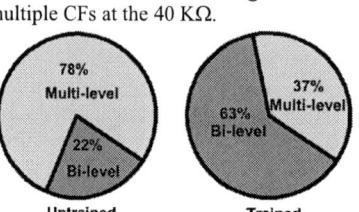

Figure 7. Temperature profiles under reset condition for cells with one single CF and multiple CFs at the 40 KΩ.

Figure 10. Percentage of cells with different RTN characteristics in the cells with and without experiencing reset trainings.

Figure 8. Read current distributions of the cells at initial, after forming process and after training operation. Training operation for minimizing unnecessary CFs is completed by applying 5 pulses at V_{WL}=1.2V, V_{SL}=2V and pulse width= 50μs.

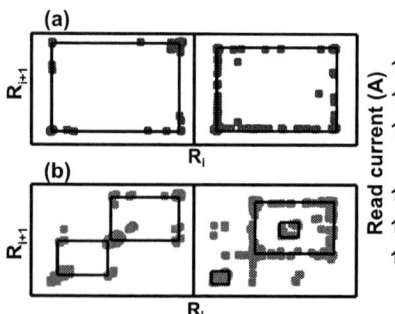

Figure 9. Lag plots, depicting resistance at certain moment versus its previous value, are used to determine (a) bi-level RTN signal (b) Cells with multi-level RTN

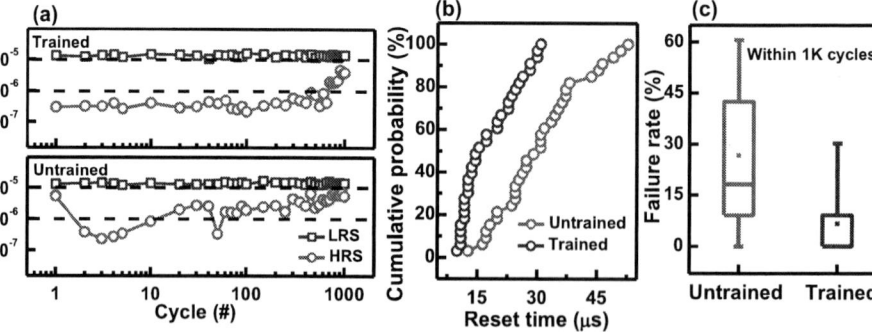

Figure 11. (a) ISPP cycle comparison between trained cells and untrained cells. More stable read current windows are found in trained cells. (b) Reset time needed within 1K ISPP cycles. Reset time is effectively shortened after minimizing redundant CFs. (c) HRS failure rate in 1000 cycles of trained/untrained cells. After training reset operation, HRS failure decreases in 1K ISPP cycle test.

978-1-7281-0943-5/19 $31.00 © 2019 IEEE

A Variability Source for Nanosheet GAA Transistors for sub-7nm Nodes

P. Harsha Vardhan, Swaroop Ganguly, and Udayan Ganguly

Department of Electrical Engineering, IIT Bombay. Email: udayan@ee.iitb.ac.in

ABSTRACT

Due to the complex process flow of the Gate All Around (GAA) stacked NSFET, new forms of process variability arise which are dominant over the conventional process variability like MGG and LER. Multi-V_T options in NSFET are achieved by gate metal thickness modulation, which can cause V_T variability if the process lacks precision. In this paper, we present a new form of process variability called Metal Thickness Variation (MTV) caused by inter-sheet spacing variation, compared to an 'ideal NSFET'. There are currently no models available in TCAD to study the thickness dependent Work-function (WF) V_T engineering. Hence, using physics based analytical model, we analyzed the variability in V_T due to this new variability source. We obtained the V_T distribution caused by MTV and estimated the standard deviation in V_T due to process variations and provided a way to suppress variability. Our results show that MTV emerges as a dominant source of variability and is insensitive for gate length scaling.

INTRODUCTION

Nanosheet transistors (NSFETs) can replace the FinFETs in future technology nodes owing to better control of SCE and high drain currents [1, 2]. Multi-V_T options in advanced technology nodes are achieved by various methods, from which metal thickness modulation is well demonstrated earlier experimentally for NSFETs [2]. Process induced variability has been a challenge for all the recent technology nodes due to decreased dimensions and complexity in the fabrication process [3]. However, process induced variability in the key transistor parameters like V_T in NSFET remains unexplored. Experimental demonstration of achieving multi-V_T options in the stacked nanosheet architecture has been reported in [2] using metal thickness modulation technique. A high sensitivity of 21mV/Å w.r.t metal thickness is also reported [2]. However, the same sensitivity might be an issue if a precise control of inter-sheet spacing is lacking in the process flow being followed. This results in the variations in WF on each sheet of each transistor which leads to a significant V_T variability. An analytical model to generate V_T distribution is proposed in this work which is adopted from a computationally efficient analytical model for FinFET and GAA NWFET variability [3, 4] developed earlier in our group.

METHODOLOGY

The entire methodology of obtaining the V_T distribution is summarized in the Fig. 2. The following are the key steps involved. They are:

1. The key dimensions of nanosheets like sheet thickness (T_{sh}), width (W_{sh}), Length (L_{sh}), inter-sheet spacing excluding IL/HK (T_{sp}) etc. are adopted from earlier experimental works [1, 2].
2. From the TEM image analysis in Matlab™ and from [2], the mean metal thicknesses are assumed as shown in the Table. I.
3. Gaussian distribution of metal thicknesses with an assumed mean ($\mu[t_a]$ of 0.75-2.25nm) and standard deviation ($\sigma[t_a]$ of 0.1-0.2nm) are generated as Metal: A thickness (t_a) values.
4. Metal: B thickness (t_b) for every t_a value is therefore obtained using the constraint that the sum of the metal thicknesses should be equal to T_{sp}.
5. For each face (gate) of each sheet of the sheet, the deviation is calculated for both the metals ($\Delta t_a = \mu[t_a] - t_a$ and $\Delta t_b = \mu[t_b] - t_b$).

6. For each gate of each sheet, the boundary condition for the Laplace is determined by multiplying the Δt_a (similarly for Δt_b) with their respective sensitivity values as shown in Fig. 2 to obtain ΔWF.
7. These boundary conditions are used to obtain electrostatic potential inside each nanosheet by solving 3D Laplace equation analytically. The details of this method were well reported in [3] for FinFETs.
8. Later, by using percolation resistance method [3], $I_{D,i}$ is obtained for each sheet and the total current from all the three sheets results in V_T of the NSFET as shown in the Fig. 3.
9. This process is repeated for 300 transistors to obtain the V_T distribution from which $\mu[V_T]$ and $\sigma[V_T]$ are obtained.

RESULTS AND DISCUSSION

Since the T_{sp} varies, metal: B thickness also varies due to its constrained dependence on t_a. Hence, a significant variability still exists even if the Metal: A is uniformly deposited. In Fig. 4, it is shown that even with constant t_a, the $\sigma[V_T]$ increases with increasing $\sigma[T_{sp}]$ for both the mean values of T_{sp}. Similarly, two studies are done to show the dependence of $\sigma[t_a]$ on $\sigma[V_T]$ in Fig. 5(a). It is evident that even if T_{sp} is maintained constant, the V_T distribution still has significant $\sigma[V_T]$ due to variation in t_a, however, it is better than the case when both t_a and T_{sp} are varying. In the Fig. 5(b), $\sigma[V_T]$ is plotted w.r.t t_a for two values of $\mu[T_{sp}]$ and 3 values of $\sigma[T_{sp}]$. The conclusion to be drawn out of this plot is that by increasing the T_{sp} and t_a mean values, the impact of MTV variability can be suppressed.

In this work, using stochastic simulations, we showed the impact of the variability of the individual single sheet ($\sigma[V_{T,ind}]$) on the mean threshold voltage of ($\mu[V_T]$) of the total stack of three sheets (defined as NSFET) in Fig. 6(a). This plot shows that $\mu[V_T]$ value drops with increasing V_T variability of individual sheet transistors which is a result of stacked architecture in which the V_T closely follows the lowest $V_{T,ind}$ of three sheet transistors. The effect of gate length scaling on variability is also studied. It is observed that the MTV variability is almost insensitive of gate length scaling as shown in the Fig. 6(b). The predicted $\sigma[V_T]$ agrees well with the experimentally observed value reported in [2] marked by green colored star.

CONCLUSION

Metal Thickness Variation (MTV) induced V_T variability is more important compared to any other sources of variability and can be suppressed by maintaining higher inter-sheet spacing. Hence, there is a need to develop more compact or analytical models to study various V_T engineering methods in NSFETs and the variability associated with each of these methods. A high sensitivity of V_T to t_a though advantageous for multi-V_T device options produces additional challenges in terms of variability.

REFERENCES

[1] N. Loubet *et al.*, *Symp. VLSI Tech.* T230, 2017.

[2] J. Zhang *et al.*, *IEDM*, p.22.1.1-4, 2017.

[3] P. H. Vardhan *et al*, *IEEE TED*, vol 64, p.3071-3076, 2017.

[4] P. H. Vardhan *et al*, *SSE*, vol 147, p.26-34, 2018.

978-1-7281-0943-5/19 $31.00 © 2019 IEEE

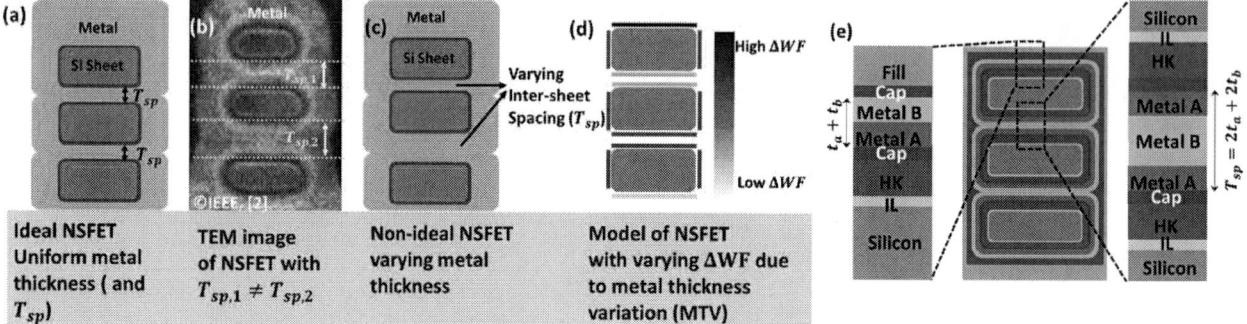

Fig.1. Motivation: (a) Schematic of the 'ideal NSFET' with uniform inter-sheet spacing; (b) TEM image of NSFET with varying inter-sheet spacing (T_{sp}); (c) Schematic of 'Non-ideal NSFET' with different T_{sp}; (d) Electrostatic equivalent model of NSFET with varying boundary conditions indicated by gray scale as per the metal thicknesses considered in figure (c) . Due to variation in inter-sheet spacing, as shown in the TEM image above, the metal thicknesses vary and hence its Work-Function varies which ultimately results in V_T variability; (e) Structure of the NSFET showing a detailed gate stack indicating two types of metals (A & B) typically used to tune the effective WF for multi-V_T options. It can be observed that the metal thicknesses are constrained by T_{sp} while the top gate metal is not constrained.

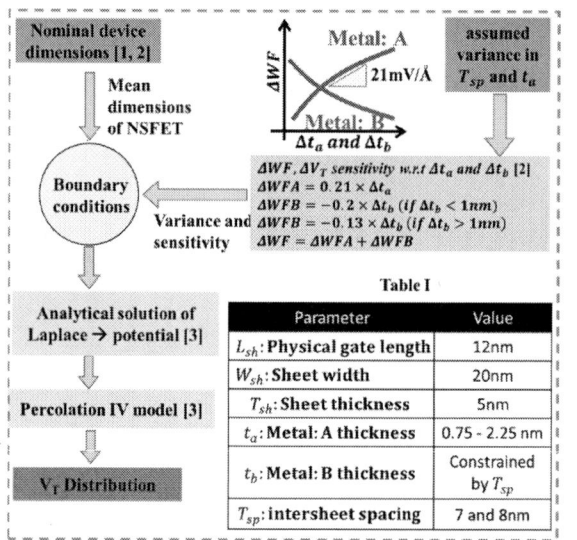

Fig.2. Flow chart of the model used to generate V_T distribution. Nominal dimensions taken from experimental works [1, 2]. On top of that, variance in inter-sheet spacing and metal thicknesses are assumed. The resulting metal thicknesses gives boundary conditions to Laplace equation. From potential profile, V_T is obtained using percolation method.

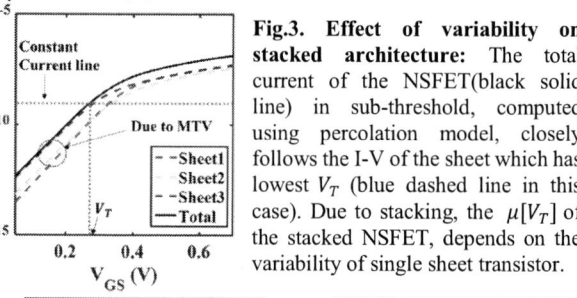

Fig.3. Effect of variability on stacked architecture: The total current of the NSFET(black solid line) in sub-threshold, computed using percolation model, closely follows the I-V of the sheet which has lowest V_T (blue dashed line in this case). Due to stacking, the $\mu[V_T]$ of the stacked NSFET, depends on the variability of single sheet transistor.

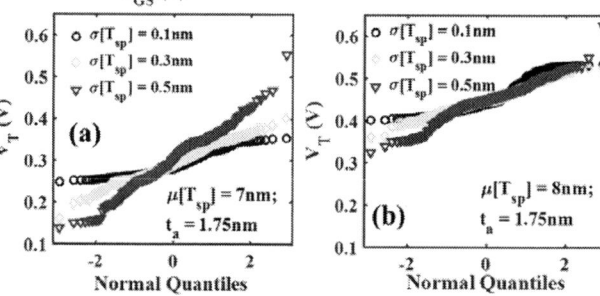

Fig.4. Quantile plots of the V_T values due to the random variation in T_{sp} for three values of $\sigma[T_{sp}]$ assumed with (a) mean T_{sp} of 7nm; (b) mean T_{sp} of 8nm. For both the plots, Metal: A thickness of 1.75nm is assumed constant.

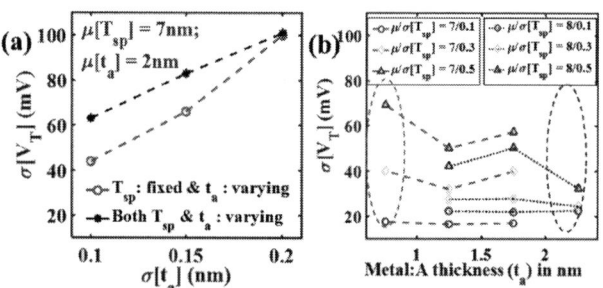

Fig.5. (a) Plot showing the sensitivity of Metal: A thickness variation in two scenarios where inter-sheet spacing (T_{sp}) is fixed (red circles) and T_{sp} is also varied along with t_a (black) using Gaussian distribution; (b) $\sigma[V_T]$ w.r.t Metal: A thickness for two mean values of T_{sp} viz. 7nm and 8nm for varied standard deviations in T_{sp}. V_T Variability due to MTV can be reduced by increasing T_{sp} and t_a.

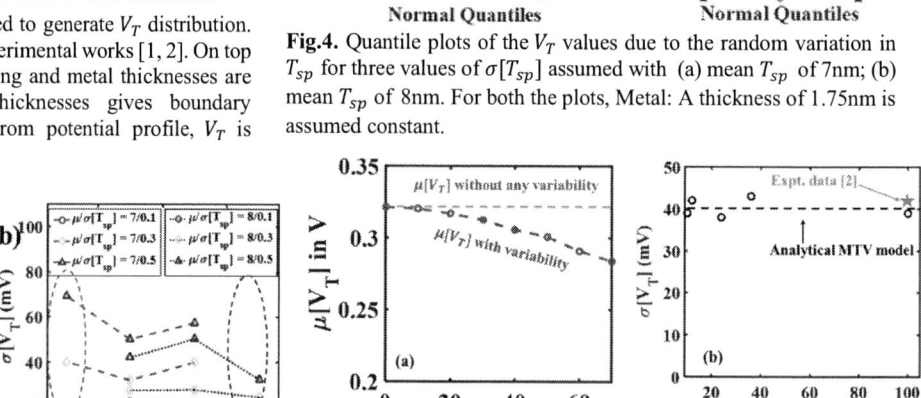

Fig.6. (a) Plot showing the dependence of $\mu[V_T]$ w.r.t variation in a single sheet transistor. The fall of the mean V_T can be attributed to the averaging effect caused by stacked Nanosheets; (b) V_T variability for the MTV is plotted here w.r.t Gate length (L_{sh}). MTV is dominant for V_T variability and scaling does not influence $\sigma[V_T]$ significantly. **The predicted $\sigma[V_T]$ matches with the experimental data given in [2] for 100nm long channel device for the assumed T_{sp} variation.**

978-1-7281-0943-5/19 $31.00 © 2019 IEEE   36

Impact of Multi-Domain Interaction on ON-State Characteristics of MFIS-Type 2D Negative-Capacitance FETs

Po-Sheng Lu, Chia-Chen Lin, and Pin Su

Department of Electronics Engineering & Institute of Electronics, National Chiao Tung University, Taiwan
E-mail: rsps950103.ee05g@g2.nctu.edu.tw, pinsu@faculty.nctu.edu.tw

ABSTRACT

In this work, with the aid of segmented SPICE simulation, we investigate the impact of multi-domain interaction on MFIS-type 2D Negative-Capacitance FETs with emphasis on the ON-state characteristics. Our study indicates that the multi-domain interaction enhances the lateral electric field for the MFIS device, leading to a higher ON-current. In addition, the multi-domain interaction increases the saturation drain voltage of the MFIS device due to the rise of the internal voltage near the drain-side. Our study also suggests that the negative differential resistance (NDR) effect present in MFIS devices may result from the strong domain interaction in addition to negative DIBL.

INTRODUCTION

Negative-Capacitance FETs [1] have been attracting substantial interests for low-power applications such as the IoT. 2D layered transition-metal-dichalcogenides (TMDs) have been regarded as potential channel materials for future high-density integration. 2D NCFETs (Fig. 1) with sub-60 mV/decade swing are especially interesting [2]-[5]. For the MFIS-type 2D NCFETs (Fig. 1 (b)), the non-uniform charge in the channel may lead to non-uniform polarization, and the multi-domain interaction of the ferroelectric [6] should be considered. Recently, experimental data of long channel MFIS devices has shown the NDR effect [7], [8], in which the multi-domain interaction may play a role. In this work, the impact of multi-domain interaction on MFIS-type 2D NCFETs is investigated by the aid of segmented SPICE simulation. The NDR effect is also examined.

METHODOLOGY

Segmented SPICE simulation method combined with the model in [3] is used to simulate the MFIS-type 2D NCFETs (Fig. 2 (b)). We divide the channel of the MFIS-type 2D NCFET into N segments with each sub-transistor treated as the MFMIS-type (Fig. 2 (a)). By the segmented method, the distribution of internal voltage V_{mos} (x) along the channel can be demonstrated and the ferroelectric layer is divided into N domains. The distribution of polarization in each domain can also be demonstrated by locally coupling the polarization charge of each sub-transistor. In addition, the multi-domain interaction of the ferroelectric [6] can be added into 1-D static state Landau-Khalatnikov (L-K) equation because the polarization in each domain is different:

$$V_{FE,n} = T_{FE}\left(2\alpha Q_n + 4\beta Q_n^{3}\right) + k(2Q_n - Q_{n-1} - Q_{n+1})$$

where V_{FE} represents the voltage drop across the ferroelectric, T_{FE} the ferroelectric thickness, k the multi-domain interaction factor (see Fig. 2 (b)), and α, β the ferroelectric parameters. The multi-domain interaction can be viewed as a strength that uniforms the distribution of polarization in the ferroelectric. The segmented simulation also shows good agreement with the MFIS-type analytical model in [4] (for the k=0 case) as shown in Fig. 3.

RESULTS AND DISCUSSION

Fig. 4 shows the hysteresis-free I_D-V_g characteristics for the designed low- and high-P_r MFMIS and MFIS 2D NCFETs with various k value, where P_r is the remnant polarization of the ferroelectric. When considering the multi-domain interaction, the ON-current of the MFIS device is higher than the case without interaction. It is noteworthy that a large enough k value can enable the high-P_r MFIS device to approach the MFMIS device [9]. Fig. 5 shows the G_m characteristics for the low- and high-P_r MFMIS and MFIS 2D NCFETs. The multi-domain interaction enhances the G_m for both the low- and high-P_r MFIS devices.

Fig. 6 shows the distribution of the polarization charge from source to drain for the MFIS 2D NCFETs with various k value. When considering the multi-domain interaction, the charge near the source-side decreases and the charge near the drain-side increases, so the distribution of the polarization charge becomes more uniform. In order to raise the charge near drain-side, the internal voltage near drain-side needs to pull up simultaneously. The higher the k value, the higher the internal voltage near drain-side. Similarly, the internal voltage near source-side drops to decrease the charge. Consequently, for the low-P_r case, the internal voltage becomes more distributed (Fig. 7). On the contrary, for the high-P_r case, the distribution of internal voltage becomes more uniform. Under a large k value, the distribution of internal voltage of the high-P_r MFIS device can approach a constant, just like the MFMIS counterpart.

Overall, the multi-domain interaction boosts the distributed effect for the low-P_r MFIS device, and results in higher ON-current. On the other hand, for the high-P_r MFIS device, the multi-domain interaction diminishes the distributed effect, bringing the high-P_r MFIS device closer to the MFMIS device. The multi-domain interaction can enhance the lateral electric field in the channel for both the low- and high-P_r MFIS devices (Fig. 8) by increasing the internal voltage near the drain-side and decreasing it near the source-side. Due to the enhanced lateral electric field, the G_m and ON-current are increased by the multi-domain interaction.

Fig. 9 shows the output characteristics for the low- and high-P_r MFMIS and MFIS 2D NCFETs with various k value. Due to the rise of the internal voltage near the drain-side, higher drain bias is needed for channel pinch-off. In other words, the multi-domain interaction increases the saturation drain voltage ($V_{D,sat}$). For the high-P_r MFMIS, significant NDR is observed. However, due to the relative long channel, the negative DIBL does not occur in this case (the insets of Fig. 4). The NDR of the high-P_r MFMIS device is due to the deamplification of the capacitance matching caused by the drain voltage. Significant NDR of the high-P_r MFIS device may also be caused by a large multi-domain interaction as shown in Fig. 9.

REFERENCE

[1] S. Salahuddin and S. Datta, *Nano Lett.*, vol. 8, no. 2, p. 405, 2008. [2] M. Si et al., *IEDM*, p. 573, 2017. [3] W.-X. You and P. Su, *IEEE TED*, vol.64, no. 8, p. 3476, 2017. [4] W.-X. You and P. Su, *IEEE TED*, vol.65, no. 10, p. 4196, 2018. [5] A. Nourbakhsh et al., *Nanoscale*, vol. 9, p. 6122, 2017. [6] K. Jang et al., *IEEE JEDS*, vol.6, p. 346, 2018. [7] K.-T. Chen et al., *IEEE JEDS*, vol.6, p. 900, 2018. [8] Z. Yu et al., *IEDM*, p.577, 2017. [9] A. K. Saha et al., *IEDM*, p. 326, 2017.

Acknowledgment- The support from Ministry of Science and Technology, Taiwan, under MOST-106-2221-E-009-148-MY2, MOST-107-2633-E-009-003, and MOST-107-3017-F-009-002.

978-1-7281-0943-5/19 $31.00 © 2019 IEEE

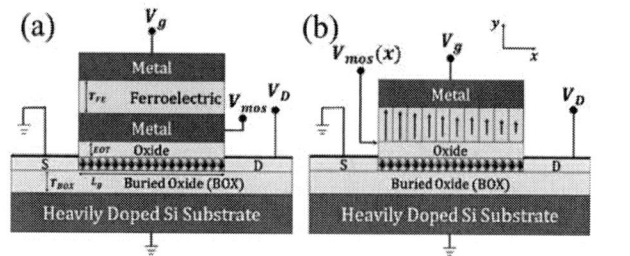

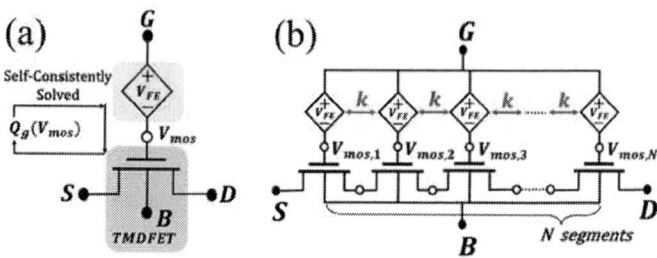

Fig.1 Schematic of the (a) MFMIS- and (b) MFIS-type 2D NCFETs. The monolayer MoS₂ is used as the channel material.

Fig. 2 Simulation framework of the (a) MFMIS- and (b) MFIS-type 2D NC FETs.

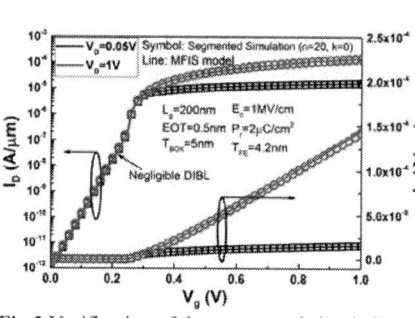

Fig.3 Verification of the segmented simulation (with k=0) using the analytical MFIS-type model in [4].

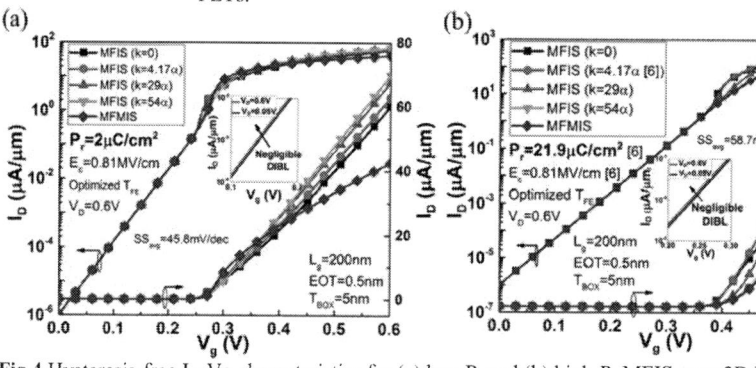

Fig.4 Hysteresis-free I_D-Vg characteristics for (a) low-P_r and (b) high-P_r MFIS-type 2D NCFETs with various k value. The insets show negligible DIBL of the MFIS-type device.

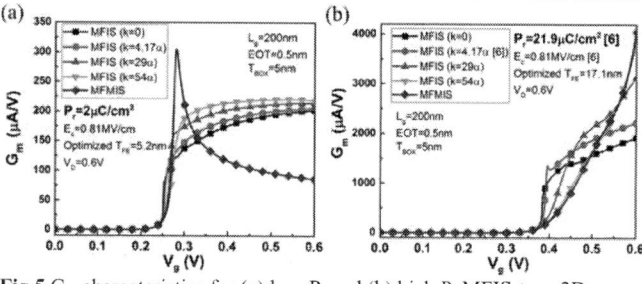

Fig.5 G_m characteristics for (a) low-P_r and (b) high-P_r MFIS-type 2D NCFETs with various k value.

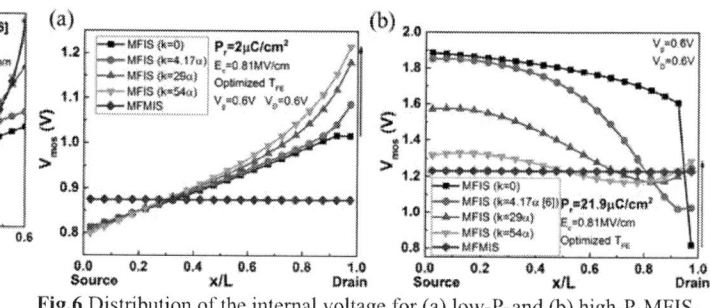

Fig.6 Distribution of the internal voltage for (a) low-P_r and (b) high-P_r MFIS-type 2D NCFETs with various k value.

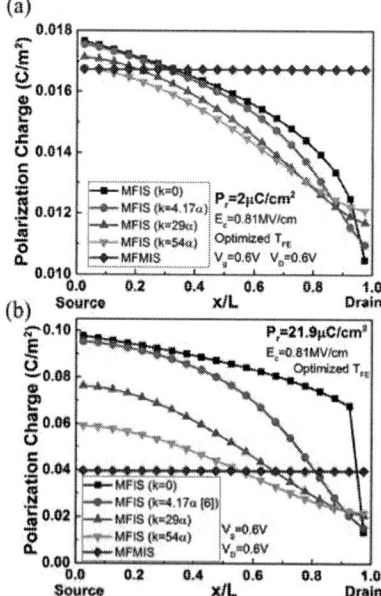

Fig.7 Distribution of the polarization charge in the ferroelectric for (a) low-P_r and (b) high-P_r MFIS-type 2D NCFETs with various k value.

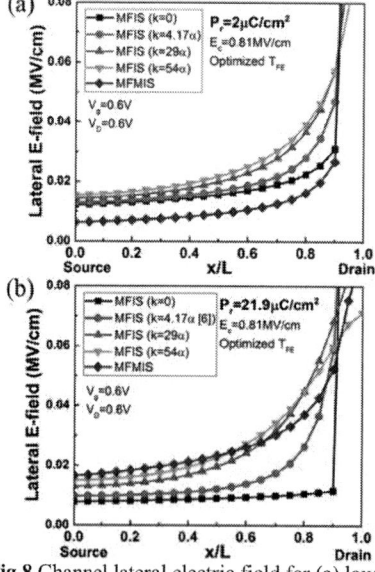

Fig.8 Channel lateral electric field for (a) low-P_r and (b) high-P_r MFIS-type 2D NCFETs with various k value.

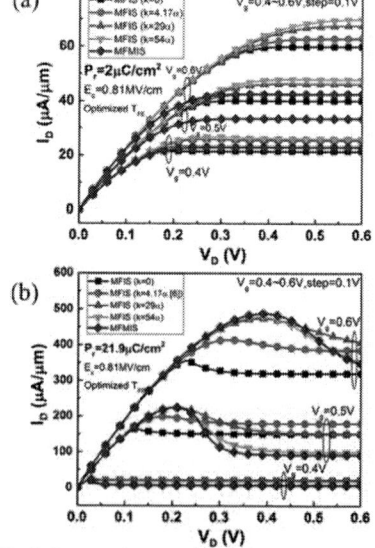

Fig.9 Output characteristics for (a) low-P_r and (b) high-P_r MFIS-type 2D NCFETs with various k value.

978-1-7281-0943-5/19 $31.00 © 2019 IEEE

Accurate Measurement of Sneak Current in ReRAM Crossbar Array with Data Storage Pattern Dependencies

Yaqi SHANG, Takashi OHSAWA

Graduate School of Information, Production and Systems, Waseda University, Japan
E-mail: edoshine@moegi.waseda.jp, takashi.ohsawa@aoni.waseda.jp

Abstract— In this paper, readout scheme in ReRAM crossbar array based on sneak current compensation is introduced. This scheme consists of two cycles. In the first measurement cycle, an accurate sneak current is measured which is dependent on the data storage patterns of the crossbar arrays in which the selector ON/OFF ratio is not large enough. In the second cycle, the measured sneak current is subtracted from the total bit line current to predict the cell current accurately. To make the measured sneak current accurate, the crossbar array is divided into many blocks only in one direction so that the impact of array size increase is minimal. The scheme is validated by using HSPICE.

Keywords—ReRAM CBA, sneak current, compensation, data storage pattern dependency

I. INTRODUCTION

Among the candidates to replace the existing memory devices facing scaling problems, ReRAM crossbar array (CBA) technology is one of the most attractive solutions for achieving the possibility of $4F^2$ cell size and neuromorphic design. However, there are several issues to be solved in this architecture. One of the severe issues is the sneak current from the unselected cells which will cause inaccurate data reading. Although some techniques have been proposed to address this problem by the sneak current compensation, these compensation strategies usually didn't cover the data storage pattern dependency (DSPD) for the compensation current under the assumption that the ON/OFF ratio of the cell selector diode is large enough for neglecting the dependency. However, the ratio of the diode in many CBAs is around the order of $10^2\sim10^3$ [1]. Therefore, it is necessary to take DSPD into consideration for these CBAs. Furthermore, the state-of-the-art technique for sneak current compensation is based on the subtraction of a large current measured in a pre-read cycle from a large read current observed in a read cycle to predict a small cell current. Therefore, it is difficult to predict an accurate cell current. In this paper, we proposed a sneak current compensation scheme in which an accurate cell current prediction can be achieved for CBAs with data-storage-pattern-dependent sneak currents.

II. SNEAK CURRENT ANALYSIS AND ISSUES

The conventional way of reading the data from ReRAM CBA is shown in the Fig. 1 (a). In this floating reading scheme, the CBA is divided into four parts as shown in Fig. 1 (b) [2-3]. One part is the selected part and the resistance is described as R_S. The other three parts constitute of the sneak path. Two of them are half-selected parts with resistances R₁ and R₂ in forward diode bias condition, and the other is an unselected part with R₃ in reverse bias condition. Thus the sneak resistance R_{sneak} can be derived by the following equation

$$R_{sneak} = R_1 + R_3 + R_2. \tag{1}$$

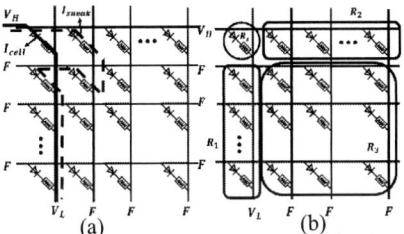

Fig.1. (a) cell current and one of sneak paths by floating reading scheme. (b) the four parts of CBA

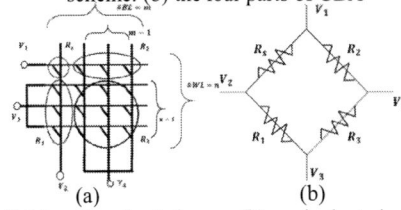

Fig. 2 (a) $m \times n$ shorted array, (b) equivalent circuit

For one-diode-one-resistor (1D1R) cell, if we denote the resistances of the forward and reverse bias diodes as R_{for} and R_{rev}, respectively, a random cell resistance R_i in a m × n size CBA can be expressed using the cell resistance R_{CELL_i} as

$$R_i = R_{CELL\_i} + R_{for}(or\ R_{rev}) \tag{2}$$

For unselected parts, since $R_{rev} \gg R_{CELL\_i}$, $R_i = R_{rev}$; for half-selected parts, since $R_{for} \ll R_{CELL}$, $R_i = R_{CELL\_i}$. So from the equivalent circuit for CBA read shown in Fig. 2 (b) which represents network of CBA at read shown in Fig. 2 (a), the equation (1) can be expressed as

$$R_{sneak} = \frac{1}{\sum_{i=1}^{n-1}\frac{1}{R_{CELL\_i}}} + \frac{R_{rev}}{(n-1)(m-1)} + \frac{1}{\sum_{i=n}^{n+m-2}\frac{1}{R_{CELL\_i}}}. \tag{3}$$

V₁, V₂, V₃ and V₄ stand for selected WL, selected BL, unselected WL and unselected BL in Fig. 2. Thus, if we can measure R_{sneak}, we can predict the actual cell current by subtracting the sneak current from a total bit line current as $I_{cell} = I_{total} - I_{sneak} = I_{total} - V/R_{sneak}$, where V stands for the voltage between of selected WL and BL.

The state-of-the-art technology assumed that $R_{rev} \gg m \times R_{CELL}$ [2] so that

$$R_{sneak} \approx \frac{R_{rev}}{(n-1)(m-1)}, \tag{4}$$

neglecting the DSPD. However, for a CBA with $\frac{R_{rev}}{R_{cell}} \sim m = 10^3$[1], DSPD cannot be neglected.

If we assume the data stored in the cells is the same, equation (3) can be transformed to

$$R_{sneak} \approx \frac{R_{CELL}}{n-1} + \frac{R_{rev}}{(n-1)(m-1)} + \frac{R_{CELL}}{m-1}$$
$$\approx \frac{2R_{CELL}}{m}\left(1 + \frac{R_{rev}/R_{CELL}}{2m}\right) \quad (m = n \gg 1)$$
$$\approx \frac{3R_{CELL}}{m} \ll R_{CELL} \qquad \left(\frac{R_{rev}}{R_{cell}} \sim m\right) \tag{5}$$

978-1-7281-0943-5/19 $31.00 © 2019 IEEE

Since the size of CBA is large ($m \gg 1$), sneak current is much larger than the cell current. Cell current prediction by $I_{cell} = I_{total} - I_{sneak} = I_{total} - V/R_{sneak}$ may be inaccurate, because the prediction of a small cell current by subtracting a large sneak current from a large bit line total current is very difficult.

III. COMPENSATION SCHEME OF SNEAK CURRENT

A. Data Storage Patterns Dependency

According to equation (3), R_1 and R_2 depend on DSPD, while R_3 does not. Our scheme is to measure R_1 and R_2 by biasing V_3 and V_4 at the same voltage V_M between V_1 (V_H) and V_2 (V_L) to inhibit the current that flows through R_3. By this bias application, the current I_1 measured at terminal V_4 corresponds to R_2 ($R_2 = (V_H - V_M)/I_1$), and the current I_2 measured at terminal V_3 corresponds to R_1 ($R_1 = (V_L - V_M)/I_2$). And R_3 can regarded as a known offset resistance.

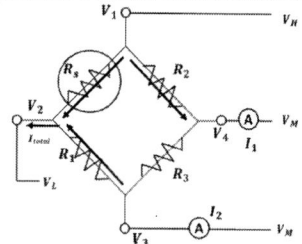

Fig. 3 Measurement bias condition for sneak current measurement

B. Reduction of Sneak Current for Accurate Prediction

In order to keep the subtraction $I_{cell} = I_{total} - I_{sneak} = I_{total} - V/R_{sneak}$ accurate enough, we only need to reduce n (m) to less than hundred so that the first (third) term in equation (3) dominates. And both I_{total} and I_{sneak} is limited within the two order larger than I_{cell} at the largest. We adopt the block selection to reduce the sneak current so that cell current can be accurately predicted. Fig. 4 shows an example of a 1Mb CBA that is divided into 64×16kb blocks.

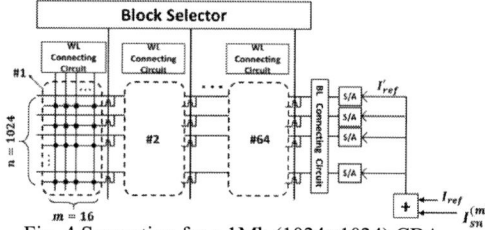

Fig. 4 Separation for a 1Mb (1024×1024) CBA

C. Compensation Circuit Design

We adopt a two-cycle read operation as shown in Fig. 5 to compensate the influence of the sneak current. In the first cycle, S_1 is turned on and S_2 is turned off. We measure the value of the sneak current as $I_{sn}^{(m)}$ by using the bias condition shown in Fig.3 and store the gate voltage V_{sn} of M_1 in a capacitor C; in the second cycle, S_1 is

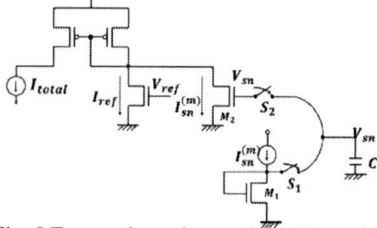

Fig. 5 Two-cycle read operation with sneak current compensation

turned off and S_2 is turned on so that V_{sn} is applied to the gate of M_2 which has the same size as M_1. The measurement sneak current is added to the reference current to compare with the total current to sense the real state of the cells.

IV. SIMULATION RESULTS

We verified our compensation method by using HSPICE simulation. The size of the block $m \times n$ is 16×1024 (16 cells on BL and 1024 cells on WL). Fig. 6 (a) and (b) show the compensation results by changing the data storage pattern for cells on the selected BL and selected WL. DSPD appears for the data storage pattern on BL, while it doesn't appear on WL. This is explained by eq. (3). They show that the compensated cell currents and various data storage patterns can be predicted which are closed to the actual cell current. And it also shows that the separated BL limits the sneak current within one order of magnitude larger than cell current. The results prove that our compensation scheme works.

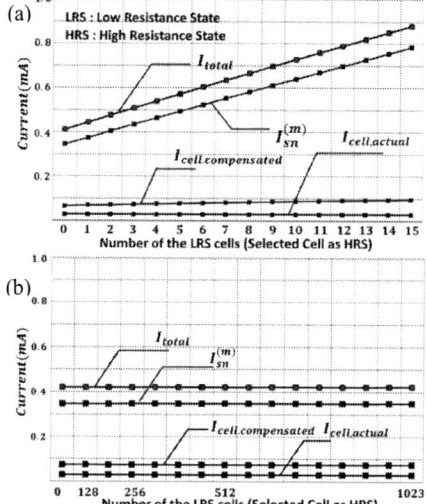

Fig. 6 Compensation results for different data storage patterns of cells changed along a selected BL (a) and WL (b).

V. CONCLUSION

We propose an accurate two-cycle sneak current compensation scheme applicable to the ReRAM CBA. This scheme can accurately predict the cell current in large size CBAs with data-storage-pattern-dependent sneak currents with minimal array size increment.

ACHNOWLEGDEMENT

This work was supported by VLSI Design and Education Center (VDEC), The University of Tokyo with collaboration with Cadence Corporation and Synopsys Corporation.

REFERENCE

[1] A. Kawaraha et al., "An 8Mb multi-layered cross-point ReRAM macro with 443 MB/s write throughput," *IEEE J. Solid-State Circuits*, vol.48, no.1, pp. 178-185, Jan. 2013.

[2] W. Bae, et al., "A crossbar resistance switching memory readout scheme with sneak current cancellation based on a two-port current-mode sensing," *Nanotechnology*, vol. 27, no. 48, 485201, Dec. 2016.

[3] M. A. Zidan, et al., "Memristor multiport readout: A closed-form solution for sneak paths," *IEEE Trans. Nanotechnology*, vol. 13, no. 2, pp. 274-283, 2014.

Selective Etching of Native Silicon Oxide in Preference to Silicon Oxide and Silicon

Christopher Ahles[†], Jong Choi[†], Raymond Hung[§], Namsung Kim[§], Srinivas Nemani[§] and Andrew Kummel[†,‡]

[†]Materials Science and Engineering Program and [‡]Department of Chemistry and Biochemistry,
University of California, San Diego, La Jolla, California 92093, United States
[§]Applied Materials, Sunnyvale, California 94085, United States
Email: cahles@ucsd.edu

ABSTRACT

An in-situ dry clean which removes native SiO_x and flowable oxide but does not etch the underlying silicon, thermal SiO_2 or SiN_x is reported. This process utilized a remote $NF_3/NH_3/Ar$ plasma, and the selectivity was studied as a function of temperature and time. Under the optimized conditions, the native SiO_x on Si was removed after ~15 seconds of plasma exposure whereas the etching of as-sputtered SiO_2 was zero within this time period. Selectivity on a nanometer scale was confirmed by TEM of a patterned Si wafer showing that the optimized dry clean removed flowable SiO_2 but does not etch Si and leaves SiN_x/thermal SiO_2 fins undamaged. Furthermore, this cleaning procedure was used to remove the native oxide on a SiGe-based patterned sample containing SiO_2/SiN_x fins in preparation for $MoSi_x$ atomic layer deposition (ALD). The selectivity between two types of silica relied on defective or weak Si-O bonds in native SiO_x compared to SiO_2.

INTRODUCTION

As devices are scaled to sub 5nm, it is critical to prepare clean and atomically flat surfaces. The traditional aqueous HF clean for removal of native Si oxide suffers from an inevitable air exposure resulting in re-oxidation of the Si surface as well as carbon contamination.[1] The Siconi™ process is a dry clean which utilizes a low temperature (<30°C) NF_3/NH_3 based plasma to selectively etch the native oxide layer on Si without significantly etching the underlying Si layer.[2] However, unlike aqueous HF, the Siconi™ process leaves behind an ammonium hexafluorosilicate salt, $(NH_4)_2SiF_6(s)$, which must be removed in a subsequent anneal. Furthermore, the selectivity of this process for various forms of SiO_2 is not known. Miki et al.[3] found that native Si oxide could be selectively etched with respect to various other silicon oxides using anhydrous HF(g). The selectivity was attributed to different oxides having different amounts of physisorbed H_2O, and this surface H_2O helped to dissociate HF(g) and promote etching. However, they found that the dry etching of native silicon oxide with HF(g) leaves the surface Si-F terminated, and this surface termination has detrimental effects on subsequent processing steps. In this work a process is reported which selectively etches native SiO_x and flowable SiO_2 in preference to Si, thermal SiO_2 and SiN_x. This process utilizes a downstream $NF_3/NH_3/Ar$ plasma which avoids the use of toxic anhydrous HF(g) and does not leave the surface Si-F terminated. The insulator selectivity is consistent with the contrast between weak bonding in native oxide and flowable oxide versus strong bonding in thermal SiO_2 and SiN_x.

RESULTS

The etching of native SiO_x and SiO_2 was studied in-situ using a pair of quartz-crystal microbalances (QCMs). A Si-sputtered quartz crystal containing native oxide and a SiO_2-sputtered quartz crystal were loaded on two different QCMs in the same chamber and subjected to the same plasma conditions ($NF_3:NH_3:Ar = 1:10:1.5$, chamber pressure of 190 mTorr, and a plasma power of 100W for 2 minutes at 45 °C; Figure 1). In this experiment, the samples were subjected to two consecutive plasma pulses separated by

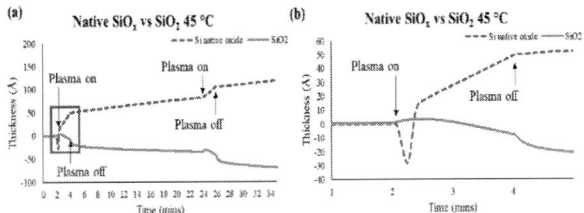

Figure 1. Thickness versus time for Si with native oxide and SiO_2 subjected to two consecutive $NF_3/NH_3/Ar$ plasma doses at 45 °C. (a) The thickness versus time data is shown for both consecutive plasmas separated by approximately 20 minutes. (b) An expansion of the region outlined by the red box in Fig. 1a is shown. This data showed that the first plasma removed the native SiO_x on Si and did not etch the underlying Si. The onset for etching of SiO_2 began at around 1 minute of plasma exposure. The process parameters were: $NF_3:NH_3:Ar = 1:10:1.5$ at a chamber pressure of 190 mTorr and a plasma power of 100W for 2 minutes.

approximately 20 minutes to observe the difference in Si etch rate with and without native oxide. The samples were not exposed to air between the first and second plasma pulses and, therefore, should not have reformed a native oxide (Fig. 1a). It was observed that the first 2-minute plasma rapidly etched the native oxide on Si while no etching of Si was observed during the second 2-minute plasma. As shown in Fig 1b, the native oxide on Si was rapidly etched during the first ~15 seconds of the plasma exposure, after which only deposition was observed.

This etching process was tested on crystalline Si (001) to determine the process parameters for selective native SiO_x etching vs crystalline Si (001). A Si coupon was degreased and loaded into the UHV chamber for XPS analysis (see Figure 2a). The degreased Si sample had 37% O and 8% C contamination. After the dry clean, all of the O was removed, and the sample surface consisted of 11% C, 43% F, 30% Si (of which 20% was Si^0 and 10% was oxidized Si) and 15% N. The XPS was consistent with a clean Si^0 surface covered with a layer of $(NH_4)_2SiF_6(s)$ salt. It is known that the $(NH_4)_2SiF_6(s)$ salt

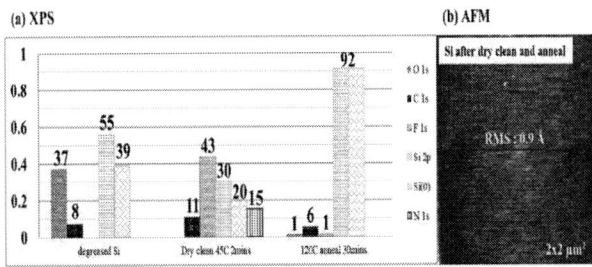

Figure 2. Chemical composition and surface topography of Si(001) subjected to the dry clean and anneal. (a) XPS of Si before and after the dry clean and a subsequent 120 °C anneal. The conditions for the dry clean were: $NF_3:NH_3:Ar = 1:10:1.5$ at a chamber pressure of 190 mTorr and a plasma power of 100W for 2 minutes at 45 °C. The 120 °C anneal was performed for 30 minutes. (b) AFM of the Si surface after the dry clean and anneal at 120 °C. The Si surface has an RMS roughness of 0.9 Å.

must be removed in a subsequent anneal step[4]; therefore, the sample was annealed at 120 °C for 30 minutes in the UHV chamber. After the anneal, XPS showed that the Si surface consisted of 92% Si (all of which was Si[0]) along with 1% O, 6% C and 1% F contamination (Figure 2a). The AFM images of dry cleaned and annealed Si sample shows that the Si surface had an RMS roughness of <1 Å (Fig. 2b).

To determine the selectivity on the nanoscale, the dry clean was performed on a nanoscale patterned sample (Figure 3). The patterned sample was a Si substrate with poly-Si fins coated with SiN_x on the top and sides, and thermal SiO_2 in between the poly-Si and Si substrate (schematic shown in Fig. 3a). The entire patterned sample was coated with a layer of flowable SiO_2. A TEM image of the patterned sample before any plasma treatment is shown in Fig. 3b. The patterned sample was subjected to two 30-second plasma pulses using the standard conditions and TEM was performed (Fig. 3c and 3d). It can be seen that the dry clean etched all of the flowable SiO_2 but did not etch the fins or the Si substrate. The collapse of three of the fins in Fig. 3c is believed to be due to mechanical damage during the sample cleaving process in preparation for TEM. Fig. 3d shows a higher magnification TEM of the region shown in Fig. 3c. Upon closer inspection it is seen that the fins remain intact and the thermal SiO_2 layer was not etched.

TEM – Patterned Sample – Before vs After Dry Clean

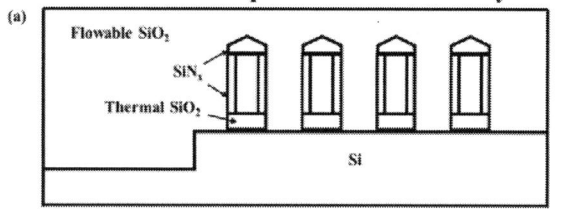

Figure 3. Selective Clean on NanoScale Patterned Sample. Two 30-second plasma pulses were employed using the standard conditions: NF_3:NH_3:Ar = 1:10:1.5 at a chamber pressure of 190 mTorr and a plasma power of 100W at 45 °C. **(a)** Schematic representation of the patterned sample. **(b)** TEM image at 13,000 x magnification of a patterned sample with no plasma treatment. **(c)** TEM image at 13,000 x magnification shows that the flowable oxide has been completely etched by the dry clean, while the fins and Si substrate remain unetched. **(d)** TEM image at 135,000 x magnification shows that the fins, including the thermal SiO_2 and SiN_x, were not etched.

This process was used to remove the native oxide from a patterned sample in preparation for atomic layer deposition (ALD) of a $MoSi_x$ film. $MoSi_x$ ALD is known to deposit selectively on Si but not SiO_2 or SiN, and it has been shown that when aqueous HF is used to remove the native SiO_x, there always exists an ~2.8 nm thick interfacial oxide layer between the Si and ALD $MoSi_x$.[5] A cross sectional STEM EELS study after ALD of a $MoSi_x$ layer on the dry cleaned pattern sample shows that the native oxide was removed while leaving the Si, SiN and SiO_2 not etched (Figure 4a). EELS elemental mapping shows that the $MoSi_x$ deposited selectively on the SiGe substrate, showing that the dry clean did not negatively affect the inherent selectivity of this process. Figure 4c shows an overlay of the elemental mapping. It can be seen that there is no oxygen at the

STEM/EELS – Dry cleaned patterned sample with MoSi$_x$ ALD

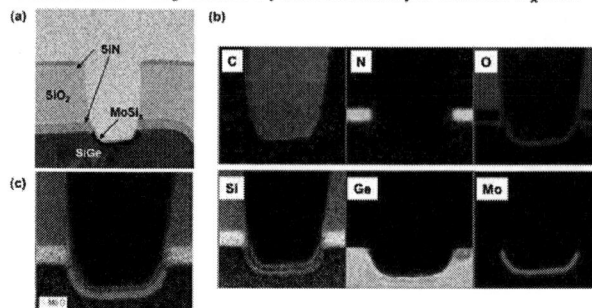

Figure 4. TEM images of a patterned sample subjected to a 1-minute plasma clean followed by MoSi$_x$ ALD. The dry clean was performed under the standard conditions: NF_3:NH_3:Ar = 1:10:1.5 at a chamber pressure of 190 mTorr and a plasma power of 100W at 45 °C for 1 minute. **(a)** TEM shows the structure of the patterned sample. **(b)** Elemental mapping showing the distribution of C, N, O, Si, Ge and Mo. **(c)** An overlay of the Si, Mo and O distribution shows that there is no interfacial oxide in between the SiGe and $MoSi_x$.

$MoSi_x$/SiGe interface, showing that the dry clean is superior to the traditional aqueous HF clean in that it does not lead to an interfacial oxide layer in this ALD process.

CONCLUSION

In conclusion, an in-situ dry clean has been developed which removed the native oxide from silicon and etched flowable SiO_2 but did not etch the underlying Si, bulk SiO_2 or SiN_x. It was found that careful control of the temperature was crucial in order to control the selectivity, and at 45 °C the native SiO_x on Si was removed with no etching of the underlying Si, and no etching of sputtered SiO_2 in this time period. XPS showed that the dry clean produced a very clean Si surface with only 6% carbon, <1% oxygen and <1% fluorine contamination. AFM showed that the dry-cleaned Si surface was atomically flat with an RMS roughness of ~1 Å. TEM images showed that the dry clean did not damage thermal SiO_2 or SiN_x features, however flowable SiO_2 was rapidly etched under these same conditions. This shows that this plasma process may be used to selectively etch flowable SiO_2 in the presence of Si, thermal SiO_2 and SiN_x. TEM and EELS measurements showed that the dry clean produced a cleaner Si surface than ex-situ HF(aq) because it eliminated the interfacial SiO_x layer in between Si and ALD $MoSi_x$. This showed that this dry clean should find applications in the preparation of patterned Si samples for selective ALD.

ACKNOWLEDGEMENTS

This work was supported in part by Semiconductor Research Corporation (SRC).

REFERENCES

1. J. Lei et al., 2006 IEEE International Symposium on Semiconductor Manufacturing, Tokyo, 2006, pp. 393-396.
2. Tang et al., US 8,501,629 B2.
3. Miki et al., IEEE TRANSACTIONS ON ELECTRON DEVICES. VOL. 37. NO. I, JANUARY 1990.
4. Nishino et al., J. Appl. Phys. 74 (2), 15 July 1993.
5. J. Choi et al., Applied Surface Science 462 (2018) 1008-1016.

Enhancing IGZO Thin Film Transistor Scalability Through Tunneling Contact

Zichao Ma[1], Xintong Zhang[1], Clarissa Prawoto[1], Lining Zhang[2],
Longyan Wang[3] and Mansun Chan[1]

[1]The Hong Kong University of Science and Technology, Hong Kong
[2]College of Electronic Science and Technology, Shenzhen University, Shenzhen, 518060, P. R. China
[3]Visionox, Beijing, China
E-mail: zmaaa@connect.ust.hk

ABSTRACT

By inserting a single layer of graphene in between the metal to In-Ga-Zn-O (IGZO) contact, the channel carrier injection mechanism of an IGZO Thin-Film-Transistor (TFT) can be converted from thermionic emission to tunneling. Experimental results show that the tunneling-TFT can be scaled to a small dimension without compromising leakage current. A $2\mu m$ tunneling-TFT shows early saturation and similar subthreshold slope as the long channel devices, while the conventional TFTs can hardly be switched off. Further reduction of the IGZO film thickness can better suppress the short channel effect of the TC-TFTs.

INTRODUCTION

Amorphous metal-oxide thin-film transistors (TFT) based on indium-gallium-zinc oxide (IGZO) are commonly used in the driving circuits of advanced flat-panel displays. High carrier mobility and low leakage current in IGZO TFTs result in superior performance than amorphous silicon and organic based counterparts [1,2]. To maximize the fill factor in an LED pixel, it is desirable to minimize the area of the driver circuits, while maximizing the light emitting area. This drives the necessity for IGZO TFT scaling. However, it has been reported that short channel effects start to dominate in normal IGZO TFTs at a channel length of ~3μm, making the scaling of IGZO TFTs challenging [3-5]. In our earlier work, we have demonstrated that tunneling contact TFT (TC-TFT) [6] can be used to improve the on-off current ratio and the steepness of the subthreshold slope. In this work, we show advantages in the scaling of TC-TFTs compared to normal-TFTs (n-TFT).

FABRICATION AND OPERATION OF TC-TFT

Fig. 1 shows the device structure of the proposed TC-TFT. Indium-tin oxide (ITO) is used as the bottom-gate electrode on a thermally oxidized silicon wafer substrate. Gate dielectric of 100 nm SiO_2 is deposited by at 300°C. IGZO active layer with thickness of 20 nm is sputter-deposited at a molar ratio of $In_2O_3:Ga_2O_3:ZnO = 1:1:1$. This IGZO film is then patterned and etched to define the channel region. Graphene is then transferred on the substrate for TC-TFT, while this is not performed for the normal TFT. Source and drain regions are metalized with a stack of 20 nm titanium and 30 nm gold. All the structures are patterned by i-line photolithography with a resolution of 1 micrometer. A 300 nm SiO_x passivation is subsequently deposited. The devices are annealed at 300°C for 1 hour in oxygen ambient for thermal "activation" [2]. All TFTs have a channel width of 5 μm.

The operating mechanism of TC-TFT is explained in Fig. 2. The channel region is defined as the area between source and drain, represented by a variable resistor controlled by the gate and drain voltages. In the source/drain junction, instead of an ohmic contact as in n-TFT, the TC-TFT has a Schottky barrier between the source/drain and the channel. Thus, the source contact in TC-TFT, that behaves like a reverse-biased diode in response to drain voltage (V_D), controls the current by quantum tunneling when the device is turned on. As a result, TC-TFT shows better immunity to V_D (Fig. 2(b)) compared to n-TFT that is resistively controlled by V_D (Fig. 2(a)).

PERFORMANCE OF SCALED TC-TFT

The scalability of TC-TFT compared to n-TFT is evaluated using long channel (7 μm) and short channel (2 μm) devices. The transfer curves in Fig. 3 shows subthreshold slope (SS) is degraded in n-TFT with shorter channel length. On the other hand, no significant degradation of SS is observed in TC-TFT. Moreover, the output performance in Fig. 4 suggests the short channel n-TFT devices fail to saturate. In contrast, TC-TFT of the same channel length has already reached saturation, showing a good immunity of TC-TFT to short channel effect in a general view. Extracting the subthreshold slope in Fig. 3, we quantify the degradation of SS with respect to scaling in Fig. 5. The TC-TFT shows maintained SS at 2-μm channel length, whereas the SS degrades by more than 2 times in n-TFT, which indicates an advantage in scaling of the TC-TFT.

To investigate the scaling performance following a constant field scaling rule, we compare TC-TFTs and n-TFTs with gate oxide of 300 nm and IGZO thickness ranging from 40 nm to 65 nm. The channel length of the devices is kept at 2 μm, which equivalent to a 0.66 μm channel length at 100 nm gate oxide thickness and 20 nm IGZO film thickness. The transfer curves TC-TFT with different IGZO layer thicknesses are plotted in Fig. 6. It is shown that TC-TFT with 65 nm IGZO film, maintained a large on-off current ratio at around 10^5. As a comparison, the transfer curve of n-TFT is also plotted by the dashed line in Fig. 6, which shows performance degradation due to the absence of a tunneling contact. In addition, further reducing IGZO film thickness can effectively reduce the subthreshold slope and off-state current. As shown in Fig. 6, by reducing the channel thickness to 2/3 of its initial value, the SS is significantly reduced and on-off current ratio is increased by one order of magnitude.

978-1-7281-0943-5/19 $31.00 © 2019 IEEE

CONCLUSION

We have shown with tunneling contact, thin-film transistors (TFT) can be scaled down to 2 μm, while maintaining a large on-off current ratio and steep subthreshold slope. In contrast to normal TFT, the Schottky barrier contact at source region reduces the dependence of source current to the drain voltage. This suggests promising scalability of TC-TFTs towards sub-micrometer channel length

ACKNOWLEDGEMENT

This work is supported by an AoE fund from the Research Grant Council of Hong Kong under project number AoE/P-04/08-1. The fabrication work was done at the Nanosystem fabrication facility with equipment supported by various funds such as CRF C6007-15E.

REFERENCES

[1] J. H. Choi, J. Pi, et al., "Toward sub-micron oxide thin-film transistors for digital holography", Journal of the Society for Information Display, vol. 25, no. 2, pp. 126-135, Feb. 2017.

[2] L. Lu, J. Li, et al, "High-performance and reliable elevated-metal metal-oxide thin-film transistor for high-resolution displays", In Electron Devices Meeting (IEDM), IEEE, pp. 32, Dec. 2016.

[3] P. G. Bahubalindruni, A. Kiazadeh, et al., "Influence of Channel Length Scaling on InGaZnO TFTs Characteristics: Unity Current-Gain Cutoff Frequency, Intrinsic Voltage-Gain, and On-Resistance", Journal of Display Technology, vol. 12, no. 6, 2016.

[4] C. Yang, T. Chang, et al., "Drain-Induced-Barrier-Lowing-Like Effect Induced by Oxygen-Vacancy in Scaling-Down via-Contact Type Amorphous InGaZnO Thin-Film Transistors", Journal of Electron Devices Society, vol. 6, 2018.

[5] S. Lee, Y. Song, H. Park, et al., "Channel scaling and field-effect mobility extraction in amorphous InZnO thin film transistors", Solid-State Electronics, vol. 135, pp. 94-99, 2017.

[6] L. Wang, Y. Sun, et al., "Tunneling contact IGZO TFTs with reduced saturation voltages", Applied Physics Letters 110, 152105, 2017.

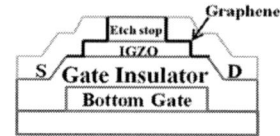

Fig. 1 device structures of tunneling contact TFT

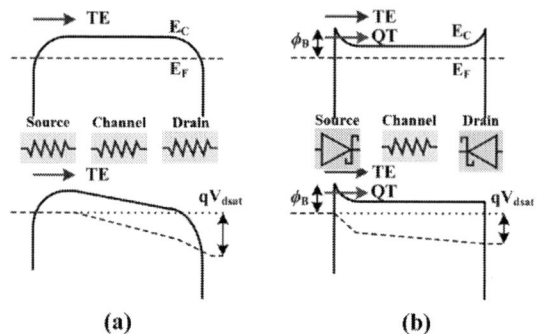

Fig. 2 Operation mechanisms of normal TFT(a) and tunneling contact TFT (b)

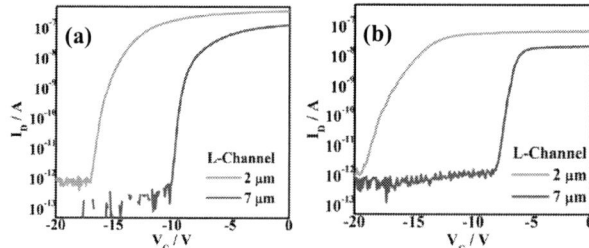

Fig. 3. Transfer performance of tunneling contact TFT (a) and normal TFT (b).

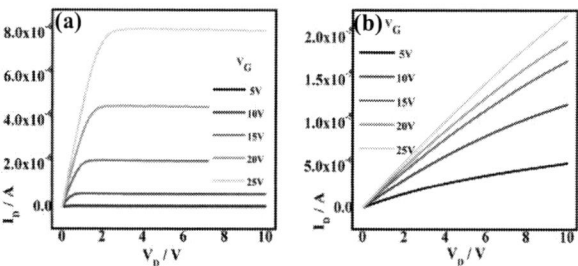

Fig. 4. The output characteristics of tunneling contact TFT (a) and normal TFT (b) with the channel length of 2 μm.

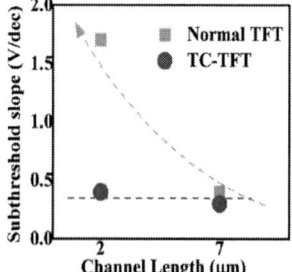

Fig. 5 Subthreshold slope of tunneling contact TFT and normal TFT with 2 μm and 7 μm channel length.

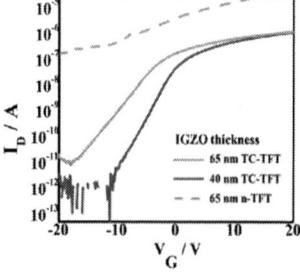

Fig. 6 Transfer characteristics of 2 μm channel length tunneling contact TFT and normal TFT with different IGZO film thickness.

Cell Array Design with Row-Driven Source Line in Block Shunt Architecture Applicable to Future 6F^2 1T1MTJ Memory

Tongshuang Huang, Takashi Ohsawa

Graduate School of Information, Production and Systems, Waseda University, Japan

E-mail: tylerhuang@akane.waseda.jp, takashi.ohsawa@aoni.waseda.jp

ABSTRACT

In this paper, we propose a new 1T1MTJ cell array architecture with SL parallel to WL to achieve a small cell size, in which page mode write can be realized without performance degradation. We propose a row-driven source line (RSL) 1T1MTJ memory cell array architecture for minimizing the cell size and a corresponding operational waveform. A block shunt architecture (BSA) that shunts lower source line (LSL) and upper source line (USL) is proposed to make page mode write possible. Size of 1T1MTJ cell can be shrunk to 6F^2 when the state-of-the-art design rules are applied.

INTRODUCTION

In conventional 1T1MTJ cell array architecture, source lines (SLs) are placed perpendicular to the word lines (WLs). This causes cell size to be large, because the SLs must run next to bit lines (BLs) instead of above the BLs. To solve this problem, architectures that run SL parallel to WL have been proposed [1]. However, disturb problem of writing remains unsolved. One architecture can avoid this problem but cannot be applied to a page mode write.

In this paper, we propose a new 1T1MTJ cell array architecture with SL parallel to WL to minimize the cell size. By adopting a new write operation and a new SL shunt scheme, disturb problem is avoided with the page mode write. It is shown that this new 1T1MTJ cell array architecture makes a 6F^2 STT-MTJ cell feasible.

ROW-DRIVEN SOURCE LINE ARRAY AND OPERATIONAL WAVEFORM

Fig.1 (a) shows the conventional architecture of 1T1MTJ cell array. A SL runs perpendicular to WLs and makes a pair with a BL. Referring to Fig.1 (b), we see that SL (M1) runs next to BL (M2) to avoid the M1 buffer layer between MTJ and BL. This makes the cell size large.

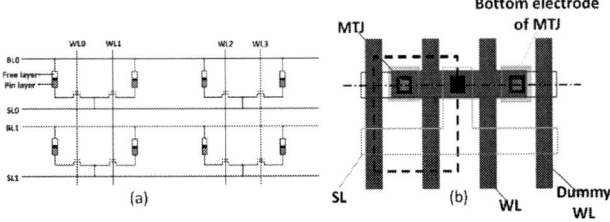

Figure 1 (a) Equivalent circuit and (b) layout of the conventional 1T1MTJ cell array

Fig.2 (a) is an equivalent circuit of the cell array architecture we propose named row-driven source line (RSL) architecture, for the SL is driven by a row driver instead of a column driver. Fig.2 (b) is the corresponding layout design. The individual cell size shrinks drastically, because BLs can run in a minimum pitch. One SL is no longer corresponding to one BL anymore. On the contrary, one SL can connect to all cells along one WL. In order to write data into these cells, the source line voltage should be driven low and high in one write cycle so that the shared SL can flow current to BL in different directions, ensuring that MTJ can be switched correctly in two phases as shown in Fig.2 (c).

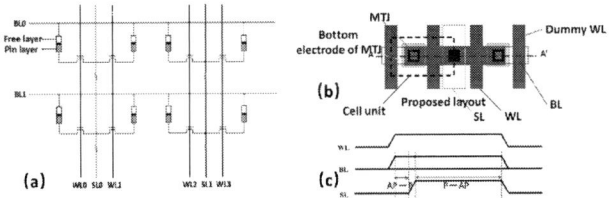

Figure 2 (a) Equivalent circuit, (b) layout of proposed 1T1MTJ cell array and (c) operational waveform

BLOCK SHUNT ARCHITECTURE

In the proposed RSL array, one SL is shared by a large number of cells. Since large current flows through one SL in writing operation, the parasitic resistance cannot be neglected. A large voltage drop along SL will induce writing errors for the cells far from the driver of SL. Writing errors occur when the current cannot reach the critical current of MTJ to switch. To address this issue, block shunt architecture (BSA) is proposed. Fig.3 shows the structure of BSA. The array is divided into four corners. Every corner consists of 256 lower source lines (LSLs) and 512 BLs. The LSL is the same with the conventional SL made by M1 layer. However, to mitigate the impact on parasitic resistance, we divide the source line into several parts. Different from the conventional hierarchical cell array design, the LSLs will not be driven by specific local source line drivers, but just shunted to upper source lines (USLs). USLs are made by M3 above BL (M2) and the width of USL is designed to be the maximum width of that metal layer. In Fig .3, one USL is directly connected to four LSLs, and the width of USL is 3 μm in 90nm technology. Due to the large value of the width, the resistance of USL is decreased considerably. In Fig.3, the LSL is divided into two parts and these parts shunt to the USL. This hierarchical structure electrically shortens the LSL length to half of the previous design and eases the severe situation of parasitic resistance of LSL. In Fig.3, M2 buffer layers used are placed in the middle of every part. And these connections serve to shunt four LSLs to one USL. To clarify BSA more explicitly, the layout design of BSA and cross-section views are shown in Fig.4.

978-1-7281-0943-5/19 $31.00 © 2019 IEEE

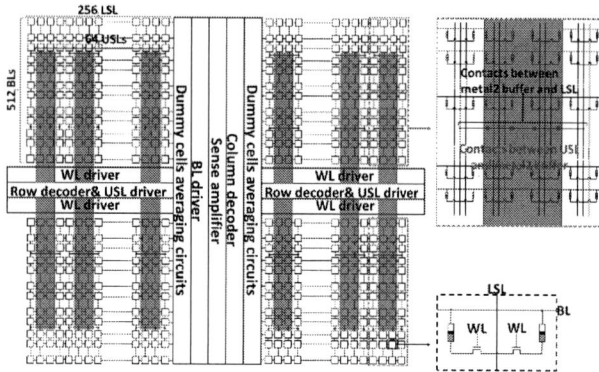

Figure 3 Block shunt architecture in two divisions of 1Mb 1T1MTJ array

The cross-section view of AA' and BB' are shown in Fig.4 (b) and (c). As we can see in Fig. 4(c), BB' cross-section view shows the connection between LSLs and buffer M2 and USL. Four LSLs (M1) connection to one USL (M3) is the most appropriate architecture in 90nm technology, because the width of M2 exceeds the maximum width of M3 (3 μm) when eight LSLs are connected to one USL. So only four LSLs are connected to one USL.

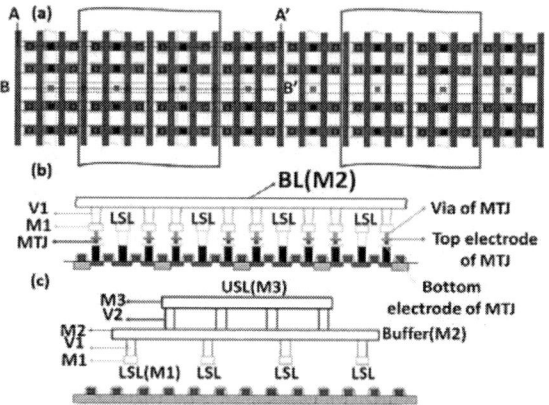

Figure 4 (a) Block shunt architecture layout design (b) AA' cross-section view and (c) BB' cross-section view

SIMULATION RESULTS

Since this design is based on 90 nm MOSFET design rules matched with 100 nm MTJ, the parameters for MTJ are: R_P=1kΩ, TMR =100%, critical current I_{P-AP}= 400μA, critical current I_{AP-P} = 200μA [2]. We use 1.8V as the supply voltage of this cell array. In Fig. 5, the number of divisions in this architecture is eight. The entire simulation shows two 75 ns writing cycles. Fig. 5 (a) shows the current flowing through MTJs which are the most far from the USL drivers and BL drivers. In other words, these cells are suffering the worst situation of parasitic resistance and capacitance among all cells in this cell array. All cells switch successfully. In Fig. 5(b), the MTJ resistance rises from R_P state to R_{AP} state (The medium resistance level is R_{AP} state of MTJ under large voltage bias, because R_{AP} value of MTJ model changes under different voltage). And in the second write cycle, MTJs switch from AP state to P state. We can draw the conclusion that all cells can be switched successfully in this case. By applying more divisions into BSA, the performance can improve further. Fig. 6 shows the improved performance versus array area increase. According to this graph, the performance will be improved sharply when the number of divisions is small. Yet, as the number of divisions increases, the performance improvement is saturated. Since the shunt design requires no local drivers for hierarchical design, the dead space of this array is small.

Meanwhile, the increasing area of BSA owing to increase in the number of divisions is almost linear. Considering all factors, eight divisions architecture is selected as the most appropriate number of divisions of this array.

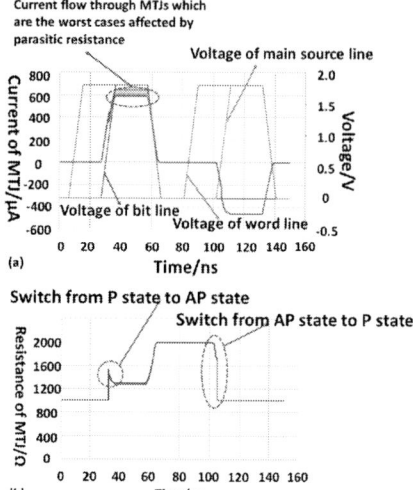

Figure 5 (a) Simulation results of writing operation of block shunt architecture with eight divisions, (b) resistance change of MTJ in simulation

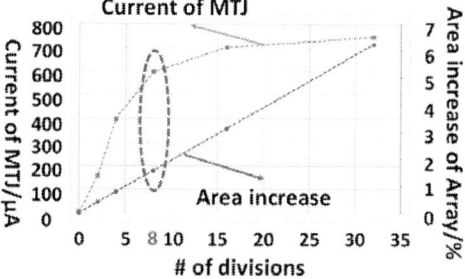

Figure 6. Trade-off between the number of divisions and array area increase

CONCLUSION

A row-driven source line array architecture is proposed with a new source line operational waveform. To address the voltage drop and current crowding issues brought by the proposed row-driven design, we propose block shunt architecture to make the source line hierarchical. By considering the trade-off between performance and chip size, the number of divisions is optimized to eight in 90nm technology node.

ACKNOWLEDGEMENT

This work was supported by VLSI Design and Education Center (VDEC), The University of Tokyo with collaboration with Cadence Corporation and Synopsys Corporation.

REFERENCES

[1] DeBrosse, John K. "STT MRAM source line configuration." U.S. Patent No. 9,875,780, 23 Jan. 2018

[2] T. Ohsawa et al. , "High-density and low-power nonvolatile static random access memory using spin-transfer-torque magnetic tunnel junction." *Jpn. J. Appl. Phys.,* 51 (2012) 02BD01.

978-1-7281-0943-5/19 $31.00 © 2019 IEEE

Evaluation of 2D Negative-Capacitance FETs for Low-Voltage SRAM Applications

Kuei-Yang Tseng, Wei-Xiang You, and Pin Su

Department of Electronics Engineering & Institute of Electronics, National Chiao Tung University, Taiwan

E-mail: crayon.pn02@nctu.edu.tw, pinsu@faculty.nctu.edu.tw

ABSTRACT

In this work, we comprehensively evaluate and analyze the stability and performance of 6T SRAM cells using 2D MFIS-type negative capacitance FETs (2D-NCFETs) based on the IRDS 2030 node with 10-nm gate length. Our results indicate that 2D-NCFETs possess better RSNM than the 2D-FET counterpart under low supply voltages. Our study also shows that 2D-NCFETs have better WSNM except for V_{DD} = 0.2V due to the existence of hysteresis loop in write curve during write operation. By using write-assist circuits or back-gating techniques, we demonstrate that the WSNM of 2D-NCFETs can be significantly improved. We further analyze the performance of read and write operations, and 2D-NCFETs have been found to possess better performance than 2D-FETs.

INTRODUCTION

For future need of low power applications such as portable devices and Internet-of-Things, it is crucial to minimize the power consumption. Negative-capacitance FET (NCFET) [1] is one of the most promising steep-slope devices that can enable low supply voltage and low power consumption. Studies have shown that transition metal dichalcogenide (TMD) field-effect transistors (2D-FET) exhibit superior immunity to short-channel effects that are very attractive for future extremely-dense low-voltage SRAM arrays [2] [3]. Combining 2D TMD channel and NCFET is especially interesting [4] [5]. In this work, through SPICE simulation, we investigate the stability, performance and the impacts of NC effect for 6T MFIS-type 2D-NCFET SRAM cells based on IRDS 2030 technology node.

METHODOLOGY

The schematic of the 2D-NCFET used in this study is shown in Fig. 1. The ferroelectric layer is placed on top of the insulator forming an MFIS-type 2D-NCFET structure, which is technologically more viable. In this work, 2D-NCFETs and 2D-FETs are designed to achieve same OFF-current (I_{OFF}) of 10 pA/μm as shown in Fig. 2. The channel length used in this work is 10 nm based on the IRDS 2030 node. Fig. 2 shows significant improvement in the subthreshold slope for the 2D-NCFET with thin BOX structure. Pertinent parameters of the HfO_2-based ferroelectric and channel materials of monolayer MoS_2-n and WSe_2-p are listed in Table 1 and Table 2.

To evaluate the stability and performance of the 2D-NCFET based SRAM cell, an accurate short-channel MFIS-type 2D-NCFET model is used [6] [7]. We extract model parameters for the baseline 2D-FET from TCAD numerical simulation. To capture the impact of drain coupling on the NC effect, the short-channel effects are considered through physical modeling of the subthreshold swing (SS) and V_T shift [8] based on a scalable short-channel polarization charge model in [6].

RESULTS AND DISCUSSION

Fig. 3 compares the read static noise margin (RSNM) for 6T 2D-NCFET and 2D-FET MoS_2-n/WSe_2-p SRAM cells at various supply voltages (V_{DD}). It is shown that, for a given V_{DD}, 2D-NCFETs possess better RSNM than 2D-FETs. In addition, the RSNM of 2D-NCFETs saturates at higher V_{DD}. This results from the lower current ratio of pull-down transistor to pass-gate transistor due to the NC effect at ON-state. Therefore, the read disturb voltage ($V_{read,0}$) keeps increasing with V_{DD} (Fig. 6 (a)) and there is no further improvement with higher supply voltage for the 2D-NCFET.

Fig. 4 compares the write static noise margin (WSNM) for 6T 2D-NCFET and 2D-FET SRAM cells. The WSNM for 2D-NCFETs is in general larger than the 2D-FET. Due to the negative differential resistance effect in the saturation region (Fig. 4 inset) and boosted drive current in the linear region, pull-up transistors become weaker and pass-gate transistors become stronger. Thus, 2D-NCFETs exhibit lower write disturb voltage ($V_{write,0}$) as shown in Fig. 6 (b) However, the hysteresis loop occurs in write curve at V_{DD} = 0.2V as shown in Fig. 5, making it not favorable for write operation.

The WSNM at low V_{DD} can be improved by using write-assist circuit techniques, such as negative bit-line and boosted word-line. Fig. 7 shows the boosted word-line voltage can increase the strength of the pass-gate transistor and the hysteresis loop can be mitigated. Thus at V_{DD} = 0.2V, the WSNM of 2D-NCFETs can be improved from 5.8mV to 110mV. In addition, it can be clearly seen that the technique is more effective for 2D-NCFETs as compared with the 2D-FET. This can be explained by the fact that for a given ΔV, the change in current for 2D-NCFETs is larger due to its steep subthreshold slope. Fig. 8 shows that back-gating techniques can also be used to improve the WSNM at V_{DD} = 0.2V for 2D-NCFETs. Due to the negative body-effect coefficient in 2D-NCFETs [9], a negative back-gate bias needs to be applied to reduce the V_T of the pass-gate transistor. The impact of hysteresis loop during write operation can be suppressed in a similar way to using write assist circuits. The WSNM of 2D-NCFETs can become better than the control cell under back-gating.

To investigate the performance of 6T SRAM cells with 2D-NCFETs, a capacitive load is added onto each bit-line to account for the capacitance of wires and the connected devices. Fig. 9 shows the cell read access time and time-to-write at various V_{DD}. The cell read access time is mainly determined by the strength of the read current through pass-gate and pull-down transistors. For read operation, with higher current in 2D-NCFETs, the cell access time can be improved more than one order of magnitude as compared with the 2D-FET. For write operation, even though 2D-NCFETs have larger gate capacitance due to voltage gain, the substantial improvement in the drive current still benefits and dominates the write performance compared with the 2D-FET counterpart.

REFERENCES

[1] S. Salahuddin and S. Datta, *Nano Lett.* 8, p. 405 2008.
[2] C.-H. Yu *et al.*, *IEEE TED*, vol. 63, no. 2, p.625, 2016.
[3] C.-H. Yu *et al.*, *IEEE TED*, vol. 64, no. 5, p. 2445, 2017.
[4] M. Si *et al.*, *Nature Nanotechnol.*, vol.13, p.24, Dec. 2017.
[5] A. Nourbakhsh *et al.*, *Nanoscale*, vol.9, no.18, p.6122, 2017.
[6] W. X. You *et al.*, *IEEE TED*, vol. 65, no. 4, p. 1604, 2018.
[7] W. X. You and P. Su, *IEEE TED*, vol. 65, no. 10, p. 4196, 2018.
[8] W. X. You *et al.*, *IEEE S3S Conf.*, San Francisco, Oct. 2018.
[9] W. X. You and P. Su, *IEEE TED*, vol. 64, no. 8, p.3476, 2017.
[10] M. Kobayashi and T. Hiramoto, *VLSI Symp.*, June 2015.
[11] P. Su and W. X. You, *IEEE S3S Conf.*, San Francisco, Oct. 2018.

Acknowledgements- The support from the Ministry of Science and Technology, Taiwan, under MOST-107-2221-E-009-090-MY2, MOST-107-3017-F-009-002 and MOST-107-2633-E-009-003.

Table 1. Pertinent parameters of the HfO_2-based ferroelectric layer [10] and the gate stack. The Al_2O_3 is used for the gate oxide.

Parameters	Monolayer WSe₂	Monolayer MoS₂
Band Gap [eV]	1.64	1.8
Electron Affinity [eV]	3.53	4.28
Dielectric Constant [e_0]	4.5	4.8
Effective Mass [m_0]	m_e=0.345 m_h=0.344	m_e=0.573 m_h=0.659

Table 2. Pertinent material parameters of monolayer MoS_2-n and WSe_2-p used in this work [2] [3].

α [m/F]	$-4.23e9$
β [m^5/FC^2]	$4.07e12$
γ [m^9/FC^4]	0
EOT [nm]	0.65
T_{FE} [nm]	3.4

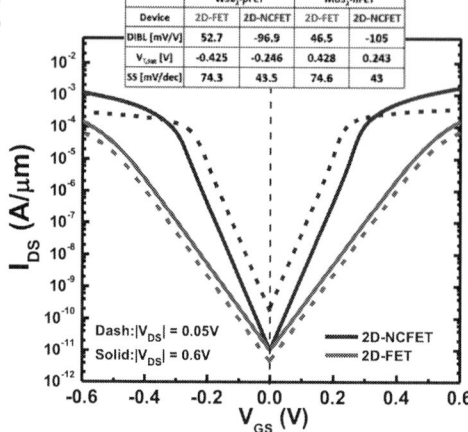

Fig. 2. Hysteresis-free I_{DS}-V_{GS} characteristics of 2D-NCFETs and 2D-FETs with equal $I_{OFF,Sat}$ at $V_{DS} = V_{DD}$. The table shows the electrical parameters of devices. The mobilities of monolayer MoS_2 and WSe_2 used in this work are 45 cm²/V-s and 38 cm²/V-s, respectively [2].

Device	WSe₂-pFET 2D-FET	WSe₂-pFET 2D-NCFET	MoS₂-nFET 2D-FET	MoS₂-nFET 2D-NCFET
DIBL [mV/V]	52.7	-96.9	46.5	-105
V_{Cite} [V]	-0.425	-0.246	0.428	0.243
SS [mV/dec]	74.3	43.5	74.6	43

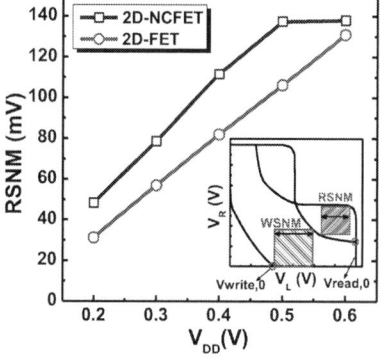

Fig. 1. Schematic of the 2D-NCFET (MoS_2-n and WSe_2-p) in this work. The BOX thickness is 5 nm. A thin BOX is crucial to the 2D-NCFET [11].

Fig. 3. The 2D-NCFET possesses larger read static noise margin (RSNM) of 6T SRAM cell than the 2D-FET counterpart under low supply voltages.

Fig. 4. Write static noise margin (WSNM) comparison of 6T SRAM for 2D-NCFET and 2D-FET. The inset shows the intersection of two load lines for MoS_2-n and WSe_2-p at V_{DD} = 0.2V.

Fig. 5. At V_{DD} = 0.2V, 2D-NCFET 6T SRAM cell exhibits a hysteresis loop in write curve.

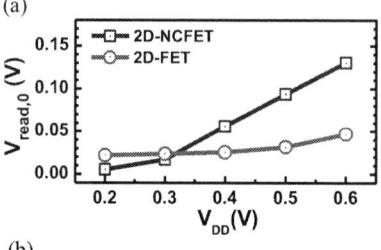

(a)

(b)

Fig. 6 . (a) The read disturb voltage of 2D-NCFETs increases with V_{DD} due to the increasing current ratio of pass-gate to pull-down devices induced by the NC effect. (b) At V_{DD} = 0.2V, a larger write disturb voltage for the 2D-NCFET can be seen.

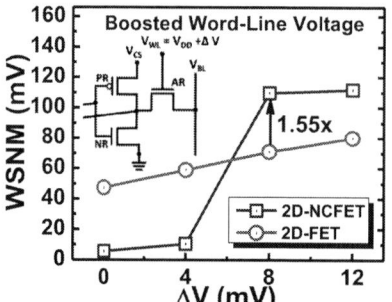

Fig. 7. The boosted word-line technique can be used to increase the WSNM for the 2D-NCFET at V_{DD} = 0.2V.

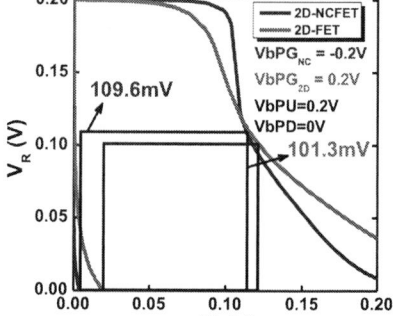

Fig. 8. Back-gating can also be used to improve the WSNM at V_{DD} = 0.2V. Note that a negative back-gate bias is needed for the 2D-NCFET case.

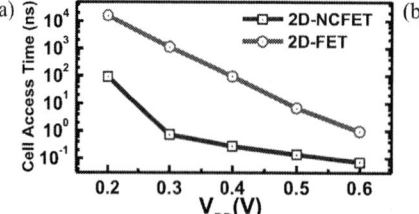

(a)

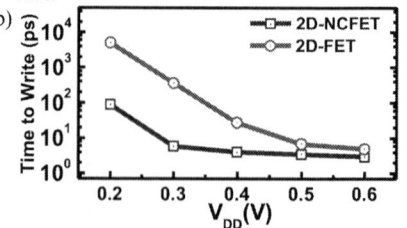

(b)

Fig. 9. Comparison of (a) cell read access time and (b) cell time-to-write for 2D-NCFETs and 2D-FETs under various V_{DD}. SRAM cells with 2D-NCFETs exhibit better performance than 2D-FETs.

978-1-7281-0943-5/19 $31.00 © 2019 IEEE

Analysis of Transient Effect on Super-Steep SS PN-Body Tied SOI-FET

Takayuki Mori, Jiro Ida, Hiroki Endo

Division of Electrical Engineering, Kanazawa Institute of Technology, Japan
E-mail: t_mori@neptune.kanazawa-it.ac.jp

Abstract

In this paper, transient effects of the super-steep subthreshold slope (SS) "PN-body tied (PNBT) silicon-on-insulator field-effect transistor (SOI-FET)" were investigated by using technology computer-aided design simulation. It was founded that the PNBT SOI-FET has switching delay time due to the carrier charging and discharging affected by the Shockley–Read–Hall (SRH) recombination on the W_b region (Base of Bipolar). It is a specific mechanism of the PNBT SOI-FET, which results in the delay time from OFF to ON state and the long time of the leakage current from ON to OFF state.

Introduction

Steep subthreshold slope (SS) devices (< 60 mV/dec), such as tunnel FETs and negative capacitance FETs have been researched for ultralow power CMOS. Instead of those, we have proposed the PN-body tied (PNBT) silicon-on-insulator field-effect transistor (SOI-FET) which has a super-steep SS (<1 mV/dec) with low drain voltage V_d (0.1 V) [1]–[3]. We consider that the super-steep SS is induced by SOI floating body effects (FBEs). Additionally, the PNBT structure has an inherent thyristor and it plays key role of inducing FBEs [2]. Therefore, we expect that the PNBT SOI-FET has switching delay time due to the carrier charging effect and/or the carrier lifetime (the Shockley–Read–Hall (SRH) recombination). This is an important issue for the high-speed operation.

Recently, we have confirmed the switching delay time in the measurement results of the PNBT SOI-FET [4]. However, the analysis of the mechanism is insufficient. In this study, we investigated the transient characteristics of the PNBT SOI-FET in more detail by using technology computer-aided design (TCAD) simulation.

Simulation Condition and Results

Fig. 1 (a) shows a bird-eye view of the PNBT SOI-FET made by 3D device simulator "HyENEXSS" [5]. We used the SRH recombination, auger recombination, and band-to-band tunneling model. The carrier lifetime parameters in the SRH model were modified (default lifetime × 0.02) to fit the measurement results [3]. The device parameters are shown in Table I.

Fig. 2 shows the simulated DC drain current I_d and the body current I_b vs. the gate voltage V_g characteristics. Super-steep SS characteristics appear at $V_g = 0.19$ V. It is noted that the hysteresis width is small and I_b is lower than I_d. We consider that the inherent thyristor turns on by the positive feedback and provides carriers under the channel [2]. It induces SOI FBEs and the super-steep SS.

We tried the transient simulation of the PNBT-SOI-FET. V_g swept as shown in Fig. 3 for the transient analysis on the "OFF to ON state" and the "ON to OFF state". Fig. 4 shows the simulated transient I_d and I_b characteristics dependence on V_g. As the "OFF to ON state", I_d and I_b flow just after $V_g = V_{gON}$; however, after a while, those shift to the larger currents flowing mode. In other words, the device switches via two-step condition change from OFF to ON state. Especially, these large currents will determine the delay time of the circuits. However, the delay time on the start of the large currents decreases when V_g increases. It is a promising future for reducing the delay. It was also observed with the measurements [4]. As the "ON to OFF state", I_d and I_b decrease just after $V_g = 0$ V, and shows the long time of the leakage current, which was also observed with the measurements [4]. With all V_g condition, the approximately same currents flow at 100 μs, and it takes 1 s until the complete OFF state is reached. It is noted that I_d once increases after $V_g = 0$ V. The two-step condition change and the increasing I_d just after $V_g = 0$ V are confirmed only in this simulation results. It is considered that these characteristics had been hid in the measurement limit of the equipment.

Next, the carrier concentrations in the device are shown for the detailed transient analysis. Figs. 5 (a)–(d) show the transient I_d and I_b characteristics, the electron and hole concentration, and the SRH G-R rate when $V_{gON} = 0.4$ V on the "OFF to ON state". The electron conduction mode changes from surface to entire SOI region as shown in Fig. 5 (b). The holes also increase abruptly as shown in Fig. 5 (c). It seems to be SOI FBEs. In contrast, the SRH recombination rate on the W_b region (Base of Bipolar) gradually increases as shown in Fig. 5 (d). Just after ON state, the conventional subthreshold current flows between the source and the drain. Additionally, the electrons diffuse from the source to the W_b region and it induces the positive feedback same as a thyristor, which results in the large current. It is considered to be the two-step condition change. We consider that the origin of the delay time come from the hole arrival time from the body to the channel region, affected by the SRH recombination on the W_b region. When the SRH recombination rate reaches until some limit, some holes do not recombine on the W_b region and begin to diffuse from the body to the channel region. Figs. 6 (a)–(d) show the transient I_d and I_b characteristics, and the electron and hole concentration when $V_{gON} = 0.4$ V on the "ON to OFF state". There are only few electrons on the surface at $V_g = 0$ V point as shown in Fig. 6 (b). However, some holes still accumulate under the channel. The holes also exist the W_b region as shown in Fig. 6 (c). We assume that the holes recombine on the W_b region, even ejecting to the channel direction. Therefore, I_d once increases just after $V_g = 0$ V because the ejected holes to the channel region increase the SOI body potential. The remaining holes under the channel eject to the source; however, the ejection speed is slow as shown in Fig. 6 (d). The potential under the channel logarithmically decreases because of the effect which is known as the negative feedback on SOI body potential. This is same as a conventional partially depleted SOI-MOSFET [6].

Conclusion

We investigated the transient characteristics of the PNBT SOI-FET by using TCAD simulation. The device has the switching delay time at the "OFF to ON state" due to the carrier charging and discharging affected by Shockley–Read–Hall (SRH) recombination on the W_b region. When it is true, it will open the way achieving the high-speed switching on the PNBT SOI-FET by changing the impurity concentration design on the W_b region and also the channel-region.

ACKNOWLEDGEMENT

This work was partially supported by JST-CREST Grand Number JPMJCR16Q1, Japan.

REFERENCES

[1]J. Ida, et al., in *IEDM Tech. Dig.*, 2015, pp. 624–627.[2]T. Mori, et al., in *EDTM Tech. Dig.*, 2018, pp190-192.[3]T. Mori, et al., in *Proc. SSDM*, 2018, pp. 839–840.[4]H. Endo, et al., *EDTM* 2019 submitted.[5]HyENEXSS™ ver. 5.5, Selete, 2011.[6]G. G. Shahidi, et al., in *ISSCC Tech. Dig.* 1999, pp. 426–427.

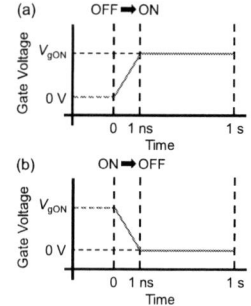

Fig. 1. Device structure of PNBT SOI-FET. (a) bird-eye view, (b) front view, (c) top view.

Table I Simulation device parameters.

Parameter	Value
Gate Oxide T_{ox}	5 nm
SOI Thickness T_{Si}	50 nm
Buried Oxide Thickness T_{BOX}	200 nm
Gate Length L_g	200 nm
Gate Width W_g	1 µm
N-region Width W_b	1.2 µm

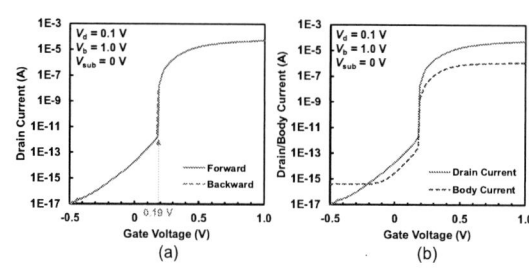

Fig. 2. DC I–V_g characteristics of the PNBT SOI-FET. (a) hysteresis characteristics. (b) I_d and I_b–V_g characteristics.

Fig. 3. V_g pulse setup in this simulation. (a) OFF to ON state. (b) ON to OFF state.

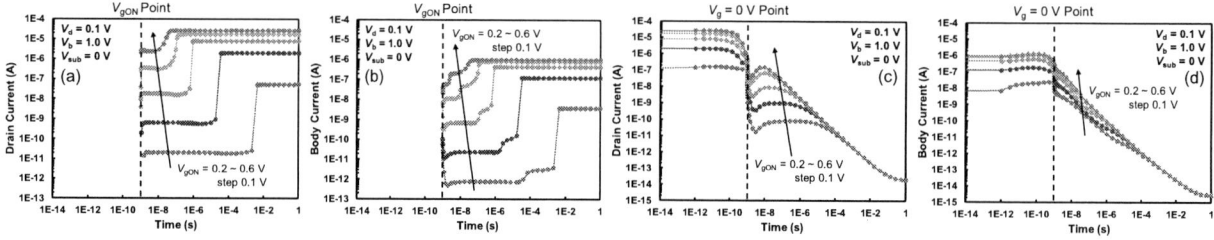

Fig. 4. Transient characteristics of the PNBT SOI-FET. (a) I_d characteristics on the "OFF to ON state". (b) I_b characteristics on the "OFF to ON state". (c) I_d characteristics on the "ON to OFF state". (d) I_b characteristics on the "ON to OFF state".

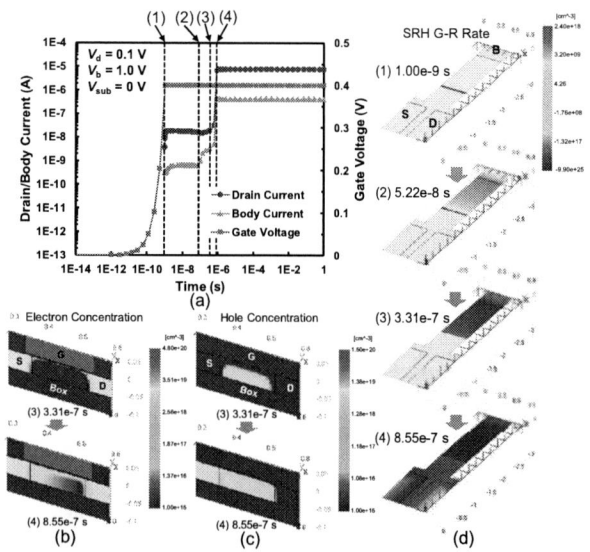

Fig. 5. Transient (a) I_d and I_b characteristics, (b) electron concentration at forefront, (c) hole concentration at forefront, and (d) SRH G-R rate at center of SOI when V_{gON} = 0.4 V on the "OFF to ON state".

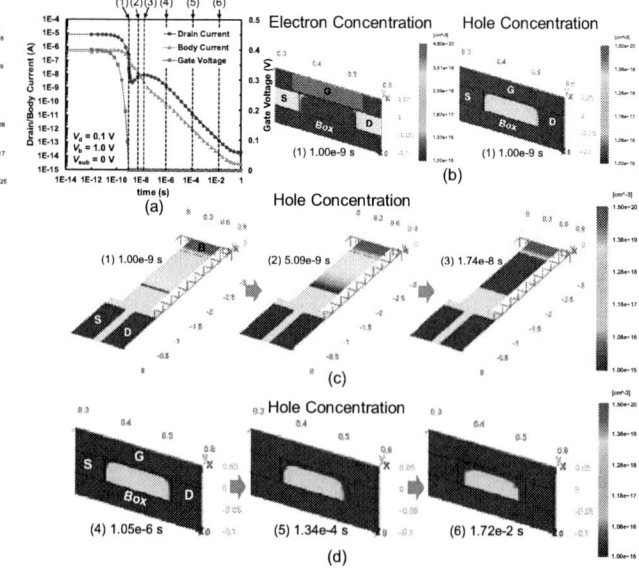

Fig. 6. Transient (a) I_d and I_b characteristics, (b) electron and hole concentration at forefront, (c) hole concentration at center of SOI, and (d) hole concentration at forefront from (4) to (6) state when V_{gON} = 0.4 V on the "ON to OFF state".

978-1-7281-0943-5/19 $31.00 © 2019 IEEE

A Large Dynamic Range Current Sensor Using Magnetic Tunnel Junction on 8" Si process line

D. Y. Wang*, I. J. Wang, C. S. Lin, J. W. Su, H. H. Lee, Y. C. Hsin, S. Y. Yang, Y. J. Chang, Y. C. Kuo, Y. H. Su, S. Z. Rahaman, G. L. Chen, S. H. Li, J. H. Wei, K. C. Huang and C. I. Wu

Electronic and Optoelectronic System Research Laboratories, Industrial Technology Research Institute, Hsinchu, Taiwan

E-mail: dy_wang@itri.org.tw

ABSTRACT

A large dynamic range magnetic field sensor for the current sensing application was designed and successfully fabricated. In this paper, we had tuned the MTJ film structure, cap layer and devices shape to achieve the good linearity and sensitivity. After choice the suitable magnetic free layer and large aspect ratio MTJ cell design, the linear dynamic range and sensitivity of our TMR sensor are over 300 Oe and 0.07%/Oe respectively.

INTRODUCTION

The magnetic tunnel junction (MTJ) has attracted great attention for many important applications, such as magnetic random access memory (MRAM), hard-disk read heads, and magnetic sensors.[1-4] For the magnetic sensing applications, industrial and bio applications are the hot topics. The tunneling magneto-resistance (TMR) sensor is a candidate for the industrial application because of its linear response, high sensitivity, wide dynamic range, and temperature ranges. However, fabricating a wide dynamic range for the larger sensing range is a challenging work, because it involves the fine tuning of MTJ film structure and the MTJ cell design. In this study, we fabricated a magnetic current sensor with complementary concept and characterized the sensing properties.

EXPERIMENTAL RESULTS

For the high-current level (>100A) sensing, the TMR's linear dynamic range should be > +/- 100 Oe and this value is much larger than our previous TMR field sensor[5]. To increase the dynamic range, we modify the MTJ free layer first. The simplify MTJ film structure is shown in Fig. 1(a). The MgO based MTJ with NiFe or CoFeB free layer was studied and the VSM data of them are shown in Fig, 1(b) and (c). From the VSM data, NiFe and CoFeB free layer are almost identical and we cannot know which one is better from the film data only. To fabricate the TMR sensor, the MTJ films are grown on the 8" Si/SiO2 substrate. To achieve the high dynamic range and the low hysteresis curves, the MTJ cell is designed as the high aspect ratio shape. The design of the MTJ cell and the direction of free layer (FL) and pinned layer (PL) are shown in Fig.2(a). Fig. 2(b) shows the TEM image of the MTJ's structure and the MgO layer is clear and well-aligned. It means that our MTJ's process is good and not process induced drawback in this sensor. The magnetic moment of pinned layer is aligned on short axis by annealing the wafer with high magnetic field and the free layers moment is aligned on the long axis due to the shape's anisotropy. When the current-induced magnetic field is applied on the sensor, the moment of PL and FL rotated with field as shown in Fig 2(c) and they will return to the original 90 degree interlacing as the magnetic field is removed.

Figs. 3(a) and (b) show the measurement results of CoFeB and NiFe free layer sensors. Although the VSM data of them are similar, the R-H loop of them are totally different. For the CoFeB-based devices, we got an evident hysteresis loop with larger MR ratio and this device should be classified as the memory, not sensor. For the NiFe-based devices, the dynamic range is > +/- 100 Oe, MR ratio is 30~35% with the slight hysteresis and the 99% linearity across the whole 8" wafer. The measured data is identical to Fig2(c) and shows a good sensor behavior. Because the NiFe is softer magnetic material, the free layer's moment can be continuous rotated from 0° to 180° and R-H curve shows the good linear behavior.

To study the MTJ's shape effect, we design the MTJ cell with different AR ratio. The length of MTJ's long axis is fixed at 16um and the length of short axis is ranged from1.5 to 0.4 um. The dynamic range of devices are plotted in Fig.4(a) and the range is increased with reduce the short axis and reach to 330 Oe with the 0.4-um short axis. The device's sensitivity is 0.07%/Oe. Fig 4(b) counts the linear range from the 10% to 90% of the transition curve only and separated the dynamic range into positive and negative magnetic field part. We can see the dynamic range of each parts is still reach 150Oe at 0.4um case. In Figs. 3(b), the R-H curve of the device is not perfectly pass the zero point and the dynamic range of negative branch is larger than the one of positive branch. It means that the magnetic moments of free layer and pinned layer are not well aligned in 90°. It will be improved by the new design of MTJ films and process flow.

Finally, we use our TMR sensor to construct a current sensing module and the system diagram is plotted in Fig. 5(a). We use an Iron to collect the magnetic field which is induced by the wire's current. We cut a small open in the Iron ring and insert the TMR sensor sub-chip into the gap to sense the magnetic field. The TMR sub-chip connects with the control chip and the system is shown in Fig.5(b). Figs. 5(c) and (d) show the current source signal and the output data of current sensor module. The output signal is will fitted with the input current signal and response time of this module and sensor is sub-ms level. Although the demo data shown in Figs. 5(c)and (d) is less 10A, according our TMR sensor's dynamic range, 100A current inside the Iron ring can be precisely sensed by our module.

CONCLUSION

In this study, a large dynamic range magnetic sensor is successfully fabricated by an MTJ with NiFe free layer. This magnetic sensor shows the dynamic range over 300 Oe and the sensitivity is 0.07%/Oe.. Using this MTJ chip to construct the current sensor module, we can get the real-time response of the input current. The response time is sub-ms level and the current sensing range is over 100A.

REFERENCES

[1] .Z. Diao et al., *Appl. Phys. Lett.*, 2010, 96, p. 202506.
[2] S. Yussa et al., *J. Phys. D*, 2007, 40, R337.
[3] C.-W. Chien et al., *IEEE Elec. Dev. Lett.*, 2014, 35, p. 738.
[4] L. Jogschies et al., Sensor 2015 15, p28665.
[5] D. Y. Wang et al., *VLSI TSA*, 2015.

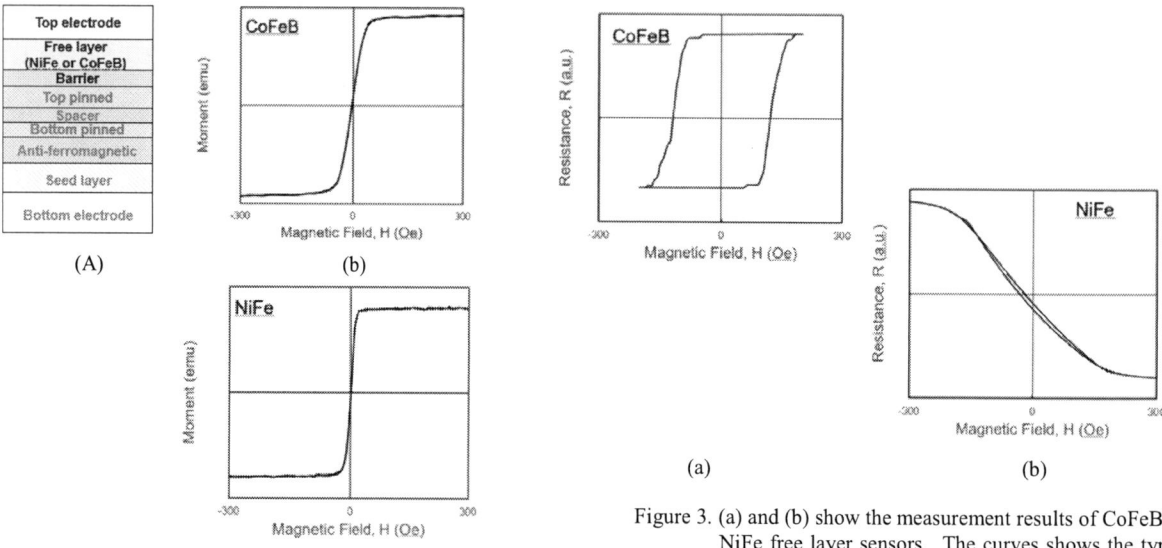

Figure 1. (a) The simplify MTJ film structure and the MTJ'sVSM data for (b) CoFeB and (c) NiFe free layer.

Figure 3. (a) and (b) show the measurement results of CoFeB and NiFe free layer sensors. The curves shows the typical memory-like" and "sensor-like" behaviors.

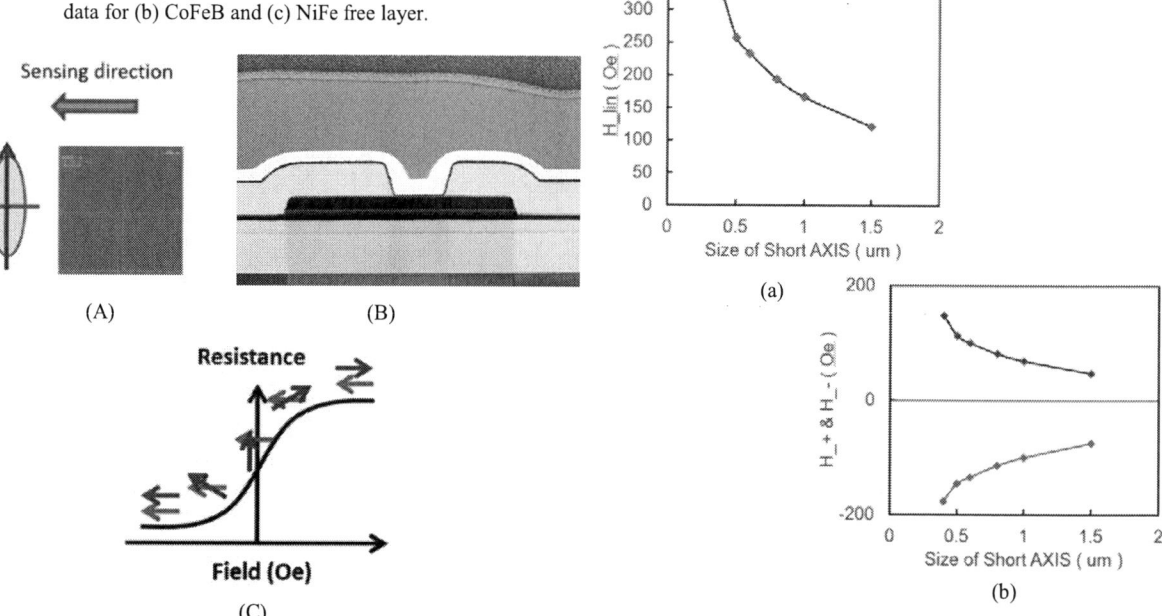

Figure 2. (a) The design of the MTJ cell and the direction of FL PL and SEM image of the MTJ cell. (b) TEM image of the MTJ's structure. (c) The moment rotation of PL and FL with field.

Figure 4. (a) The trend of dynamic range of TMR sensor, the length of short axis is ranged from1.5 to 0.4 um. (b) The dynamic range under positive (H_+) and negative (H_-)magnetic field

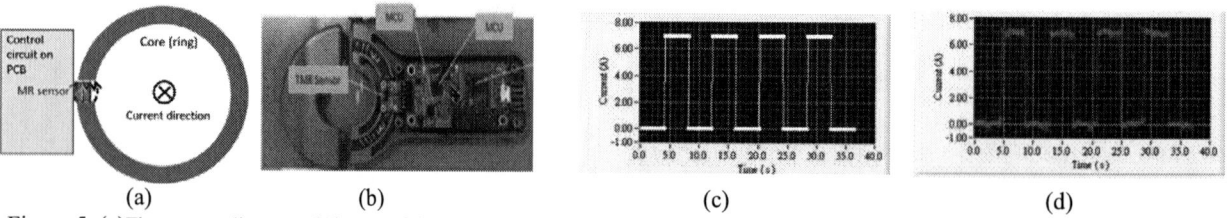

Figure 5. (a)The system diagram, (b)image of the system, (c) the current source signal and (d) the output data of current sensor module.

Architecture Evaluation for Standalone and Embedded 1T-DRAM

Md. Hasan Raza Ansari[1], Nupur Navlakha[1], Jyi-Tsong Lin[2], Abhinav Kranti[1]*

[1]Low Power Nanoelectronics Research Group, Discipline of Electrical Engineering,
Indian Institute of Technology Indore, Simrol, Indore 453552, India. *E-mail: akranti@iiti.ac.in
[2]Department of Electrical Engineering, National Sun Yat-Sen University, Kaohsiung 80424, Taiwan, R.O.C.

ABSTRACT

This paper analyzes different transistor architectures (inversion mode (IM), accumulation mode (AM) and junctionless (JL)) for standalone as well as embedded DRAM. The performance metrics (retention time (RT), sense margin (SM), current ratio (CR) and write time (WT)) of JL based 1T-DRAM can be improved through stacked JL (SJL) and core-shell (CS) topologies, which separate conduction and storage regions. Results including gate length scalability (25 nm) and high temperature (125 °C) operation indicate the preference for SJL for standalone applications while CS architecture for embedded DRAM.

INTRODUCTION

The increase in the demand of dynamic random access memory (DRAM) due to the advent of mobile and cloud computing, virtual reality, and internet of things (IoT) applications along with conventional computing requires innovative devices to provide maximum storage, high performance, high retention, and low energy at reduced size [1]. The functionality of 1T-DRAM is primarily governed by transistor architecture, storage region, bias optimization and lifetime [2]. JL transistor, with same type of dopants in source/drain (S/D) and channel regions show usability as DRAM [3], [4]. The work investigates transistor architectures to facilitate high retention along with scalability, WT, and high temperature operation in standalone and embedded 1T-DRAM. While retention and scalability is vital for standalone DRAM, a dense memory with high speed is better suited as embedded DRAM. The conventional JL transistor has been modified to separate conduction and storage regions through an oxide in SJL configuration, and using an intrinsic region in CS JL device [5]. DRAM performance metrics are also compared with those achieved through IM and AM devices.

RESULTS AND DISCUSSION

Double Gate (DG) IM (@ $N_a = 10^{15}$ cm^{-3}), AM (@ $N_d = 10^{18}$ cm^{-3}), and JL (@ 5×10^{18} cm^{-3} to $N_d = 10^{19}$ cm^{-3}) transistors (Fig. 1a) along with CS JL (Fig. 1b) and SJL (Fig. 1c) topologies were analyzed with ATLAS tool [6] with models for band-to-band tunneling (BTBT), impact ionization, temperature and concentration dependent carrier lifetime, Lombardi mobility, bandgap narrowing, Shockley Read Hall (SRH) and Auger recombination. The transfer characteristics (I_d-V_G) were well calibrated with experimental data for IM [7] (Fig. 2a) and JL [8] (Fig. 2b) devices. Independent gate operation (front gate for conduction and back gate for storage) was utilized for DRAM. Gate length (L_g) was varied from 200 nm to 25 nm with an underlap length of 10 nm, and front and back gate oxide thickness (T_{ox}) was 1 nm. Film thickness of 12 nm was used for IM, JL and AM devices. In CS architecture, shell thickness (T_{Shell}) was fixed at 2 nm and core thickness (T_{Core}) was varied from 3 nm to 8 nm while for SJL transistor, the conduction (T_{Si1}) region was 7 nm with a storage (T_{Si2}) region of 9 nm and separation oxide thickness (T_{SOX}) of 3 nm. The doping (N_a) of storage region was 10^{17} cm^{-3} for SJL topology.

Fig. 3 shows the methodology adopted to extract the performance metrics in DRAM. Figs. 4a-c show the contour plots of electrostatic potential for conventional JL, CS and SJL transistors, respectively, with ΔV indicating the potential depth along the x-direction for charge storage. Conventional JL device exhibits least ΔV due to lesser depletion of electrons. CS architecture [5] with a thin heavily doped shell and a thicker (intrinsic) core achieves a deeper potential well. However, the sharing of storage region with heavily doped n^{++} S/D regions increases generation of holes due to BTBT and recombination during Hold operation, which reduces retention. The conduction and storage regions are segregated through a physical oxide in SJL transistor and a deeper well is formed due to the depletion of holes from T_{Si2} underneath S/D region, which is based on metal-oxide-semiconductor (MOS) capacitor, and observed through variation in potential along y-direction (Fig. 4c).

Biasing and timing schemes are illustrated in Table I. Holes are generated for State "1" through BTBT mechanism (lower power consumption) and removal of holes for State "0" is through forward bias mechanism [2], [4]. The difference between State "1" and "0" current is termed as SM and RT is evaluated as the time when SM reaches to 50% of peak value [2], [4]. WT is defined as the time when SM attains constant value during Write "1" for a fixed drain voltage [2], [4]. Due to higher lifetime and deeper potential well, IM and AM devices attain high RT (Fig. 5). Conventional JL DRAM achieves RT of 700 µs at 85 °C, which is lower than 64 ms for standalone and 4~16 ms for embedded DRAM [9]. Fig. 6 shows that CS and SJL devices ($N_d = 10^{19}$ cm^{-3}, $L_g = 200$ nm) can attain a maximum RT of ~15 ms and ~1.2 s, respectively, at 85 °C.

In JL devices, the potential depth (ΔV) is governed by doping and device parameters, which controls BTBT and recombination of holes, and thus, the DRAM performance [4]. Figs. 7a-b show that with a reduction in T_{Si}, the state currents (I_1 and I_0) and SM are lower in CS device (@ T_{Shell} of 2 nm) as compared to JL topology due to an enhanced depletion of electrons. Read sensitivity is measured through ($CR = I_1/I_0$) and ($SM = I_1-I_0$) [2]. CS topology exhibits higher CR as compared to JL DRAM. In terms RT, a maximum RT of ~110 ms and ~28 ms for CS and JL devices, respectively, is obtained at T_{Si} of 7 nm, which reduces to ~85 ms and ~2.5 ms, respectively, at T_{Si} of 12 nm. Fig. 8a shows the variation in ΔV and RT with shell doping (N_d). An increase in N_d reduces ΔV due to lesser depletion and results in a lower RT. However, vertical stacking of alternately doped layers i.e. n-p-n (Fig. 8b) can enhance RT as the relative work function difference between n- and p-regions increases, which depletes electrons and enhances ΔV. Consequently, RT increases from ~15 ms for an intrinsic core to ~33 ms (@ $N_a = 5\times10^{18}$ cm^{-3}). 25 nm SJL based DRAM shows better scalability over IM device as RT of ~125 ms (> 64ms) can be achieved (Fig. 9), thus exhibiting potential for standalone DRAM. SJL based DRAM achieves maximum RT of ~95 ms (> 64ms) at 125 °C with $L_g = 200$ nm. As shown in Fig. 11, CS architecture with better SM, CR and $RT > 4$ ms at 85 °C [9] shows suitability for eDRAM in which speed can be improved through optimization of bias and shell thickness.

CONCLUSION

An assessment of DRAM performance metrics indicate that separation of conduction and storage regions can yield very high retention values (better than IM) at lower gate lengths and elevated

978-1-7281-0943-5/19 $31.00 © 2019 IEEE

temperatures in a stacked junction transistor, which is more appropriate for standalone applications. Core shell architecture is more suitable for embedded DRAM with an appropriate balance between sense margin, current ratio and retention time. The work provides a comprehensive assessment of device architectures and new viewpoints for next generation DRAM.

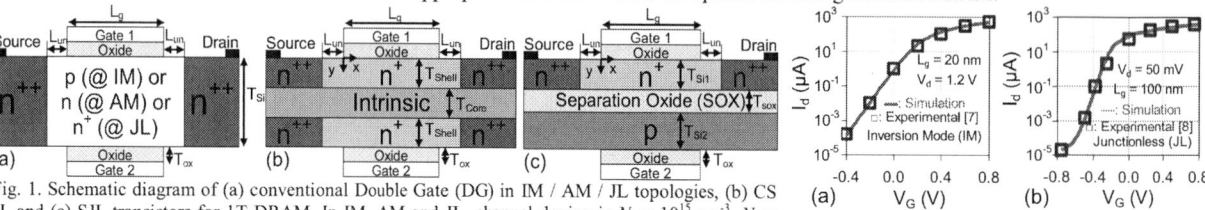

Fig. 1. Schematic diagram of (a) conventional Double Gate (DG) in IM / AM / JL topologies, (b) CS JL and (c) SJL transistors for 1T-DRAM. In IM, AM and JL, channel doping is $N_a = 10^{15}$ cm^{-3}, $N_d = 10^{18}$ cm^{-3} and $N_d = 5 \times 10^{18}$ cm^{-3} – 10^{19} cm^{-3}, respectively. In CS and SJL architectures, doping for channel and storage regions are $N_d = 10^{18}$ cm^{-3} – 10^{19} cm^{-3} and $N_a = 10^{17}$ cm^{-3}, respectively.

Fig. 2. Comparison of transfer characteristics with published experimental data of (a) IM [7] and (b) JL [8] devices.

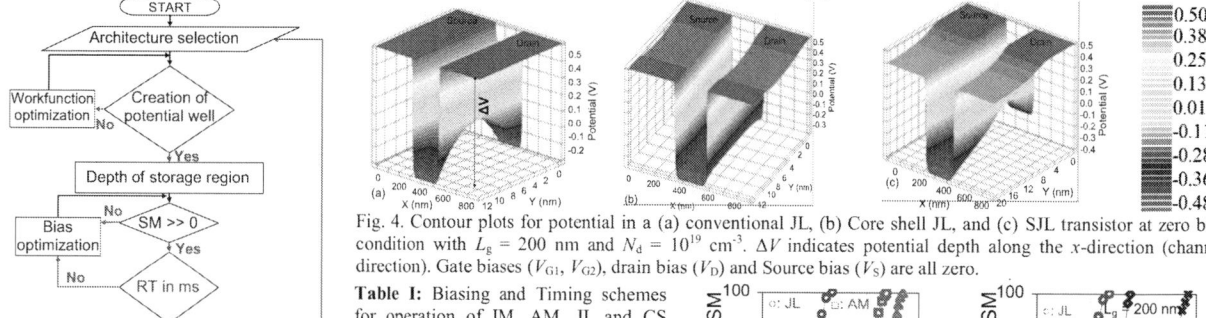

Fig. 4. Contour plots for potential in a (a) conventional JL, (b) Core shell JL, and (c) SJL transistor at zero bias condition with $L_g = 200$ nm and $N_d = 10^{19}$ cm^{-3}. ΔV indicates potential depth along the x-direction (channel direction). Gate biases (V_{G1}, V_{G2}), drain bias (V_D) and Source bias (V_S) are all zero.

Fig. 3. Flow chart for assessing DRAM functionality.

Table I: Biasing and Timing schemes for operation of IM, AM, JL and CS based 1T-DRAM

Operation	V_{G1}^* (V)	V_{G2}^* (V)	V_S^* (V)	V_D^* (V)	Time (ns)
Write 1 (W1)	1.0	-1.6	0.0	1.5	50
Hold (H)	0.0	-0.1	0.1	0.1	--
Read (R)	1.0	0.1	0.0	0.1	100
Write 0 (W0)	1.5	1.5	0.0	0.0	50

*For SJL: $V_{G2\_W1} = -1.5$ V, $V_{S/D\_H} = 0$ V, $V_{G2\_H} = -0.2$ V, $V_{G1\_R} = 0.9$ V, $V_{G1\_W0} = 0$ V, $V_{G1\_W0} = 1.0$ V and $WT = 100$ ns.

Fig. 5. Variation in percentage change in SM with hold time at $L_g = 200$ nm and 85 °C.

Fig. 6. Variation in percentage change in SM of JL, CS and SJL based DRAM at 85 °C.

Fig. 7. Variation in (a) state currents (I_1 and I_0) and (b) SM and RT with T_{Si} (= $2T_{Shell} + T_{Core}$) in JL and CS JL DRAM for $N_d = 5 \times 10^{18}$ cm^{-3}, $L_g = 200$ nm and $T_{Shell} = 2$ nm (fixed for CS) at 85 °C.

Fig. 8. Variation of ΔV and RT with (a) shell doping (N_d) and (b) core region doping (N_a) in n-p-n vertical structure with $N_d = 10^{19}$ cm^{-3}, $L_g = 200$ nm at 85 °C.

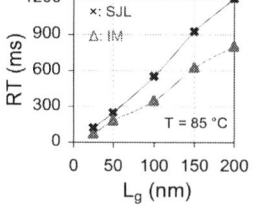

Fig. 9. Variation in RT with L_g of IM and SJL based DRAM at 85 °C with $N_d = 10^{19}$ cm^{-3} (@ SJL).

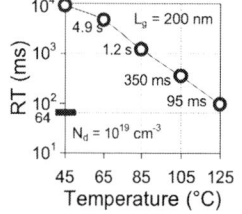

Fig. 10. Variation in RT with temperature for SJL based DRAM with L_g of 200 nm.

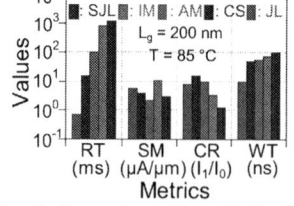

Fig. 11. Comparison of IM, AM, JL, CS and SJL based 1T-DRAM metrics with $L_g = 200$ nm at 85 °C.

ACKNOWLEDGMENT

This work was supported by Department of Science and Technology, Government of India, through Global Innovation and Technology Alliance under grant GITA/DST/TWN/P-70/2015, and by the Ministry of Science and Technology of Taiwan, R.O.C., under contract 104-2923-E-110-001-MY3.

REFERENCES

[1] K. Kim, *Symp. VLSI Tech.*, pp. 5, 2008.

[2] N. Navlakha *et al.*, *J. Appl. Phys.*, pp. 214501, 2016.

[3] J.-P. Colinge *et al.*, *Nat. Nanotechnol.*, pp. 225, 2010.

[4] M. H. R. Ansari *et al.*, *IEEE Trans. Electron Devices*, pp. 1205, 2018.

[5] Y.-J. Lee *et al.*, *IEEE International Electron Devices Meeting*, pp. 32.7.1, 2014.

[6] Atlas User's Manual. Silvaco Int., 2015.

[7] M. Vinet *et al.*, *IEEE Electron Device Lett.*, pp. 319, 2005.

[8] D.-Y. Jeon *et al.*, *Solid. State. Electron.*, pp. 86, 2013.

[9] H.-Y. Yang *et al.*, *IEEE Trans. Very Large Scale Integr. Syst.*, pp. 1715, 2012.

TIPS-Pentacene:PS Blend Organic Field-Effect Transistors with Hybrid Gate Dielectric on Paper Substrate

Vivek Raghuwanshi, Deepak Bharti, Ajay Kumar Mahato, Ishan Varun, and Shree Prakash Tiwari

Department of Electrical Engineering, Indian Institute of Technology Jodhpur,
Jodhpur, Rajasthan 342037, India
E-mail: sptiwari@iitj.ac.in

ABSTRACT

We report on the fabrication of bottom gate top contact organic field effect transistors with hybrid gate dielectric and TIPS-Pentacene:PS blend as active layer on PowerCoat™ HD 230 paper substrate. The hybrid dielectric layer leads the devices operated at relatively low voltage of -10 V with avg. and maximum field effect mobility of 0.52(±0.16) and 0.78 cm^2 V^{-1} s^{-1} respectively, with near zero threshold voltage and on-off current ratio approaching ~10^4. The devices have shown excellent operational stability when tested with DC bias stress of $V_{DS} = V_{GS} = $ -10 V for 1 h and minuscule changes in electrical characteristics were observed with continuous multiple transfer cycle scans. In addition, we have demonstrated a resistive load inverter circuit with varying the load resistance with these paper OFETs.

INTRODUCTION

Paper electronics is an emerging field aimed at the reusable and renewable devices, which can mitigate the solid waste problems by lesser environmental footprint at the end of the device life span. Organic field effect transistors which are intensively studied in past few decades due to their low cost and large area processability are typically fabricated on silicon, glass and plastic substrates [1-3]. However, fabrication on paper substrate broadens their application area and also allows the development of new electronic devices, which will have lesser impact on the environment. Paper is made from natural wood sources and has various advantages as it is the most ubiquitous substance used in day to day life with benefits of low cost, bio-degradability, foldability, low printing cost etc. But despite of these advantages the higher surface roughness and its porous nature, restrict the fabrication of high performance OFET devices on it. Thus, a barrier layer is required to overcome the aforementioned constrains. There are few reports in the literature that have used either thermally coated layer or a solution processed coating to planarize and protect the paper surface from further chemical processing and provide a smooth surface for viable high-performance device fabrication.

Here, in this report, we have used solution processed polyvinyl alcohol (PVA) layer to planarize the rough paper surface, and operationally stable organic field effect transistors with hybrid gate dielectric and 6,13-bis(triisopropylsilylethynyl)pentacene:Polystyrene (TIPS-Pentacene:PS) blend as active layer were fabricated on it. The devices have shown highly stable electrical characteristics with avg. and maximum field effect mobility of 0.52(±0.16) and 0.78 cm^2 V^{-1} s^{-1}, with near zero threshold voltage and on-off current ratio approaching ~10^4 at -10 V operation.

EXPERIMENTATION

Devices in the bottom gate top contact structure were fabricated on PowerCoat™ HD 230 paper substrate. The substrates were cleaned with heavy blow of nitrogen and spin coated at 1000 rpm for 60 s with 8 wt. % PVA solution in water followed by heating at 60 °C for 10

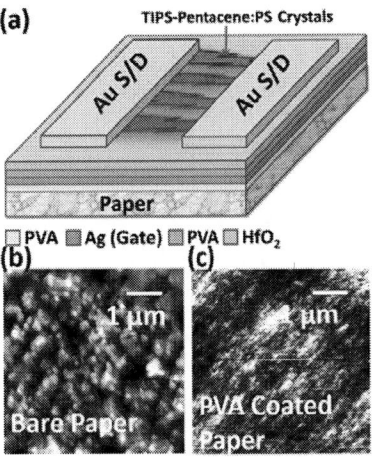

FIGURE 1. (a) Schematic of BGTC OFET device. AFM images of (b) bare, and (c) PVA coated paper substrate.

minutes to achieve a smooth barrier layer. A 100 nm thick gate (Ag) electrode was deposited by thermal evaporation. Further a 4 wt. % PVA solution was spin coated at 2000 rpm for 60 s to deposit the polymer gate dielectric, later a 40 nm HfO$_2$ layer was deposited over PVA layer in a savannah S-200 ALD system from Cambridge nanotech to complete the bilayer dielectric. Further semiconductor processing and deposition steps are similar to our previous reports [2]. Fig. 1(a) shows the schematic of the fabricated device.

RESULTS AND DISCUSSION

Fig. 1(b) and (c) show the AFM images of the PowerCoat™ HD 230 bare paper and PVA coated paper surface respectively. The bare paper was found to have an avg. surface roughness (Rq) of 11.9(±4.0) nm, which is too high for further processing as the same roughness could be replicated to the other device layers if this couldn't be resolved. An 800 nm thick layer of PVA was thus deposited to planarize the surface and to have a barrier layer. The PVA layer was found to have the Rq of 3.3(±1.1) nm, thus it suppressed the surface roughness and make the surface viable for further device fabrication.

The output and transfer characteristics of the typical paper based representative OFET are shown in Fig. 2(a) and (b) respectively. The devices have shown excellent p-type characteristics with clear linear and saturation region under -10 V operation. Despite of having high surface roughness of the paper substrate, the devices yielded high performance with maximum filed effect mobility of 0.78 cm^2 V^{-1} s^{-1} and an average mobility of 0.52 cm^2 V^{-1} s^{-1} with standard deviation of 0.16, low threshold voltage of -0.56(±0.24) and an on-off current ratio approaching 10^4 was found for the set of 10 devices. The high-performance in these OFETs can be attributed to the smooth PVA

978-1-7281-0943-5/19 $31.00 © 2019 IEEE

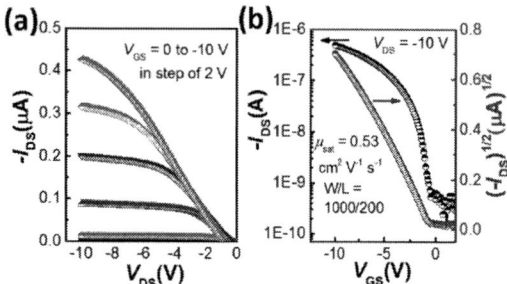

FIGURE 2. (a) Output and (b) Transfer characteristcis of the representative paper OFET.

planarization layer that made the paper surface viable for OFET fabrication, which is further planarize by the gate (Ag), PVA and HfO$_2$ dielectric layers, thus providing a smooth surface for active layer deposition. Another reason to high performance can be the high degree of dielectric-semiconductor interface which is achieved through the vertical phase separation in TIPS-Pentacene:PS blend during the process of slow solvent evaporation in drop casting from the blend solution.

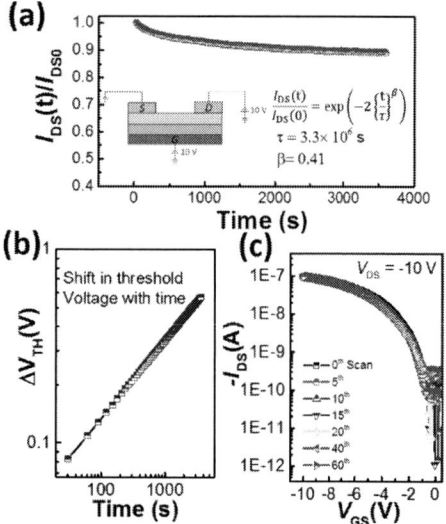

FIGURE 3. (a) Effect of DC bias stress on (a) normalized drain current and (b) threshold volatage with stress time. (c) Multiple transfer scans.

Fig. 3(a) shows the time dependent variation in normalized drain current, when the DC bias stress of $V_{DS} = V_{GS} = -10$ V is applied for 1 h in ambient. The device has shown the drain current decay of $\sim$ 10 % after 1 h, which is due to the charge trapping at various locations due to the applied stress for example at the grain boundaries, in the bulk of the semiconductor, at the dielectric-semiconductor interface etc. The experimental data was fitted to the normalized drain current equation as shown in inset of Fig. 3(a) and the values of τ (which reflects the typical trapping time) and β (which indicates the width of the trap distribution) were extracted [4]. The high value of τ (3.3×10^6) indicates lesser trapping sites, which indicates a smoother dielectric-semiconductor interface. The shift in threshold voltage was obtained by stretched exponential function and the corresponding shift is represented in Fig. 3(b) [5]. Further to test the operational stability of the devices, the transfer characteristics were captured continuously for more than 60 scans. The devices have shown stable characteristics with almost complete overlap of each scan. Which again indicates less

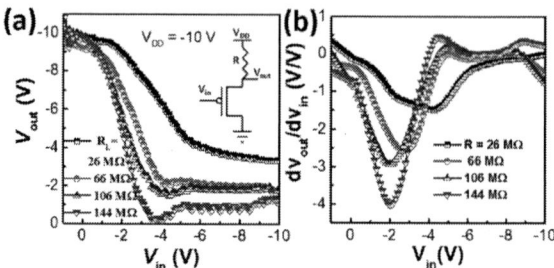

FIGURE 4. (a) Voltage transfer characteristics of the resistive load inverter circuit with paper OFET. (b) Static gain with different resistive load.

degree of charge trapping and an operationally stable device for circuit applications. In addition, to check the potential of these paper OFETs towards circuit application, we have demonstrated a resistive load inverter circuit by connecting external loads to paper OFET, with varying resistance. Inset of Fig. 4(a) shows the schematic of the external load inverter circuit. The circuit was operated well below -10 V. Fig. 4(a) shows the static transfer characteristic of the inverter circuit, it is observed from the graph that the circuit works correctly as inverter with a clear output high or the logic 1 at low input voltage and a clear output low or logic 0 at high input voltage. However, with increasing the load resistance the transfer characteristics were found to improve because of the higher voltage drop across higher resistance. Fig. 4(b) shows the inverter static gain at different load resistance, and was found to have the maximum value of 4 for 144 MΩ load.

CONCLUSION

In this work, the OFETs with bilayer dielectric (HfO$_2$/PVA) were successfully fabricated on low cost paper substrate. The devices were found to be highly operationally stable with performance comparable to those on plastic and glass substrate with the max. and avg field effect mobility of 0.78 and 0.52($\pm$0.16) cm^2 V^{-1} s^{-1} respectively with avg. threshold voltage of -0.56($\pm$0.24) V. In addition, a resistive load inverter circuit was successfully demonstrated with these OFETs.

REFERENCES

[1] Yi, H.T., M.M. Payne, J.E. Anthony, and V. Podzorov, *Ultra-Flexible Solution-Processed Organic Field-Effect Transistors*. Nature Communications, 2012. **3**: p. 1259.

[2] Raghuwanshi, V., D. Bharti, I. Varun, A.K. Mahato, and S.P. Tiwari, *Performance Enhancement in Mechanically Stable Flexible Organic-Field Effect Transistors with Tips-Pentacene:Polymer Blend*. Organic Electronics, 2016. **34**: p. 284-288.

[3] Yadav, S. and S. Ghosh, *Amorphous Strontium Titanate Film as Gate Dielectric for Higher Performance and Low Voltage Operation of Transparent and Flexible Organic Field Effect Transistor*. ACS Applied Materials & Interfaces, 2016. **8**(16): p. 10436-10442.

[4] Tiwari, S.P., X.-H. Zhang, W.J.P. Jr., and B. Kippelen, *Study of Electrical Performance and Stability of Solution-Processed N-Channel Organic Field-Effect Transistors*. Journal of Applied Physics, 2009. **106**(5): p. 054504.

[5] Mathijssen, S.G.J., M. Cölle, H. Gomes, E.C.P. Smits, B. de Boer, I. McCulloch, P.A. Bobbert, and D.M. de Leeuw, *Dynamics of Threshold Voltage Shifts in Organic and Amorphous Silicon Field-Effect Transistors*. Advanced Materials, 2007. **19**(19): p. 2785-2789.

978-1-7281-0943-5/19 $31.00 © 2019 IEEE

Origin of fixed charges and dipole in GeO$_x$/Al$_2$O$_3$ gate stack based on Ge

Lixing Zhou[1,2], Xiaolei Wang[1], Xueli Ma[1], Jinjuan Xiang[1], Chao Zhao[1,2], Tianchun Ye[1,2], and Wenwu Wang[1,2]

[1]Key Laboratory of Microelectronics Devices & Integrated Technology, Institute of Microelectronics, Chinese Academy of Sciences, Beijing 100029, China
[2]University of Chinese Academy of Sciences, Beijing 100049, China
E-mail: wangxiaolei@ime.ac.cn

ABSTRACT

The origin of fixed charges and dipole in GeO$_x$/Al$_2$O$_3$ gate stack deposited by atomic layer deposition was experimentally investigated. Three kinds of charges in the gate stacks: the Ge/GeO$_x$ interface charges Q_1, GeO$_x$/Al$_2$O$_3$ interface charge Q_2, and GeO$_x$/Al$_2$O$_3$ interface dipole Q_{dipole} were obtained by C-V characteristics. The XPS was used to explore the interfacial defect at both the Ge/GeO$_x$ and GeO$_x$/Al$_2$O$_3$ interfaces. It is found that positive charges Q_1 at the Ge/GeO$_x$ interface is due to the oxygen vacancy, negative charges Q_2 at the GeO$_x$/Al$_2$O$_3$ interface is due to the oxygen dangling bonds, Q_{dipole} is due to the band alignment at the GeO$_x$/Al$_2$O$_3$ hetero-interface, respectively.

INTRODUCTION

Germanium is a very promising material because of its higher mobility for the application of CMOS circuit [1]. Flat band voltage (V_{FB}) or threshold voltage (V_{TH}) shift due to fixed charges and dipole have always been a critical issue affecting the performance of MOS devices [2-4]. It is necessary to investigate the defect in the gate stack based on Ge. In this study, we explored the fixed charges and dipole in GeO$_x$/Al$_2$O$_3$ gate stack by C-V measurement and analysis their origin using XPS.

EXPERIMENT

To extract the gate charges, capacitors with two gate structures: terraced GeO$_x$ structure with different GeO$_x$ thicknesses (0.7-1.23 nm) and the same Al$_2$O$_3$ thickness (10 nm) and terraced Al$_2$O$_3$ structure with the same GeO$_x$ thickness (0.7 nm) and different Al$_2$O$_3$ thicknesses (5-15 nm), were fabricated. GeO$_x$ and Al$_2$O$_3$ was grown in ALD chamber. GeO$_x$ was oxidized by ozone. Al was as the gate electrode. In addition, capacitors with terraced GeO$_x$ and Al$_2$O$_3$ structures annealed in N$_2$, O$_2$, and H$_2$ were also fabricated. Three series of samples with Ge/GeO$_x$/Al$_2$O$_3$ structure was used for the XPS measurement. The first is with GeO$_x$ thickness varying from 0.22 to 0.7 nm and 2 nm Al$_2$O$_3$ on GeO$_x$ for protection. The second is with 0.7 nm GeO$_x$ and Al$_2$O$_3$ thickness varying from 1.5 to 4 nm. The third is with 0.7 nm GeO$_x$ and 2 nm Al$_2$O$_3$ annealed in N$_2$, O$_2$, and H$_2$ ambiences. The XPS measurement was performed using Thermo Scientific ESCALAB 250xi equipped with a monochromatic Al Kα radiation source. The data were collected at a take-off angle of 90° relative to the sample surface.

RESULTS AND DISCUSSION

Firstly, we extracted the charge distribution in gate stacks using the relationship of V_{FB} and EOT of the capacitors with terraced GeO$_x$ and Al$_2$O$_3$ structures shown as Eq. (1) and Eq. (2) in Table 1. Fig. 1 (a) and (b) give the V_{FB}-EOT plots of the terraced GeO$_x$ and Al$_2$O$_3$ structures. The good linear relation indicates the bulk charge in the GeO$_x$ and Al$_2$O$_3$ can be negligible. Three kinds of charges exist in the gate stacks: the Ge/GeO$_x$ interface charges Q_1, GeO$_x$/Al$_2$O$_3$

interface charge Q_2, and GeO$_x$/Al$_2$O$_3$ interface dipole Q_{dipole}. Their distribution is shown in Fig. 2. It is can be seen that the dipole is one order of magnitude larger than Q_1 and Q_2. Fig. 3(a) is the Ge 3d spectra of 0.7 nm GeO$_x$. Ge^{1+} appears which indicates an insufficient oxidation process. Fig. 3(b) is the the areal intensity ratio of Ge(GeO$_x$)/O(GeO$_x$) for different GeO$_x$ thicknesses. The value of Ge(GeO$_x$)/O(GeO$_x$) decreases with GeO$_x$ thickness increment. It suggests that oxygen content in GeO$_x$ increases as the GeO$_x$ thickness increases. Oxygen is relevantly deficient in the beginning of ozone oxidation and this causes oxygen vacancy. Oxygen vacancy with a +2 charge state has the lowest formation energy [5], which is consistent with the positive charges at the Ge/GeO$_x$ interface in Fig. 2. To investigate the origin of defect at the GeO$_x$/Al$_2$O$_3$ interface, we analysis the ratio of Al 2p (Al$_2$O$_3$) vs O 1s (Al$_2$O$_3$). Fig. 4 shows the value of Al(Al$_2$O$_3$)/O(Al$_2$O$_3$) increases with Al$_2$O$_3$ thickness increment. This result demonstrates that oxygen is relatively sufficient compared to Al at the Al$_2$O$_3$ initial growth stage and causes oxygen dangling bonds at the GeO$_x$/Al$_2$O$_3$ interface. The transition level of oxygen dangling bonds is localized below the Ge valence band maximum (VBM), which indicates that oxygen dangling bonds are negative charged [6], consist with the negative charges at the GeO$_x$/Al$_2$O$_3$ interface as shown in Fig. 2. Fig. 5 depicts the value of dipole at the GeO$_x$/Al$_2$O$_3$ interface increases but the polarity become opposite. Fig. 6 shows the Ge 3d spectra with and without PDA. It can be seen that N$_2$, O$_2$, and H$_2$ annealing makes the peak position of Ge 3d^{GeO_x} shift to a lower binding energy. These results can be explained by the gap states and charge neutrality level (CNL) theory. When the CNL in Al$_2$O$_3$ is higher than that in GeO$_x$, electron transfers from Al$_2$O$_3$ to GeO$_x$ to form positive dipole as shown in Fig. 7(a). PDA in different ambiences will change the relative position of CNL in Al$_2$O$_3$ and GeO$_x$. When the CNL in Al$_2$O$_3$ is lower than that in GeO$_x$, electron transfers from GeO$_x$ to Al$_2$O$_3$ to form negative dipole as shown in Fig. 7(c). Fig. 7(b) and (d) show that the occurrence of positive and negative dipole at the Al$_2$O$_3$/GeO$_x$ interface causes band bending downward and upward of GeO$_x$, respectively, contributing to the shift of GeO$_x$ core level.

CONCLUSION

We experimentally investigated the origin of interfacial defect in GeO$_x$/Al$_2$O$_3$ gate stack. There are positive charges at the Ge/GeO$_x$ interface due to oxygen vacancy, negative charges at the GeO$_x$/Al$_2$O$_3$ interface due to oxygen dangling bond, and dipole at the the GeO$_x$/Al$_2$O$_3$ interface due to the band alignment.

ACKNOWLEDGEMENTS

This work was financially supported by National Natural Science of China (Nos. 61574168 and 61504163).

REFERENCES

[1] S. H. Shin, et al., IEEE Transactions on Electron Devices, vol. 65, no. 5, 2018, pp. 1679-1684.

[2] John Robertson, Report on Progress in Physics, vol. 69, no. 2, 2005, pp. 327–396.

[3] X. Wang, *el al.*, Japanese Journal of Applied Physics, vol. 50, no. 10F, 2011, pp. 10PF02.

[4] Q. Zhang, *et al.*, IEEE Electron Device Letters, vol. 27, no. 9, 2006, pp. 728-730.

[5] H. C. Chang, *et al.*, Journal of Applied Physics, vol. 111, no. 7, 2012, pp. 076105.

[6] M. Choi, *et al.*, Journal of Applied Physics, vol. 113, no. 4, 2013, pp. 044501.

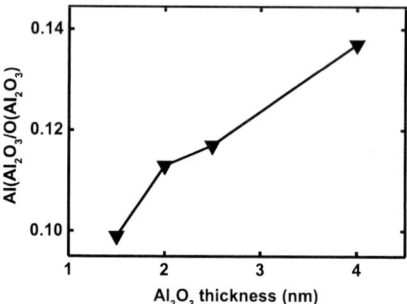

Fig. 4 Areal intensity ratio of $Al(Al_2O_3)/O(Al_2O_3)$ for different Al_2O_3 thicknesses.

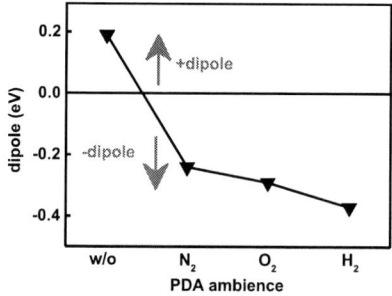

Fig. 5 Dipole distribution formed at the GeO_x/Al_2O_3 interface for the samples without and with PDA in N_2, O_2, and H_2 ambiences.

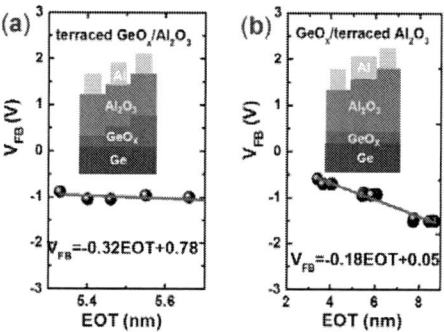

Fig. 1 V_{FB} versus EOT plots of (a) Ge/terraced GeO_x/Al_2O_3/Al and (b) Ge/GeO_x/terraced Al_2O_3/Al structure of the sample without annealing.

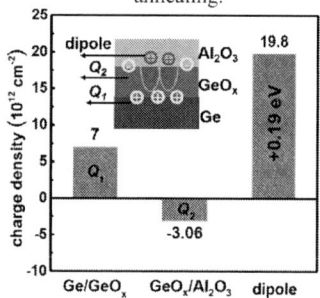

Fig. 2. Charge distribution in the Ge/GeO_x/Al_2O_3 gate stacks.

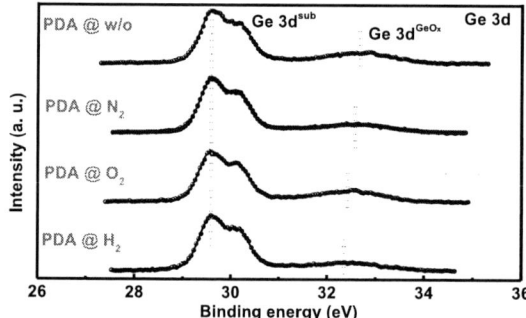

Fig. 6 The Ge 3d spectra of Ge/GeO_x/Al_2O_3 structure without and with PDA in N_2, O_2, and H_2 ambiences.

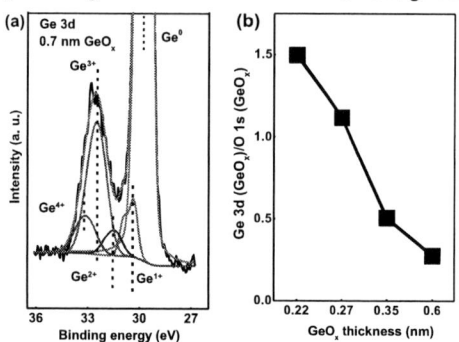

Fig. 3(a) is the Ge 3d spectrum of 0.7 nm GeO_x and (b) is areal intensity ratio of $Ge(GeO_x)/O(GeO_x)$ for different GeO_x thicknesses.

Table 1: Summary of equations to extract charge distribution in gate stacks of Ge MOSCAP.

Eq. 1 $V_{FB} = -\dfrac{Q_1}{\varepsilon_0\varepsilon_r}EOT - \dfrac{\varepsilon_1\rho_1}{2\varepsilon_0\varepsilon_r^2}EOT^2 - \dfrac{Q_2}{\varepsilon_0\varepsilon_r}EOT_2 + \dfrac{(\varepsilon_1\rho_1-\varepsilon_2\rho_2)}{2\varepsilon_0\varepsilon_r^2}EOT_2^2 + \Delta + \phi_{ms}$
for terraced GeO_x
Eq. 2 $V_{FB} = -\dfrac{Q_1+Q_2}{\varepsilon_0\varepsilon_r}EOT + \dfrac{\varepsilon_3\rho_2 EOT_1}{\varepsilon_0\varepsilon_r^2}EOT - \dfrac{\varepsilon_2\rho_2}{2\varepsilon_0\varepsilon_r^2}EOT^2 - \dfrac{Q_2 EOT_1}{\varepsilon_0\varepsilon_r} - \dfrac{\varepsilon_2\rho_2}{2\varepsilon_0\varepsilon_r^2}EOT_1^2 + \Delta + \phi_{ms}$
for terraced Al_2O_3

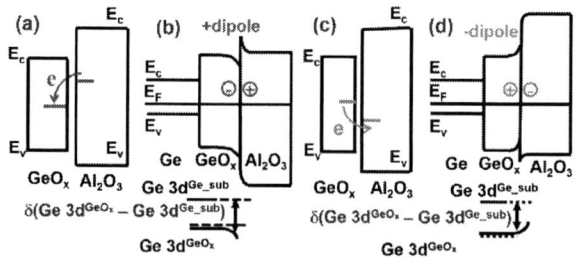

Fig. 7 Simplified band alignment diagrams of Ge/GeO_x/Al_2O_3 structure: (a) with higher CNL in Al_2O_3 and lower CNL in GeO_x before GeO_x and Al_2O_3 contact, (b) positive dipole formation after contact, (c) with lower CNL in Al_2O_3 and higher CNL in GeO_x before GeO_x and Al_2O_3 contact, and (d) negative dipole formation after contact.

978-1-7281-0943-5/19 $31.00 © 2019 IEEE

Contradiction Behaviors between I-V and C-V Curves after Self-Heating Stress in a-IGZO TFT with Triple-Stacked Channel Layers

Yu-Ching Tsao[1], Mao Chou Tai[2], Ting-Chang Chang[1*]

[1]Department of Physics, National Sun-Yat-Sen University, 70 Lien-hai Road, Kaohsiung 804, Taiwan R.O.C.
[2]Department of Photonics, National Sun-Yat-Sen University, 70 Lien-hai Road, Kaohsiung 804, Taiwan R.O.C.
Email: tcchang3708@gmail.com

ABSTRACT

In this work, an opposite sub-threshold swing trend of I-curve to C-V curve in tri-layer IGZO TFT is observed. Thermal and electrical field effects during stress duration is investigated to clarify this behavior. Results indicate a carrier distribution migration phenomenon; carriers are away from the gate insulator surface after self-heating stress which is caused by defect generations at channel/GI interface. Since carrier tends to locate at the middle layer (In-rich layer) after self-heating stress, carriers away from surface defects leads to a better sub-threshold swing in I-V curve. However, C-V curves detects the information of defects leading to S.S degradation.

INTRODUCTION

In AM displays, thin film transistors (TFT) has been playing a big role in pixel arrays. Among all materials used in the active layer of TFTs, IGZO has been the most promising candidate to be adopted in the next-generation display, due to its high electron mobility, large uniformity, room temperature fabrication process, and optical transparency [1]. However, amorphous IGZO TFT's performance is strongly dependent on the device's channel/GI interface. Accordingly, the mobility of conventional devices has a limit around $10 cm^2/V \cdot s$. Along with the development of high resolution and high frame rate LCD displays or AMOLED displays, the required mobility to drive TFT arrays has become harsher. Therefore, previous literatures have demonstrated bi- or tri-layer channels to improve carrier mobility to overcome application limits [2]. During AMOLED TFT operations, a driving TFT is constantly suffering from a high current to drive the OLED. However, a lack of work has discussed about degradation behaviors of multilayer channel TFTs under self-heating effects. In this work, a sandwich structure under self-heating effect is the first to observe a current increase and decreased subthreshold swing (S.S) in I-V curves but an increase of S.S in C-V curves. Views of thermal effects and electrical fields are discussed to clarify this phenomenon.

RESULTS AND DISCUSSION

Figure 1 illustrates the cross section of the back-channel etching type a-IGZO TFT. The channel is composed of a triple-stacked layer which could be seen from the TEM, demonstrated in Figure 2. It should be noticed that the middle layer is an In-rich layer. Figure 3 demonstrates the characteristic curves (a) and capacitance-voltage curves (b) during stress durations. Results from characteristic curves, has observed an on current increase along with a subthreshold decrease demonstrated in the voltage correction I-V curve, shown in Figure 4. On the other hand, a S.S has observed an obvious degradation in C-V curves. In common results, S.S degradations of C-V curve is related to defect generations at the channel/GI interface which is accompanied by S.S increase in I-V curves. However, in this work, contradiction results have been the first to be observed. Normally, from ideal MOSFET drain current formula, Ion increase could be explained by the shrink of effective channel length or increase in carrier mobility. However, from identical initial values of C-V curve, it has confirmed that the effective channel remains the same after self-heating stress. Furthermore, AC SHS is performed to inspect whether mobility enhance are results attributed from percolation models [3]. Figure 5 has demonstrated AC stress with a fixed cooling time but different heating times to form different Joule heat on both similar devices. Results has verified that Joule heating hasn't affected the carrier mobility. At last, from electrical properties illustrated in Figure 6, altering different drain voltages causes varies degradation amounts indicating that the phenomenon is related to strong lateral electrical field at the pinch off region. Consequently, a model is given through the results to interpret this phenomenon. After long time SHS, defect generations are caused by the strong lateral electrical field and electrons are trapped at interface defects during conduction. Therefore, an amount of negative charges causing the energy band rising at the front layer. Meanwhile, since carriers tend to conduct through low resistance paths, carriers end up passing through the middle layer (In-rich layer) which enhances the carrier mobility and carriers away from interface defects leads to a better subthreshold swing, demonstrated in Figure 7. On the other hand, C-V measurement gives information about the whole channel. Therefore, interface defects have contributed to the S.S degradation in C-V measurements.

CONCLUSION

A contradiction behavior between I-V and C-V curve is the first to be observed in a-IGZO. Both thermal and electrical propertied during stress durations were discussed. Results has shown a change in conduction path before and after self-heating stress, due to stress-induced defect generations and electron trapping at the front channel.

REFERENCE

[1] K. Nomura, H. Ohta, A. Takagi, T. Kamiya, M. Hirano, and H. Hosono, "Room-temperature fabrication of transparent flexible thin-film transistors using amorphous oxide semiconductors," Nature, vol. 432, no. 7016, pp. 488-492, Nov. 2004.
[2] H. Xie, Q. Wu, L. Xu, L. Zhang, G. Liu, C. Dong (2016). Nitrogen-doped amorphous oxide semiconductor thin film transistors with double-stacked channel layers. *Applied Surface Science*, *387*, 237-243.
[3] T. Kamiya, and H. Hosono, "Material characteristics and applications of transparent amorphous oxide semiconductors," NPG Asia Mater., vol. 2, no. 1, pp. 15-22, Jan. 2010.

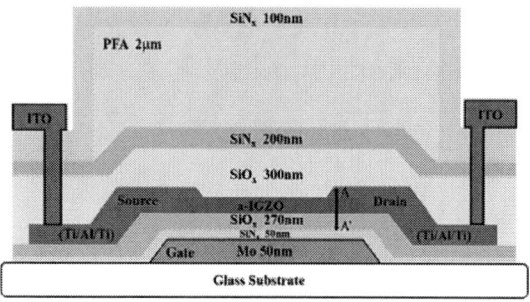

Figure 1. Cross section of the back-channel etching tripled-stacked channel layer a-IGZO TFT. More information of the tri-channel layer could be seen in Figure. 2

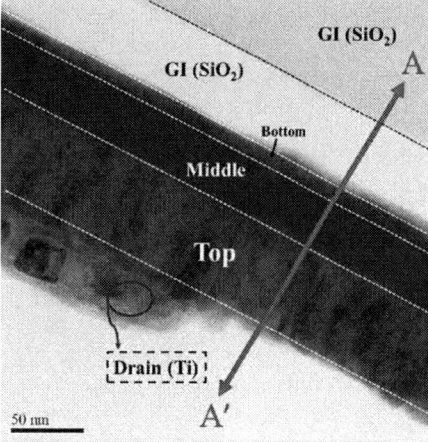

Figure 2. TEM figure from drain to gate insulator, illustrating a sandwich structure channel.

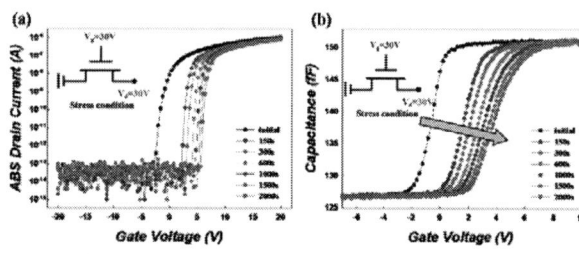

Figure 3. (a) I-V curve and (b) C-V curve under self-heating stress.

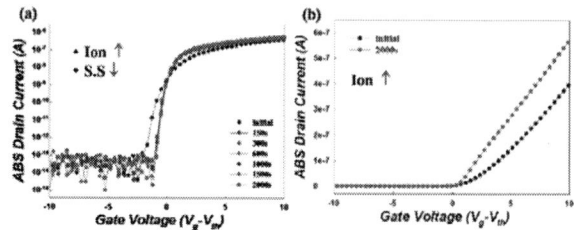

Figure 4. Gate voltage correction of the I-V curve with (a) log scale and (b) linear scale.

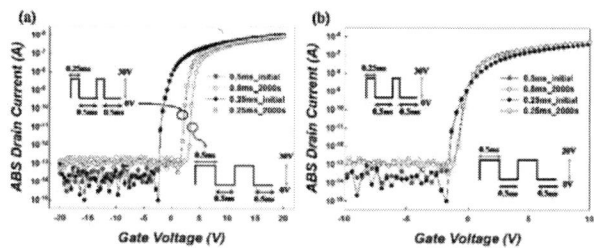

Figure 5. (a) Different heating time of self-heating stress in two similar structures. The black(red) line with a pulse of 0.25ms(0.5ms) heating time and 0.5 cooling time. (b) Gate voltage correction of both SHS stress.

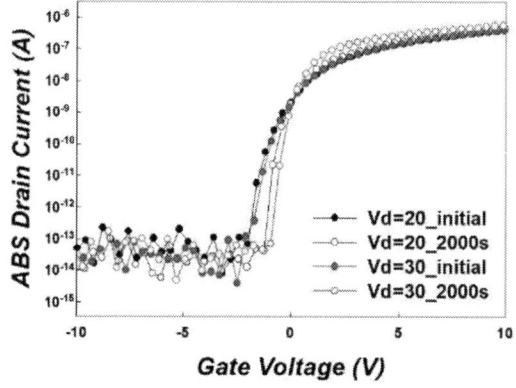

Figure 6. Gate voltage correction characteristic curve of self-heating stress with different drain voltages. The black(red) curve are applied with V_d=30V and V_d=30V(20V).

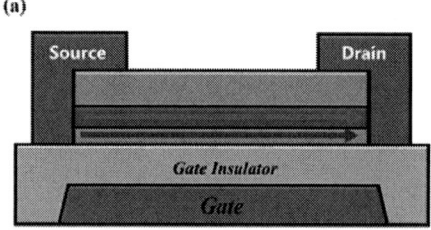

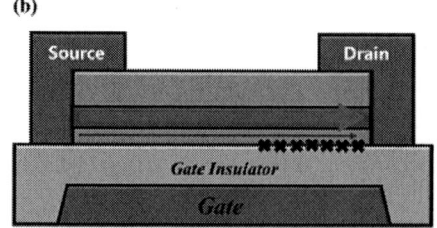

Figure 7. Illustration of carrier conduction before(a) and after(b) 2000s SHS. The thickness of each line indicates the carrier density at each layer.

978-1-7281-0943-5/19 $31.00 © 2019 IEEE   60

Defect Localization and Electrical Fault Isolation for Metal Connection using Helium Ion Microscope

Deying Xia, Shawn McVey and Wilhelm Kuehn

Carl Zeiss SMT Inc
One Corporation Way, Peabody, MA 01960, USA
E-mail: deying.xia@zeiss.com

INTRODUCTION

As semiconductor process technology moves further to smaller feature sizes, novel materials and processes are introduced to increase functional device density. [1] It is challenging to have facile approaches to evaluate the new processes for metal connections, including defect localization and electrical fault isolation. Passive voltage contrast imaging is an easy way to localize defects in conductive structures. Charge accumulation with primary electron or ion beam irradiation will cause the image darkening in floating structure regions revealing defects. In general, Ga focused ion beam (Ga-FIB) and scanning electron microscope (SEM) have been used for defect localization with voltage contrast in the semiconductor industry. [2] Helium ion microscope (HIM) has been developed as light ion beam technology for many applications in imaging, nanofabrication and elemental analysis. [3] Voltage contrast images for defect localization and verification of electrical fault isolation using HIM have advantages such as lower sputtering and higher resolution compared to Ga-FIB and greater voltage contrast compared to SEM.

INSTRUMENTS AND TEST SAMPLES

Carl Zeiss Orion NanoFab was used to provide HIM images with passive voltage contrast to examine defect locations in metal connections. NanoFab is a multi-ion beam system (Ga, He and Ne) optionally equipped with gas injection system (GIS). XeF_2 gas in GIS was used to enhance etching of metal connection in some tests.

Two kinds of test samples (TC1 and TC2) were used here to demonstrate applications of voltage contrast in HIM images for metal connection. 25-nm patterned Ti_3N_4 structures are on 20nm silicon oxide on Si substrate for TC1 sample while copper metal lines are on thick silicon oxide on Si for TC2 sample.

RESULTS AND DISCUSSIONS

HIM images could easily be examined for defect positions or disconnections in metal connections using voltage contrast. HIM images in Figure 1 show both defect-free region (Fig. 1.A) and defective region as dark stripes (Fig. 1.B). There are five long dark stripes beyond the field of view of image in Fig.1. B. Two ends of one short dark stripe in middle of image in Fig. 2.B obviously exhibit defects of metal connections. The broken gap of sub-10nm is easy to view from voltage contrast HIM images. The voltage contrast in HIM image is sufficient for quick checking in fast mode over a large area.

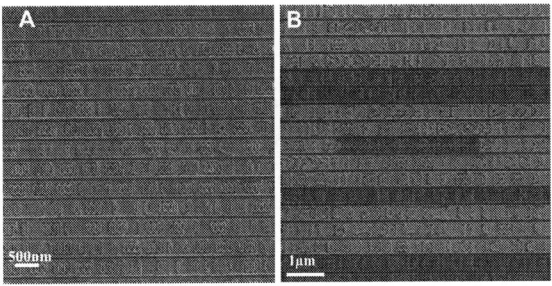

Figure 1: Helium ion microscope images for TC1: (A) defect-free region; (B) defective region showing defects responsible for electrically isolated region.

For circuit edit and failure analysis techniques, it is necessary to intentionally isolate the electrical connection. In such circumstances, passive voltage contrast can be also used to verify the successful isolation or metal cut. On TC1 sample, neither helium or neon ion beams could effectively isolate metal connections due to re-deposition issues. [2] With XeF_2 gas assistance, it is easy to form electrically isolation for metal structures using He or Ne ion beam as confirmed from voltage contrast due to formation of volatile by-products. [4] Images in Figure 2 show obvious dark voltage contrast after gas-assisted ion beam etching of metal connections. After etching both ends of segment on the middle stripe with helium ion beam and XeF_2 gas as white arrows indicated in Fig. 2.A, voltage contrast segment is formed. This approach also works for localizing the hidden defect position for one-location

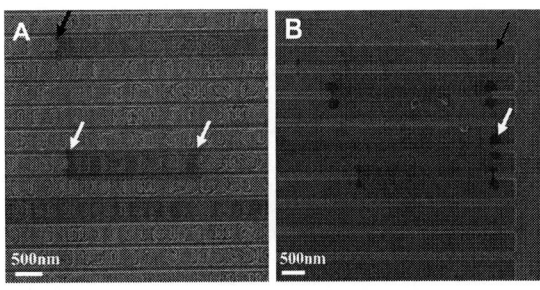

Figure 2: Helium ion microscope images with XeF_2 gas-assisted ion beam etching for TC1: (A) helium ion beam; (B) neon ion beam.

978-1-7281-0943-5/19 $31.00 © 2019 IEEE

defect situation. The upper voltage contrast segment after one end etching as indicated by black arrow in Fig. 2.A. exhibited such situation. With moving image to right, one could find another natural defect location at end of voltage contrast segment. HIM image in Fig. 2.B show the comparison with and without XeF_2 gas assistance for neon ion etching on metal line structures. The voltage contrast is obvious and the width of etched line is enhanced with XeF_2 gas as white arrow indicates in Fig. 2.B. It is not well isolated with neon ion beam only as indicated by black arrow in Fig. 2.B.

It is also possible to have a series of defects along a single long dark stripe. In that case, only the left-most and right-most defects are easily found. The neon ion beam irradiation process could be introduced to localize the intermediate defects. After neon irradiation on the P1 region, the voltage contrast segment disappeared as shown in Figure 3. If moving image to right further along P1 stripe, the closest defect location would be found as starting end of voltage contrast segment (data not showing here).

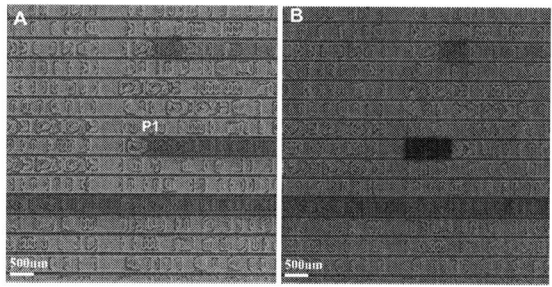

Figure 3: Helium ion microscope images for TC1: (A) before neon irradiation on P1 region; (B) after neon irradiation on P1 region.

The voltage contrast in HIM images is obvious as well for copper lines (~70nm wide) after neon cutting for TC2 sample as shown in Figure 4. Electrical measurement data also confirmed electrical isolation after neon beam cutting on vertical metal lines. More importantly, rewiring of metal connection could be carried out for further investigations in circuit edit using NanoFab system with GIS for metal deposition such as Co deposition. [5]

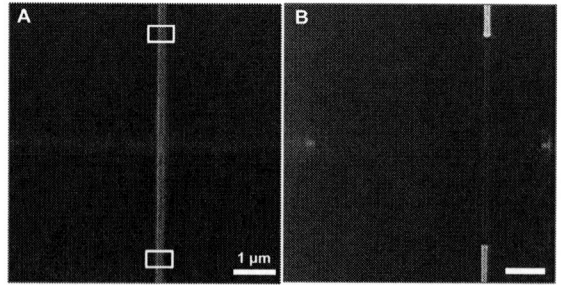

Figure 4: Helium ion microscope images for TC2: (A) before neon cutting on white box region; (B) after neon cutting.

The voltage contrast method using HIM images described here may be helpful to evaluate new semiconductor processes for nano-scale feature sizes without risk of gallium affecting the metal connection. This method is also applicable for other application fields such as nanofabrication of conductive structures, 2D materials, MEMS processing etc. [6] Active voltage contrast using HIM images could also provide the electrical potential distribution in nano-scale on capacitor structures. [7]

CONCLUSION

In summary, passive voltage contrast with helium ion microscope images for defect localization and electrical fault isolation for metal connections are demonstrated in this work. The voltage contrast using HIM images has high resolution, high contrast and less damage to sample surface. The defects in nanoscale could be easily viewed with voltage contrast and motion of imaging. With helium or neon ion beam and XeF_2 gas assistance, the electrically isolated structures were successively fabricated with voltage contrast indications. Helium ion beam etching with XeF_2 gas could be also used to diagnose the defect position for a one-location defect case. Neon ion beam irradiation on defective location was introduced to identify the intermediate defect locations in series defect cases. These results confirm that helium ion beam technique is an attractive method for circuit edit, failure analysis and process evaluations of conductive materials using voltage contrast images in semiconductors.

REFERENCES

[1] S. Tan & R. Livengood, Applications of GFIS in Semiconductors, in *Helium Ion Microscopy;* Hlawacek. G; Golzhauser, A. Springer International Publishing, Switzerland, 2016, pp.471-498.

[2] R. Rosenkranz, Failure Localization with Active and Passive Voltage Contrast in FIB and SEM. J. Mater. Sci: Mater Electron, 2011, 22, pp.1523-1535.

[3] G. Hlawacek, V. Veligura, R. Gastel, & B. Poelsema, Helium Ion Microscopy. J. Vac. Sci. Tech. 2014, B 32, 020801.

[4] M. G. Stanford, K. Mahady, B. B. Lewis, J. D. Fowlkes, S. Tan, R. H. Livengood, G. A. Magel, T. M. Moore, & P. D. Rack, Laser Assisted Focused He^+ Ion Beam Induced Etching with and without XeF_2 Gas Assist. ACS Appl. Mater. Interfaces 2016, 8, pp.29155-29162

[5] H. Wu, L. Stern, D. Xia, D. Ferranti, B. Thompson, K. L. Klein, C. M. Gonzalez, & P. R. Rack, Focused Helium Ion Beam Deposited Low Resistivity Cobalt Metal Lines with 10nm Resolution: Implications for Advanced Circuit Editing. J. Mater. Sci: Mater Electron 2014, 25, pp. 587-595.

[6] V. Oberi, I. Vlassiouk, X. G. Zhang, B. Matola, A. Linn, D. C. Joy, & A. J. Rondinone, Maskless Lithography and in situ Visulization of Conductivity of Graphene using Helium Ion Microscopy. Sci. Rep. 2015, 5, 11952.

[7] C. Sakai, N. Ishita, H. Masuda, S. Nagano, K. Onishi, & D. Fujita, In situ Voltage-application System for Active Voltage Contrast Imaging in Helium Ion Microscope. J. Vac. Sci. Technol. B, 2018, 36, 042903.

Thermal Atomic Layer Etching of Amorphous and Crystalline Hafnium Oxide, Zirconium Oxide, and Hafnium Zirconium Oxide

Jessica A. Murdzek and Steven M. George

Dept. of Chemistry, University of Colorado
Boulder, CO 80309-0215
E-mail: jessica.murdzek@colorado.edu

ABSTRACT

Thermal atomic layer etching (ALE) using the fluorination and ligand-exchange mechanism was employed to etch amorphous and crystalline films of hafnium oxide, zirconium oxide, and hafnium zirconium oxide. HF was the fluorination reactant and dimethylaluminum chloride (DMAC) or titanium tetrachloride was the metal precursor for ligand-exchange. The amorphous films etched faster than the crystalline films. The differences were most pronounced for hafnium oxide. At 250 °C, the etch rate was 0.03-0.08 Å/cycle for crystalline HfO_2 and 0.68 Å/cycle for amorphous HfO_2.

INTRODUCTION

ALE is a method to remove thin films with Ångstrom level precision using sequential, self-limiting surface reactions [1]. ALE can be accomplished with either plasma [1] or thermal [2] ALE methods. Plasma ALE is anisotropic and involves energetic ions or neutrals to remove material. Thermal ALE is isotropic and is viewed as the reverse of atomic layer deposition (ALD).

Thermal ALE can be performed using the fluorination and ligand-exchange mechanism [2]. For metal oxides, the fluorination reaction converts the surface of the metal oxide to a surface fluoride layer. The ligand-exchange reaction can then volatilize the metal fluoride layer. Many metal oxides have been etched including Al_2O_3, ZrO_2, and HfO_2 [3]. Metal nitrides such as AlN can also be etched using a similar approach [4]. Other mechanisms for thermal ALE also are possible including oxidation and fluorination to a volatile fluoride. This mechanism has been demonstrated for TiN ALE [5].

The thermal ALE of amorphous ZrO_2 and HfO_2 has been demonstrated using HF for fluorination and dimethyl aluminum chloride (DMAC) or $SiCl_4$ as metal precursors for ligand-exchange [3]. In contrast, there have been no reports for the thermal ALE of crystalline ZrO_2 and HfO_2. The thermal ALE of crystalline AlN has been demonstrated using HF for fluorination and $Sn(acac)_2$ as the metal precursor for ligand-exchange [4]. There have been no reports for the thermal ALE of amorphous AlN.

This study focuses on the thermal ALE of amorphous and crystalline phases of ZrO_2 and HfO_2. The amorphous films were deposited using ALD. The crystalline films were obtained by annealing. The temperature of crystallization for thin films of ZrO_2 is between 200-550 °C and for thin films of HfO_2 is between 300-800 °C. The etching of crystalline materials is important because etching is required to obtain ultrathin crystalline films. Films may not crystallize easily when they are too thin. Consequently, films may have to be grown thicker, crystallized, and then etched back to obtain the desired ultrathin thickness [6]. A schematic illustrating this processing sequence is shown in Figure 1.

Differences between the thermal ALE of amorphous and crystalline films may also be important for selective ALE. Selectivity is obtained when two different materials have different etch rates under the same conditions or when one material etches while the other does not etch [3]. Selective thermal ALE has been observed for a variety of metal precursors and materials [3]. Selectivity could also be observed between amorphous and crystalline phases of the same material.

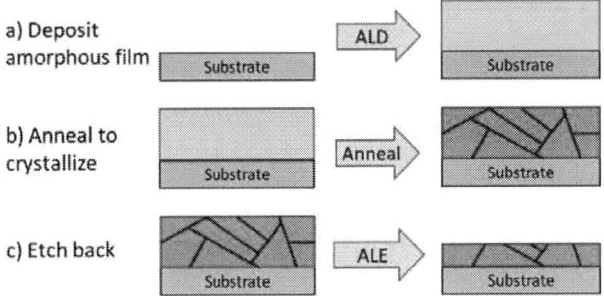

FIGURE 1. Schematic detailing processing sequence to obtain an ultrathin crystalline film.

This work used ZrO_2, HfO_2, and $HfZrO_4$ thin films on silicon coupons provided by the Tokyo Electron Limited (TEL) Technology Center, America, LLC in Albany, NY. The as deposited films were deposited using ALD methods at 250 °C. These as deposited, amorphous films were used without further modification. These films had a thickness of approximately 100 Å. The anneal 1 and anneal 2 films were both annealed at 800 °C in an N_2 environment. The anneal 1 films had a polycrystalline morphology. The anneal 2 films were fiber textured. The annealed ZrO_2 materials were tetragonal. The annealed HfO_2 materials were monoclinic. The annealed $HfZrO_4$ materials were a mixture of monoclinic and tetragonal. The crystal structures were verified using grazing incidence x-ray diffraction (GIXRD).

Thermal ALE experiments were performed in a viscous flow reactor. Each thermal ALE cycle included one exposure for 1 second of HF and one exposure for 2 seconds of DMAC or $TiCl_4$. There was a 30 second purge between each reactant exposure. N_2 was employed as the viscous flow carrier gas. The reactor pressure with flowing N_2 carrier gas was 1 Torr. The reported film thicknesses were measured using ex situ spectroscopic ellipsometry (SE) measurements.

RESULTS

Hafnium Oxide: Figure 2 shows the results for the hafnium oxide films. SE measurements were used to measure HfO_2 film thicknesses over 400 thermal ALE cycles using HF and DMAC at 250 °C. The amorphous HfO_2 film was etched completely after 150 cycles. The amorphous HfO_2 film had an etch rate of 0.68 Å/cycle. In contrast, the anneal 1 and anneal 2 HfO_2 films had much lower etch rates of 0.03 and 0.08 Å/cycle, respectively.

Results with HF and $TiCl_4$ as the reactants at 250 °C showed similar differences between amorphous and crystalline HfO_2. The etch rate for the amorphous HfO_2 film was 0.36 Å/cycle with HF and $TiCl_4$.

978-1-7281-0943-5/19 $31.00 © 2019 IEEE

The etch rate of the amorphous HfO$_2$ film was 14-24 times higher than the etch rate for crystalline HfO$_2$ films.

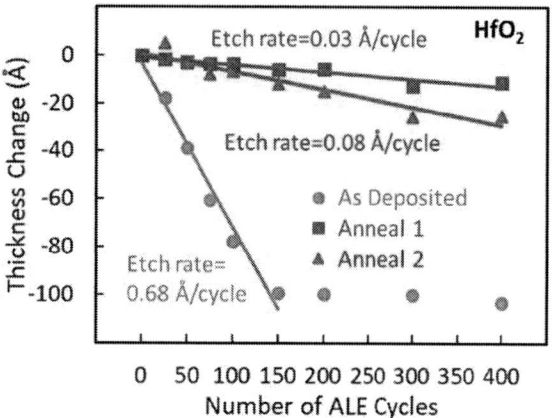

FIGURE 2. ALE for three different hafnium oxide films using HF and DMAC at 250 °C.

Zirconium Oxide: Figure 3 displays the results for the zirconium oxide film*s*. The amorphous ZrO$_2$ films and the crystalline ZrO$_2$ films were etched with HF and DMAC at 250 °C. Under the same reaction conditions, all three ZrO$_2$ films etched faster than any of the HfO$_2$ films. The etch rates were 1.11, 0.74, and 0.82 Å/cycle for the amorphous, anneal 1, and anneal 2 ZrO$_2$ films, respectively. In contrast to the results for the HfO$_2$ films, the etch rates for the amorphous and crystalline ZrO$_2$ films were similar. The ZrO$_2$ films were completely etched away after 75 cycles for the amorphous ZrO$_2$ films and after 100 cycles for the anneal 1 and anneal 2 ZrO$_2$ films.

The ZrO$_2$ films were also etched with HF and TiCl$_4$ as the reactants at 250 °C. The etch rates using HF and TiCl$_4$ were lower than the etch rates using HF and DMAC. The amorphous ZrO$_2$ film had an etch rate of 0.68 Å/cycle. The crystalline anneal 1 and anneal 2 ZrO$_2$ films had etch rates around 0.2 Å/cycle.

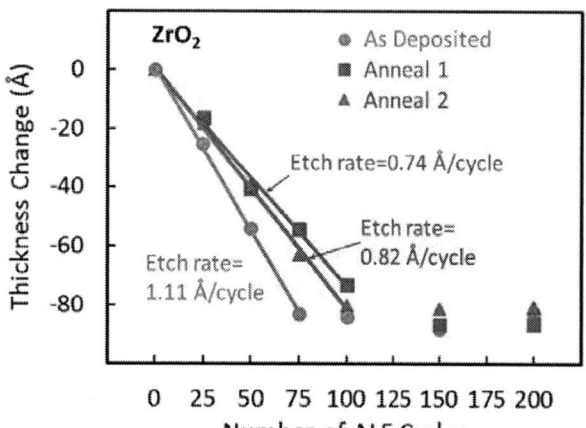

FIGURE 3. ALE for three different zirconium oxide films using HF and DMAC at 250 °C.

Hafnium Zirconium Oxide: Figure 4 shows the results for the composite HfZrO$_4$ films. The amorphous HfZrO$_4$ films and the crystalline HfZrO$_4$ films were etched with HF and DMAC at 250 °C. These measured etch rates were between the HfO$_2$ and ZrO$_2$ etch rates. The amorphous HfZrO$_4$ film was etched completely in 150 ALE cycles. The amorphous HfZrO$_4$ film had an etch rate of 0.69 Å/cycle.

The etch rates for anneal 1 and anneal 2 HfZrO$_4$ films were 0.13 and 0.16 Å/cycle, respectively.

Etch results for the HfZrO$_4$ films using HF and TiCl$_4$ as the reactants at 250 °C displayed similar results. The amorphous HfZrO$_4$ film had an etch rate that was faster than the two annealed HfZrO$_4$ films. The etch rate for the amorphous HfZrO$_4$ film was 0.35 Å/cycle. The etch rates for the two annealed HfZrO$_4$ films were 0.03-0.04 Å/cycle.

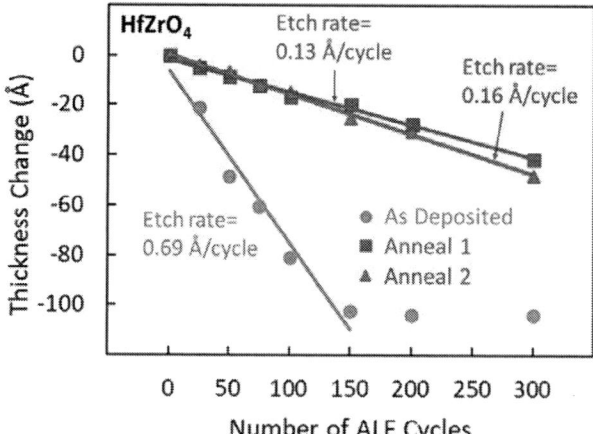

FIGURE 4. ALE for three different HfZrO$_4$ films using HF and DMAC at 250 °C

CONCLUSIONS

The etching of amorphous and crystalline HfO$_2$, ZrO$_2$, and HfZrO$_4$ films by thermal ALE was studied at 250 °C. All amorphous films etched faster than the crystalline films for each material. The difference was the most dramatic for the HfO$_2$ films where the amorphous HfO$_2$ films etched 8-22 times faster than the crystalline HfO$_2$ films. The thermal ALE of crystalline films should be useful to produce ultrathin crystalline films by depositing thicker films that can crystallize via annealing and then etching back to obtain ultrathin crystalline films.

ACKNOWLEDGEMENTS

This research was funded by the National Science Foundation through grant CHE-1609554. The authors thank Dr. Kanda Tapily and Dr. Gert Leusink from Tokyo Electron Limited for sending the HfO$_2$ and ZrO$_2$ samples. The authors also thank Prof. Andrew Kummel from the University of California San Diego for valuable discussions.

REFERENCES

[1] Kanarik, K. J.; Lill, T.; Hudson, E. A.; Sriraman, S.; Tan, S.; Marks, J.; Vahedi, V.; Gottscho, R. A. *J. Vac. Sci. Technol., A* **2015**, *33*, 020802.

[2] Lee, Y.; George, S. M. *ACS Nano* **2015**, *9*, 2061-2070.

[3] Lee, Y.; Huffman, C.; George, S. M. *Chem. Mater.* **2016**, *28*, 7657-7665.

[4] Johnson, N.R.; Sun, H; Sharma, K; George, S.M. *J. Vac. Sci. Technol., A* **2016**, *34*, 050603.

[5] Lee, Y.; George, S.M. *Chem. Mater.* **2017**, *29*, 8202-8210.

[6] George, S.M.; Lee, Y. *ACS Nano* **2016**, *10*, 4889-4894.

Thermal Stability of Shallow Ge N⁺-P Junction with Thin GeSn Top Layer

Hsiu-Hsien Liao, Yi-Ju Chen, and Bing-Yue Tsui

Institute of Electronics, National Chiao Tung University

Room ED641, No. 1001, Ta-Hsueh Road, Hsinchu, 300 Taiwan, R.O.C.

Tel: 886-3-5131570   Fax: 886-3-5724361   E-mail: bytsui@mail.nctu.edu.tw

Abstract

Thermal stability of the Ge N⁺-P junction with thin GeSn top Layer is evaluated in this work. Thin GeSn itself would not increase junction leakage current although its bandgap is narrower than Ge. However, high dose ion implantation would damage the GeSn layer and Ge substrate so that the diffusion coefficient of Sn atom in Ge is enhanced. In this case, thermal annealing higher than 500 °C would degrade junction leakage current. It is thus suggested that low defects doping technique must be developed.

Introduction

New channel materials (e.g. Ge, GaAs, and GaSb) with high carrier mobility offer nice avenues for better performance of MOSFETs. Among them, Ge is a prospective candidate f on account of its high and symmetric carrier mobility. Incorporating Sn into Ge can further boost the electron and hole mobility [1]. Several GeSn-related MOSFETs with high on-state current and mobility have been demonstrated [2-6]. However, because of the low surface energy and large covalent radius of the Sn atom, Sn has a tendency to segregate during crystal growth and post-growth processes. There are little works devoted on the shallow junction properties with GeSn surface layer [7]. Thus, a thorough investigation into the impact of GeSn film on its underneath Ge N⁺-P junction is indispensable, as it gives a guideline for the thermal budget that can be used in process integration.

Experiments

Heavily boron-doped Si (100) wafer with resistivity less than 0.1 ohm-cm was used as starting material. Ge buffer layer of several hundred nanometer thick was grown in a CVD system at 375 °C. Ultrathin $Ge_{1-x}Sn_x$ layer, 20-nm-thick for x=0.02 and 8-nm-thick for x=0.08, was grown in the same CVD system at 305 °C. On the higher Sn percentage sample, a 10-nm-thick Ge cap layer was in-situ grown at 305 °C to protect the GeSn layer. Figure 1 shows the schematic structure of these two substrates. After p-well and field implantation, a 200-nm-thick field oxide (FOX) was deposited by a PECVD system followed by a densification annealing at 400 °C for 15 min. After active region patterning, phosphorus (P) ions were implanted at 20 keV to a dose of 1×10^{15} cm⁻². Post-implantation annealing was performed at various temperatures in a rapid thermal annealing (RTA) system. Al was used to form ohmic contact at both front side and backside. The main process steps and the final structure are shown in Fig.2. Sample ID and main split conditions are listed in Table 1.

Results and Discussion

Typical I–V characteristics of the Ge, $Ge_{0.98}Sn_{0.02}$, and $Ge_{0.92}Sn_{0.08}$ N⁺-P junctions formed under 400 °C dopant activation are shown in Fig.3. The forward current/reverse current ratio exceeds 3×10^4 at room temperature. Fig.4 shows the ideality factor (η) and the leakage current density (J_r) at reverse bias at 1 V. For Ge40, GeSn240, and GeSn840, the ideality factors are equal to 1.13, 1.23, and 1.32, respectively. The higher η values for GeSn240 and GeSn840 suggest stronger contribution of recombination current in the depletion region. Apart from the ideality factor, the GeSn240 and GeSn840 diodes manifest higher J_r compared with the pure Ge junctions, which implies more e-h pair generation within the space charge region due to higher point defect density in the epitaxial GeSn and Ge layer than in the Ge bulk substrate. Figure 5 shows the area component (I_{ra}) and peripheral component (I_{rp}) of the reverse leakage current. GeSn240 and GeSn840 have significant I_{ra} component in contrast with Ge40. But, all of them have close I_{rp} values. This indicates negligible leakage contribution of the thin GeSn layer.

The leakage current density of junctions formed with different annealing temperatures are compared in Figs.6 and 7. The leakage current density shows a noticeable increase for both $Ge_{0.98}Sn_{0.02}$ and $Ge_{0.92}Sn_{0.08}$ samples when the annealing temperature increases to 550 °C. The high leakage current density of sample annealed at 350 °C may be caused by incomplete recrystallization of germanium after a high-dose P ion implantation [8].

Fig.8 shows the SIMS depth profiles of Sn and P atoms of samples annealed at 400 and 600 °C. For the 400 C annealed junction, the distribution of Sn is much shallower than that of P so that the junction exhibits low leakage current. And obviously, for the 600°C annealed junction, the low secondary ion intensity and deep distribution of Sn atoms suggests enhanced Sn diffusion in Ge which may results from the Sn-vacancy interaction [9]. This enhanced diffusion phenomenon makes Sn atoms come much closer to the metallurgical junction of N⁺-P junction when a high temperature annealing was carried out. Sn atom acts as e-h pair generation-recombination centers so that the leakage current increases

Conclusions

The effect of thermal process on Ge N⁺-P junction with GeSn top layer was studied. The leakage current exhibits a significant increase when the annealing temperature exceeds 500 °C because the diffusivity of Sn atom is greatly enhanced by implantation-induced defects and then considerably increase leakage current. Low defects doping technique must be developed otherwise the process temperature should be lower than 500 °C from the junction integrity point of view.

Acknowledgement

This work was supported by the Ministry of Science and Technology, Taiwan, R.O.C. under the contract No. MOST 107-2633-E-009-002.

References

[1] Y. Kamata, *Materials Today*, vol.11, pp.30–38, 2008.

[2] G. Han, et al., in *Tech. Dig. of IEDM*, 2011, pp. 402–406.

[3] X. Gong, et al., *IEEE Electron Device Lett.*, vol.34, pp.339–341, 2013.

[4] D. Lei, et al., *J. Appl. Phys.*, vol.119, p.024502, 2016.

[5] D. Lei, et al., in *Proc. Symp. VLSI Technol.*, 2017, pp. T198–T199.

[6] T.-H. Liu, et al., *IEEE Electron Device Lett.*, vol.39, pp.468–471, 2018.

[7] L. Wang, et al., *IEEE Electron Device Lett.*, vol.33, pp. 1529–1531, 2012.

[8] K. Benourhazi and J. P. Ponpon, *Nucl. Instrum. Methods Phys. Res. B, Beam Interact. Mater. At.*, vol.71, pp.406–411, 1992.

[9] V. P. Markevich, et al., *J. Appl. Phys.*, vol.109, p.083705, 2011.

Table 1 Sample ID and main split conditions of all samples.

Substrate	Implantation condition	Post-implantation annealing	Sample ID
Ge$_{0.98}$Sn$_{0.02}$(20 nm)/Ge	Phosphorus 20 keV 1×10^{15} cm^{-3}	350°C, 60sec	GeSn235
		400°C, 60sec	GeSn240
		450°C, 60sec	GeSn245
		500°C, 60sec	GeSn250
		550°C, 30sec	GeSn255
		600°C, 10sec	GeSn260
Ge (10 nm)/ Ge$_{0.92}$Sn$_{0.08}$ (8 nm)/Ge		350°C, 60sec	GeSn835
		400°C, 60sec	GeSn840
		450°C, 60sec	GeSn845
		500°C, 60sec	GeSn850
		550°C, 30sec	GeSn855
		600°C, 10sec	GeSn860
Ge		400°C, 60sec	Ge40
		600°C, 10sec	Ge60

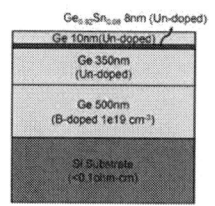

Fig.1 Schematic structure of the two substrates used in this work.

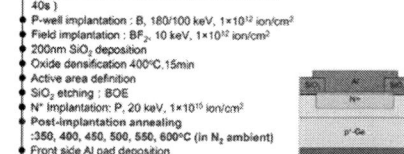

- Starting substrate clean : Acetone+ DHF (Digital etching : 10 cycles RTO, 2min + HCl 10%, 40s)
- P-well implantation : B, 180/100 keV, 1×10^{12} ion/cm^2
- Field implantation : BF$_2$, 10 keV, 1×10^{12} ion/cm^2
- 200nm SiO$_2$ deposition
- Oxide densification 400°C,15min
- Active area definition
- SiO$_2$ etching : BOE
- N$^+$ Implantation: P, 20 keV, 1×10^{15} ion/cm^2
- Post-implantation annealing :350, 400, 450, 500, 550, 600°C (in N$_2$ ambient)
- Front side Al pad deposition
- Al pad patterning
- Al pad etching
- Backside electrode Al deposition

Fig.2 Main process steps and the cross-sectional final structure.

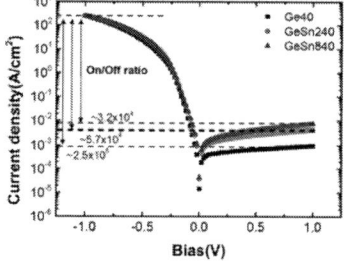

Fig.3 Typical I–V characteristics of the Ge, Ge$_{0.98}$Sn$_{0.02}$, and Ge$_{0.92}$Sn$_{0.08}$ N$^+$-P junctions formed under 400 °C dopant activation.

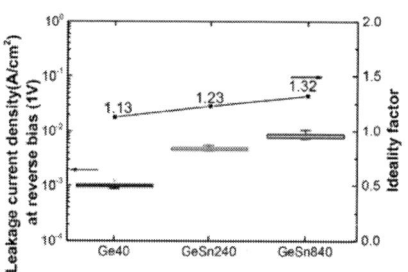

Fig.4 Ideality factor and leakage current density at reverse bias (1V) of the junctions annealed at 400 °C.

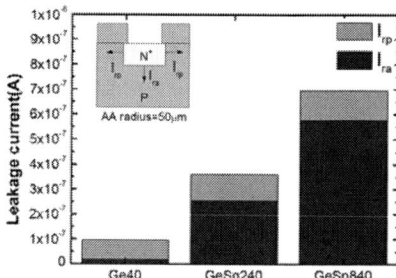

Fig.5 Area component (I$_{ra}$) and peripheral component (I$_{rp}$) of the reverse leakage current of the junctions annealed at 400 °C.

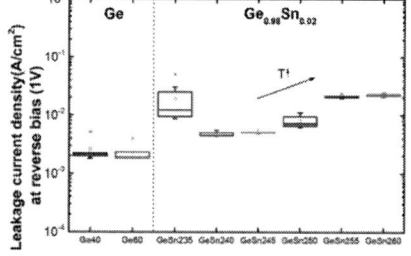

Fig.6 Leakage current density of junctions with Ge$_{0.98}$Sn$_{0.02}$ top layer after annealing at different temperatures.

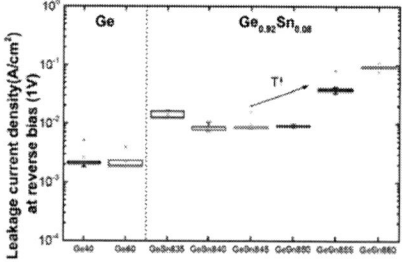

Fig.7 Leakage current density of junctions with Ge$_{0.92}$Sn$_{0.08}$ top layer after annealing at different temperatures.

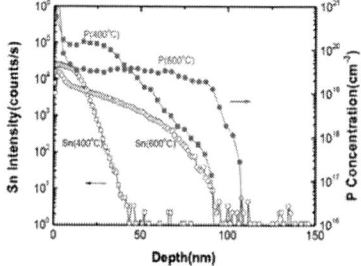

Fig.8 SIMS depth profiles of Sn and P atoms of samples annealed at 400 and 600 °C.

978-1-7281-0943-5/19 $31.00 © 2019 IEEE

Selective Atomic Layer Deposition of TiO$_2$

Christopher Ahles[†], Jong Choi[†], Keith Wong[§], Srinivas Nemani[§] and Andrew Kummel[†,‡]

[†]Materials Science and Engineering Program and [‡]Department of Chemistry and Biochemistry,
University of California, San Diego, La Jolla, California 92093, United States
[§]Applied Materials, Sunnyvale, California 94085, United States
Email: cahles@ucsd.edu

ABSTRACT

A selective atomic layer deposition of TiO$_2$ on Si and SiO$_2$ in preference to SiCOH has been developed. This process utilizes Ti(O^{i}Pr)$_4$ and acetic acid (AcOH) as the co-reactants and deposits TiO$_2$ on Si and SiO$_2$ but not on SiCOH. It is found that 3 nm of TiO$_2$ may be grown on Si and SiO$_2$ before significant deposition on SiCOH occurs and the deposited films have RMS roughnesses of 2-3 Å. The selectivity is attributed to the inherent non-reactivity of -CH$_3$ groups on SiCOH and the mechanism is shown to proceed via activation of the Si-H, Si-OH, and Ti-O^{i}Pr surfaces by reaction with AcOH.

INTRODUCTION

As lithography is reaching its fundamental scaling limit, selective chemical deposition methods are becoming a requirement for further scaling of transistors using double patterning techniques. Furthermore, water-free ALD is desirable because water can corrode metals at high temperature and damage the low-k dielectric SiCOH. Water-free atomic layer deposition (ALD) of TiO$_2$ has been studied using Ti(O^{i}Pr)$_4$ and TiCl$_4$; however, to the best of our knowledge, the selectivity of this precursor combination has not been studied.[1,2] TiO$_2$ ALD using Ti(O^{i}Pr)$_4$ and AcOH has also been studied; however, again the selectivity of this process was not investigated.[3] In this work a water-free TiO$_2$ ALD using Ti(O^{i}Pr)$_4$ and AcOH is reported which is selective for Si and SiO$_2$ in preference to -CH$_3$ terminated SiCOH.

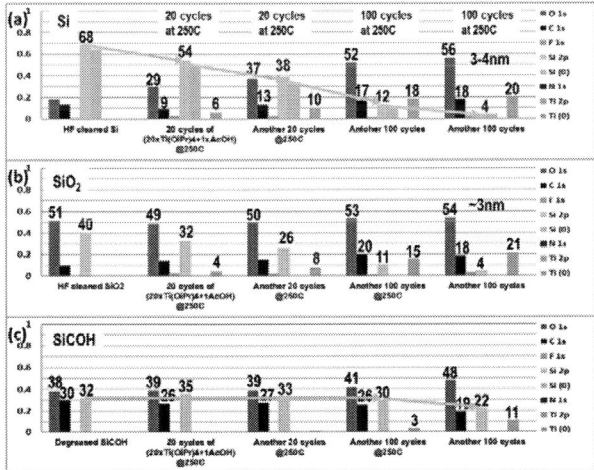

Figure 1. XPS chemical composition of (a) Si, (b) SiO$_2$ and (c) SiCOH surfaces subjected to TiO$_2$ ALD cycles using Ti(O^{i}Pr)$_4$ and AcOH at 250^{0}C. (a) After a total of 240 ALD cycles there is approximately 3-4 nm of TiO$_2$ deposited on Si (calculated based on attenuation of the Si substrate signal). **(b)** After a total of 240 ALD cycles there is approximately 3 nm of TiO$_2$ deposited on SiO$_2$ (calculated based on attenuation of the Si substrate signal). **(c)** After a total of 240 ALD cycles there is 11% Ti on the SiCOH surface. This corresponds to approximately 0.5 nm TiO$_2$, based on attenuation of the Si substrate signal. The TiO$_2$ deposition was selective for Si and SiO$_2$ for the first 140 ALD cycles which deposited ~ 2 nm TiO$_2$ on Si and SiO$_2$ while depositing only 3% Ti on SiCOH (<0.1 nm).

RESULTS

The selective ALD of TiO$_2$ was tested on Si, SiO$_2$ and SiCOH. Besides being a typical low-k dielectric, SiCOH is a model system for perfectly passivated SiO$_2$ since SiCOH is terminated in Si-O-C$_x$H$_y$ groups and should not contain any -OH groups. The Si, SiO$_2$ and SiCOH samples were loaded onto the same sample holder and subjected to the ALD together in order to ensure that all three samples received identical ALD conditions. The results of the TiO$_2$ ALD using Ti(O^{i}Pr)$_4$ and AcOH at 250^{0}C on Si, SiO$_2$ and SiCOH are shown in Figure 1. It is seen in Fig. 1a & 1b that the first 20 ALD cycles nucleated the TiO$_2$ on Si and SiO$_2$, and there was roughly the same amount of deposition on Si as on SiO$_2$. After a total of 240 ALD cycles there was 3-4 nm TiO$_2$ deposited on Si and ~ 3 nm TiO$_2$ deposited on

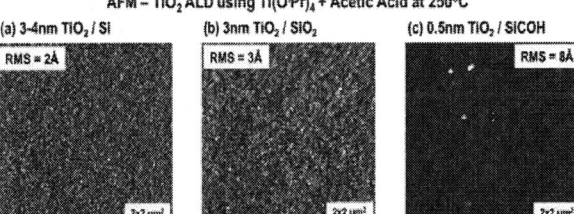

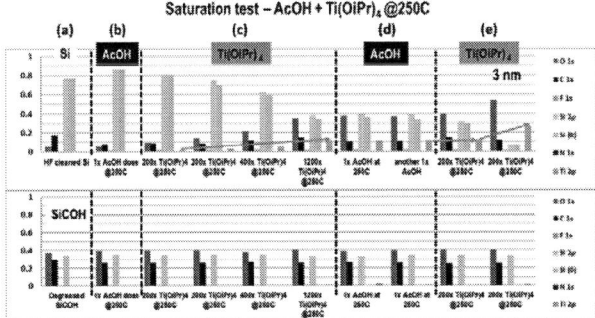

Figure 2. AFM images of the TiO$_2$ film on (a) Si, (b) SiO$_2$ and (c) SiCOH after a total of 240 ALD cycles of Ti(O^{i}Pr)$_4$ and AcOH at 250^{0}C. (a) The TiO$_2$ film on Si has an RMS roughness of 2 Å. The smoothness of this film is consistent with ALD. **(b)** The TiO$_2$ film on SiO$_2$ has an RMS roughness of 3 Å. **(c)** The RMS roughness of the SiCOH surface after the ALD cycles is 8 Å. Large (~6nm) particles are observed scattered on the SiCOH surface. These particles are too big to have been grown by ALD and therefore may indicate a slight CVD component.

Figure 3. XPS chemical composition of (a) HF cleaned Si and degreased SiCOH, (b) after dosing with AcOH at 250^{0}C, (c) after dosing with Ti(O^{i}Pr)$_4$, (d) after dosing with ACOH again and (e) after dosing with Ti(O^{i}Pr)$_4$ again. (a) HF cleaned Si has some O and C contamination. **(b)** Dosing with AcOH at 250^{0}C did little to both substrates. **(c)** Ti(O^{i}Pr)$_4$ nucleation on Si occurs via a slow CVD component at 250^{0}C. **(D)** No change in chemical composition is seen from dosing AcOH, however a change in the C oxidation state occurs (see Figure 4). **(e)** After an incubation period of 200 doses Ti(O^{i}Pr)$_4$ a second 200 doses of Ti(O^{i}Pr)$_4$ increases the Ti% from ~12 to ~28%. This shows that there is TiO$_2$ ALD in combination with a small Ti(O^{i}Pr)$_4$ CVD component.

978-1-7281-0943-5/19 $31.00 © 2019 IEEE

SiO_2. Conversely, it is seen in Fig. 1c that the first and second 20 ALD cycles deposited no Ti on SiCOH. Only after 100 ALD cycles were dosed onto SiCOH, TiO_2 nucleated on SiCOH, and after a total of 240 cycles there was only ~0.5 nm TiO_2 on SiCOH. This is consistent with Si-CH$_3$ termination prevent reaction with both precursors.

The samples were removed from the chamber and AFM was performed to quantify surface morphologies (Figure 2). In Fig. 2a, it is seen that the TiO_2 film deposited on Si had an RMS roughness of 2 Å. The high conformality of this film is consistent with ALD. In Fig. 2b, it is seen that the TiO_2 film deposited on SiO_2 had an RMS roughness of 3 Å. The high conformality of this film is also consistent with ALD. Conversely, on SiCOH, large (~ 6nm) particles are observed scattered on the surface (Fig. 2c). The films grown on Si and SiO_2 are only 3-4 nm thick and so any ALD nuclei on SiCOH should also be 3-4 nm tall. The particles on SiCOH are too tall to have been grown by ALD and are not dense enough to explain the 11% Ti on SiCOH; therefore the particles on SiCOH may be indicative of a slight CVD component which might be eliminated with better reactor design.

A saturation experiment was performed to document ALD reactions. The chemical composition of Si and SiCOH substrates subjected to saturation doses of AcOH and Ti(O^{i}Pr)$_4$ at 250^{0}C were determined by XPS. Figure 3a shows the composition of the HF cleaned Si and degreased SiCOH substrates. All of the Si was in an oxidation state of 0 which is consistent with the HF clean having removed all of the Si-O bonds. No assumptions were made about which precursor would nucleate the ALD and so the substrates were dosed first with AcOH (Fig. 3b). Only a decrease in the C on Si was observed which can be attributed to a modest desorption from the samples being heated to 250^{0}C in vacuum. If AcOH had reacted with the Si surface and left behind a surface -OAc group, a higher binding energy component in the C 1s XPS spectrum should have been observed along with increase O; however, these effects were not observed. Next the samples were subjected to doses of Ti(O^{i}Pr)$_4$ (Fig. 3c). The TiO_2 nucleation on Si occurs by a slow CVD-like dissociative chemisorption of Ti(O^{i}Pr)$_4$, as evidenced by the slow growth and lack of saturation of Ti on the surface. Next the samples were dosed with AcOH (Fig. 3d). While there is no evident change in the chemical composition of the surface, the AcOH induced formation of a high binding energy component in the C 1s XPS spectrum on Si (see Figure 4). This higher binding energy component is attributed to one of the carbons in a surface -OAc group resulting from ligand exchange of a Ti-O^{i}Pr species with Ti-OAc. Next the samples were dosed again with Ti(O^{i}Pr)$_4$ at 250^{0}C (Fig. 3e). The first 200 doses of Ti(O^{i}Pr)$_4$ did not grow any TiO_2 on Si. However, the second 200 doses of Ti(O^{i}Pr)$_4$

grew ~16% Ti on Si. This enhancement of the growth rate after dosing with AcOH shows that the Ti(O^{i}Pr)$_4$ + AcOH is an ALD process in combination with a small CVD component at 250^{0}C.

Figure 4 shows the XPS C 1s spectra on the TiO_2 film grown on Si after alternating doses of Ti(O^{i}Pr)$_4$ and AcOH at 250C. It is seen that after nucleating the ALD with a small amount of Ti(O^{i}Pr)$_4$, all of the C has a binding energy of 284-288 eV (Fig. 4a). This is consistent with surface Ti-O^{i}Pr ligands. Conversely, after dosing with AcOH, a higher binding energy C component is observed (Fig. 4b). This higher binding energy component is consistent with surface -OAc groups. It is seen that this higher binding energy C component disappears after subsequent dosing with Ti(O^{i}Pr)$_4$ (Fig. 4c). The data is consistent with the chemical mechanism shown in Figure 5. The growth is nucleated by the slow reaction of Ti(OiPr)$_4$ with the Si-H terminated surface. Si-H bonds are more reactive than the -CH$_3$ terminated SiCOH surface and so the nucleation is selective for Si over SiCOH. Dosing with AcOH replaces a Ti-O^{i}Pr group with a Ti-OAc group. This Ti-OAc group is reactive towards subsequent Ti(O^{i}Pr)$_4$ molecules. The overall mechanism is activation of the Ti-O^{i}Pr surface by reaction with AcOH.

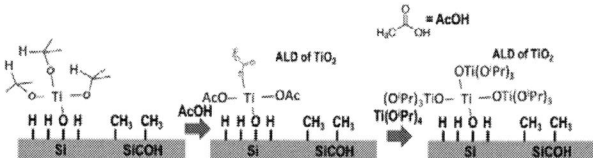

Figure 5. Mechanism of the selective ALD of TiO_2 using Ti(O^{i}Pr)$_4$ and AcOH at 250^{0}C. The mechanism involves the activation of the surface by reaction with AcOH, making it reactive towards Ti(O^{i}Pr)$_4$ again.

CONCLUSION

A selective water-free atomic layer deposition of TiO_2 using Ti(O^{i}Pr)$_4$ and AcOH at 250^{0}C has been developed. This process relies upon the reaction of AcOH with a Ti-O^{i}Pr terminated surface to give a Ti-OAc terminated surface which is reactive towards subsequent dosing of Ti(O^{i}Pr)$_4$. The origin of the selectivity is attributed to the inherent non-reactivity of -CH$_3$ terminated surfaces, such as SiCOH and other low k dielectrics. In addition, this selectivity shows that if an appropriate small molecule passivant is used on SiO_2, selectivity may be attained for deposition on Si in preference to SiO_2.

ACKNOWLEDGEMENTS

This work was supported by Applied Materials.

REFERENCES

1. V. Anderson et al., JVST A 32, 01A114 (2014).
2. R. Chaukulkar et al., JVST A 31, 031509 (2013).
3. K. Ramos et al., Chem. Mater. 2013, 25, 1706−1712.

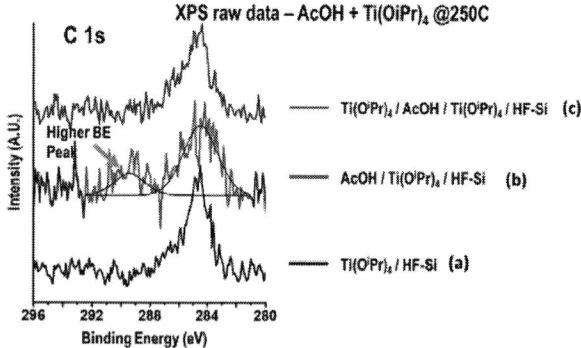

Figure 4. C 1s XPS raw data of a TiO_2 film grown on Si after Ti(O^{i}Pr)$_4$ and AcOH half-cycles at 250^{0}C. (a) The C 1s signal after a total of 2,000 doses of Ti(O^{i}Pr)$_4$ at 250C. **(b)** After dosing with AcOH a higher binding energy component is observed. This is consistent with surface –OAc groups. **(c)** Dosing with Ti(O^{i}Pr)$_4$ again removes this higher binding energy component. This is consistent with ALD via sequential ligand exchange reactions.

SWCNT and SWCNT-based heterostructures for devices

Rong Xiang,[1] Shigeo Maruyama[1,2]

[1] The University of Tokyo, Japan
[2] AIST, Japan
E-mail: xiangrong@photon.t.u-tokyo.ac.jp (RX); maruyama@photon.t.u-tokyo.ac.jp (SM)

ABSTRACT

We will present our recent efforts on controlled fabrication of single-walled carbon nanotube (SWCNT) and SWCNT-based heterostructures. In particular, we will present the full length removal of metallic SWCNTs from the as-synthesis mixtures. The obtained semi-conducting arrays are used to build high-performance SWCNT field effect transistors. In addition, we show the first demonstration of a metallic-insulator-semiconducting heterostructure nanotubes, where single- or few-walled hexagonal boron nitride nanotube (BNNT) and/or MoS2 nanotube seamlessly wrap around a SWCNT, and result in an atomically smooth coaxial nanotube consisting different materials. We name this it one-dimensional van der Waals heterostructures. Various combinations will be demonstrated and the transport properties of this new material/structure will be presented.

FULL LENGTH OF REMOVAL OF METALLIC SWCNTS

Ballistic transport and sub-10 nm channel lengths have been achieved in transistors containing one single-walled carbon nanotube (SWNT). To fill the gap between single-tube transistors and high-performance logic circuits for the replacement of silicon, large-area, high-density, and purely semiconducting (s-) SWNT arrays are highly desired.

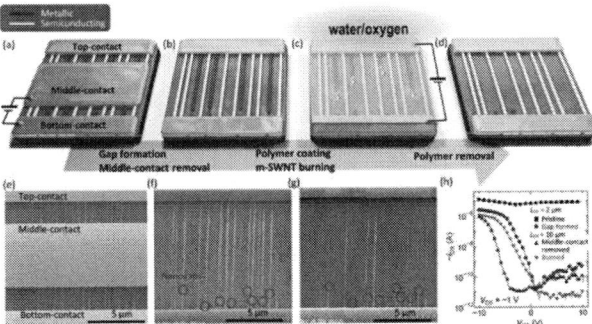

Figure 1. (a–d) Schematics for full-length removal of m-SWNTs in a controlled area.[1] **(e–g)** SEM images of SWNT arrays corresponding to (a), (b), and (d). (a,e) Initial SWNT arrays with three (top, middle, and bottom) metal contacts. A ramp voltage (VDS) was then applied between the middle and bottom contacts until all the m-SWNTs were broken down. (b,f) After electrical breakdown, the middle contacts were dissolved to leave SWNT nanogaps (circles) close to the bottom contacts. (c) After PMMA was coated onto the substrate, another ramp voltage was applied in a wet oxygen environment. All the m-SWNTs were then burned from the nanogaps in one direction. (d,g) After burning, only the s-SWNTs remained. (h) Transfer characteristics of the transistor defined by the middle and bottom contacts, before and after formation of the nanogaps (black and red, respectively), after removal of the middle contact, and after burning.

Here we demonstrate the fabrication of multiple transistors along a purely semiconducting SWNT array via an on-chip purification method. Water- and polymer-assisted burning from site-controlled nanogaps is developed for the reliable full-length removal of metallic SWNTs with the damage to s-SWNTs minimized even in high-density arrays. All the transistors with various channel lengths show large on-state current and excellent switching behavior in the off-state. Since our method potentially provides pure s-SWNT arrays over a large area with negligible damage, numerous transistors with arbitrary dimensions could be fabricated using a conventional semiconductor process, leading to SWNT-based logic, high-speed communication, and other next-generation electronic devices.

Figure 1a–d shows the purification procedures for the full-length burning of m- SWNTs. Third (middle) Au contacts, as well as two (top and bottom) contacts, were patterned on the SWNT arrays (Figure 1a) to form nanogaps in the vicinity of the bottom metal contacts, to which a lower potential was applied (cathode). A corresponding scanning electron microscopy (SEM) image is shown in Figure 1e. Electrical breakdown of the SWNTs was performed to create nanogaps between the middle and bottom contacts. After the middle metal contacts were selectively etched away, eight nanogaps in the m-SWNTs were produced beside the bottom contacts, as indicated by the circles in Figure 1f. The transfer characteristics of the transistor defined by the middle and bottom (source and drain) contacts indicate that all the m-SWNTs were broken down, as shown by the curve in Figure 1h. The reduction of the on-state current after the removal of middle contacts (Figure 1h, blue curve) is simply attributed to the change in the channel length LCH, from ~2 to ~10 μm.

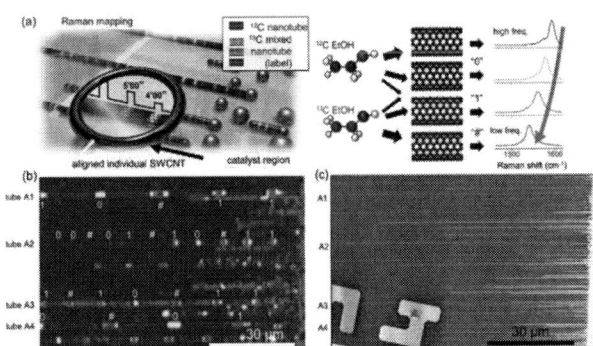

Figure 2. Digitally coded isotope labeling of individual SWCNTs.[2] (a) Schematic of isotope labeling to trace the growth of individual SWCNTs. Isotope labels with one-third and two-thirds 13C are defined as label "0" and label "1", respectively, while SWCNTs grown from 100% 13C are denoted "#", signaling the start of new binaries. Raman spectra of SWCNTs grown from ethanol with four different fractions of 13C are shown on the right. The excitation wavelength was 532 nm. (b, c) Raman mapping image (b) and SEM image (c) of the same SWCNT arrays.

978-1-7281-0943-5/19 $31.00 © 2019 IEEE

In addition, we developed a method for tracing the diverse growth profiles of individual SWCNTs by embedding digitally coded isotope labels (Figure 2a). Raman mapping showed that, after various incubation times, SWCNTs elongated monotonically until their abrupt termination (Figure 2b and c). Ex situ analysis offered an opportunity to capture rare chirality changes along the SWCNTs, which resulted in sudden acceleration/deceleration of the growth rate. Dependence on growth parameters, such as temperature and carbon concentration, was also traced along individual SWCNTs, which could provide clues to chirality control. Systematic growth studies with a variety of catalysts and conditions, which combine the presented method with other characterization techniques, will lead to further understanding and control of chirality, length, and density of SWCNTs. [2]

SWCNT-BASED HETEROSTRUCTURES

Property by design is one appealing idea in material synthesis but hard to achieve in practice. A recent successful example is the demonstration of van der Waals (vdW) heterostructures,[3-5] in which atomic layers are stacked on each other and different ingredients can be combined beyond symmetry and lattice matching. This concept, usually described as a nanoscale Lego blocks (Figure 3a), allows to build sophisticated structures layer by layer. However, this concept has been so far limited in two dimensional (2D) materials. Here we show a class of new material where different layers are coaxially (instead of planarly) stacked. As the structure is in one dimensional (1D) form, we name it "1D vdW heterostructures" (Figure 3b). We demonstrate a 5 nm diameter nanotube consisting of three different materials: an inner conductive carbon nanotube (CNT), a middle insulating hexagonal boron nitride nanotube (BNNT) and an outside semiconducting MoS2 nanotube. As the technique is highly applicable to other materials in the current 2D libraries, we anticipate our strategy to be a starting point for discovering a class of new semiconducting nanotube materials. A plethora of function-designable 1D heterostructures will appear after the combination of CNTs, BNNTs and semiconducting nanotubes.

First, we present a SWCNT-BNNT 1D heterostructure before MoS2 and other structures. We utilized SWCNTs as a template and synthesized additional hexagonal BN layers by chemical vapor deposition (CVD). Figure 3c shows a representative high resolution transmission electron microscope (HRTEM) image of this coaxial heterostructure and more images are provided as Figure S1a. From a conventional HRTEM image, this nanotube looks not distinguishable from a triple-walled pure carbon nanotube. The aberration-corrected HRTEM image of a similar tube reveals a contrast of stacking of two perfect nanotubes (Figure 3d, e). However, as the starting material is purely single-walled before we performed a post BN coating, it is reasonable to expect that the outer wall(s) are BN. This point is evidenced by electron energy loss spectroscopic (EELS) mapping shown in Figure 3f. As the reaction occurs on the outer surface, different from previous attempts inside a nanotube,[9,10] continuous coating and perfectly crystalized outer BNNTs are achieved in this work. The number of outer BNNT walls can be adjusted from 1-2 to 5-8, depending on the time of BN CVD

BNNT is an insulator with a large band-gap of 6.0 eV. Therefore, it may be used to gate the inner SWCNTs, or protect SWCNTs from the surrounding environment. Meanwhile, semiconducting MoS2 can also be coated on SWCNT and form a seamless nanotube structure. We fabricate a ternary, SWCNT-BN-MoS2 coaxial nanotube. It is 5 nm in diameter but consisting of three different materials: an inner one layer of conducting carbon nanotube, a middle three layers of insulating hexagonal BN and an outside one layer of semiconducting MoS2. This structure can be clearly visualized by the EELS mapping.

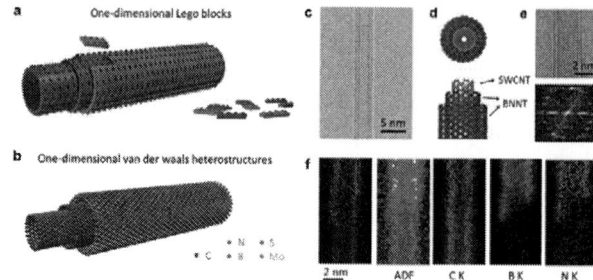

Figure 3 Overview of SWCNT based 1D van der Waals heterostructure nanotubes.[6] (a) A one dimensional Lego block showing the concept of this work; (b) the atomic model of one material built in this study: metal-insulator-semiconductor (M+I+S) 1D vdW heterostructures; (c-f) TEM image and EELS mapping of a SWCNT-BNNT 1D vdW heterostructure.

Such a structure can probably be the smallest metal-insulator-semiconductor device demonstrated so far. Field effect of the outside MoS2 may be measured using the inner SWCNT gate, and photo-response, photovoltaic effect may also be expected in such 1D vdW heterostructures.

ACKNOWLEDGMENT

Part of this work is financially supported by JSPS KAKENHI Grant Numbers JP25107002, JP15H05760, JP15K17984, JP16K05948, JP16H06333, 17K14601, 18H05329, IRENA Project by JST-EC DG RTD, Strategic International Collaborative Research Program, SICORP, and by the "Nanotechnology Platform" (project No. 12024046) of the Ministry of Education, Culture, Sports, Science and Technology (MEXT), Japan.

REFERENCES

[1] K. Otsuka, T. Inoue, E. Maeda, R. Kometani, S. Chiashi, S. Maruyama, On-chip sorting of long semiconducting carbon nanotubes for multiple transistors along an identical array, ACS Nano, (2017), 11-11, pp. 11497-11504.

[2] K. Otsuka, S. Yamamoto, T. Inoue, B. Koyano, H. Ukai, R.Yoshikawa, R. Xiang, S. Chiashi, S. Maruyama, Digital isotope coding to trace growth process of individual single-walled carbon nanotubes, ACS Nano, (2018), 12, 3994-4001.

[3] Geim, A. K. & Grigorieva, I. V. Van der Waals heterostructures. Nature 499, 419-425 (2013).

[4] Novoselov, K. S., Mishchenko, A., Carvalho, A. & Neto, A. H. C. 2D materials and van der Waals heterostructures. Science 353 (2016).

[5] Liu, Y. et al. Van der Waals heterostructures and devices. Nat Rev Mater 1 (2016).

[6] R. Xiang*, T. Inoue, Y. Zheng, A. Kumamoto, Y. Qian, Y. Sato, M. Liu, D. Gokhale, J. Guo, K. Hisama, S. Yotsumoto, T. Ogamoto, H. Arai, Y. Kobayashi, H. Zhang, B. Hou, A. Anissimov, Y. Miyata, S. Okada, S. Chiashi, Y. Li, E. I. Kauppinen, Y. Ikuhara, K. Suenaga, S. Maruyama*, One-dimensional van der Waals heterostructures, (2019), submitted. Cite: https://arxiv.org/abs/1807.06154

Toward high-mobility and low-power 2D MoS$_2$ field-effect transistors

Zhihao Yu[1], Ying Zhu[1], Weisheng Li[1], Yi Shi[1], Gang Zhang[2], Yang Chai[3] and Xinran Wang[1]*

[1]National Laboratory of Solid State Microstructures, School of Electronic Science and Engineering, and Collaborate Innovation Center of Advanced Microstructures, Nanjing University, Nanjing 210093, China, *email: xrwang@nju.edu.cn
[2]Institute of High Performance Computing, 1 Fusionopolis Way, 138632, Singapore
[3]Department of Applied Physics, The Hong Kong Polytechnic University, Hong Kong, China

ABSTRACT

2D transition-metal dichalcogenides are promising candidates for sub-5nm logic devices owing to their immunity to short-channel effects (SCE), but many issues regarding mobility, contact and interface still remain. We develop a series of interface engineering strategies using high-k dielectric and thiol chemical treatment to achieve high carrier mobility of 150 cm^2/Vs and saturation current over 450 µA/µm in long-channel monolayer MoS$_2$ FETs. Towards low power applications, we develop means to integrate ultrathin oxides with ~1nm EOT on 2D materials, which gives near ideal SS~60mV/dec and excellent leakage properties. We further use ferroelectric HfZrO$_x$ as gate dielectric and demonstrate hysteresis-free MoS$_2$ negative capacitance FETs, with sub-60mV/dec SS over 6 orders of I$_D$, minimum SS of 24 mV/dec and 10^7 on/off ratio under V_{dd}=0.5V.

INTRODUCTION

As the scaling of Si CMOS down to 10 nm node, challenges to reduce device static power consumption is demanding. 2D semiconductors represented by transition-metal dichalcogenides are promising alternative channel materials for their large bandgap, atomic channel thickness, environmental stability and possible large-area synthesis. In spite of these advantages, currently the performance of 2D FETs is still far behind theoretical limit and does not meet the requirement of the roadmap. For example, the reported mobility is much lower than theoretical limit [1], and saturation current density is only 830 µA/µm even for short-channel devices [2]. On the other hand, ITRS 2023 requires ultralow V_{dd}=0.62V for high-performance applications, so strategies to overcome the Boltzmann limit of 60 mV/dec are actively pursued. Here, taking MoS$_2$ as a prototypical example, we present a systematic study on 2D FETs, with particular attention to interface properties, mobility, ultra-thin high-κ dielectric and negative capacitance effects.

INTERFACE ENGINEERING AND MOBILITY BOOSTING IN MoS$_2$

In this section we focus on monolayer MoS$_2$, which represents the 2D limit of channel thickness. It came to our attention that most monolayer MoS$_2$ FETs reported so far have relatively low mobility ~1-100 cm^2/Vs. [3, 4] To improve mobility, we first identify the major scattering sources in the channel of a typical backgated FET (Fig. 1) and calculate the scattering rate using Boltzmann transport equation and the relaxation time approximation. Fig. 1 c-d show the calculated intrinsic and surface optical (SO) phonon-limited, CI-limited mobility as a function of temperature for several common dielectrics. µ$_{SO}$ is lower on high-κ substrate like HfO$_2$ due to lower phonon energy. Our calculations show that phonon scattering is not important except near RT, but even at RT, mobility below 100 cm^2/Vs is not limited by phonons. Rather, the mobility in most experiments is limited by Coulomb impurity (CI) scattering, which refer to charges residing at

the oxide interface or in the channel. µ$_{CI}$ depends on impurity density n$_{CI}$, oxide dielectric constant, carrier density n and temperature T. Since screening effects can reduce Coulomb scattering, low n$_{CI}$, high κ and n are desirable.

Guided by these calculations, we experimentally use thiol chemical treatment [5] and high κ substrate [1] to reduce the CI scattering and improve the mobility of monolayer MoS$_2$. Fig. 1f shows that these interface engineering approaches can significantly boost the four-probe mobility (removing contact effects) compared to bare SiO$_2$ substrate, reaching 150 cm^2/Vs at RT. Fig. 1a and b show the transfer and output characteristics of a typical 1.8 µm-long device on Al$_2$O$_3$ high κ dielectric, with two-probe mobility of 41 cm^2/Vs and saturation current over 450 µA/µm. We note that these MoS$_2$ FETs still have large contact resistance R$_c$ over 1 kΩ·µm even at on-state, which is an order larger than typical value in Si CMOS (100 Ω·µm). Achieving good electrical contacts for MoS$_2$ FETs is a prerequisite to pursue high current at ultimate scaling.

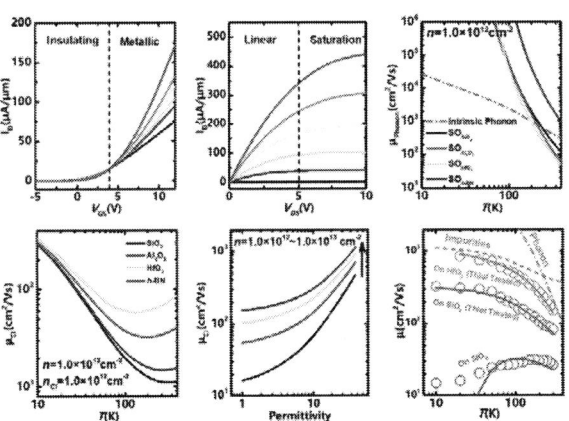

Figure 1. a) A typical I$_D$-V$_{GS}$ curves of MoS$_2$ FET on 30nm Al$_2$O$_3$ as gate dielectric, which is measured under 20K~300K. b) A typical I$_D$-V$_{DS}$ curves of device at 300K with the largest current density of over 450µA/µm. c) and d) phonon-limited and CI-limited mobility as a function of temperature on different substrates. e) CI-limited mobility as a function of permittivity at different carrier density. f) mobility as a function of temperature for bare SiO$_2$, Thoil-treated SiO$_2$, and Thoil-treated HfO$_2$ substrate, together with the best theoretical fittings, the calculated CI-limited mobility, and the calculated phonon-limited mobility

Furthermore, we need to account for the localized trap states (n$_{it}$) manifested in the commonly observed metal-insulator-transition (MIT) behavior and capacitance measurements. [6] Our model can also provide insights on the main factors leading to the low mobility. In addition, we apply our model to analyze the literature mobility data for monolayer MoS$_2$ and reveal a universal trend even though the devices

978-1-7281-0943-5/19 $31.00 © 2019 IEEE

come from several different groups. We conclude that interface engineering can effectively reduce n_{Cl} and n_{it} compared to SiO_2, and the best interface so far is achieved by BN encapsulation. Currently we still need to find a scalable process to achieve high interface quality.

LOW-POWER MoS₂ NCFETs

We need more efficient gate modulation to immunize SCE even using 2D semiconductors as channel for low-power logic devices. The excellent electrical properties of 2D materials largely originate from their perfect interface without dangling bonds. Thus, deposition of high quality ultra-thin high-κ oxides without damaging the interface of 2D materials is a necessary for electronic application. Here, we have developed a method for depositing high-κ dielectric on 2D materials by van der Waals epitaxial monolayer organic crystals as seed layers. For the first time, we achieved HfO_2 deposition by ALD with EOT of only 1 nm, breakdown field of over 15 MV/cm and no degradation at high frequencies on various 2D materials including graphene, MoS_2, WSe_2 and h-BN using a seeding layer of only 0.3 nm. Thanks to the ultra-thin gate oxide, we achieved subthreshold slope of 60mV/dec closed to the room temperature limitation in P-type WSe_2 and N-type MoS_2 MOSFETs. Furthermore, low-voltage operation was realized in a 20nm short-channel planar MoS_2 MOSFETs with a subthreshold swing as low as 75mV/dec, comparable to SOI devices at the same scale.

Our MoS_2 NCFETs show steep switching characteristics at RT. Fig. 2d compares the RT transfer curves of NCFETs of over 30 devices with different HZO thickness (t_{FE}) and a regular FET with Al_2O_3 dielectric. The best NCFET shows sub-60mV/dec over 5 orders of I_D, small V_{th} (-0.25 V and -0.19 V) and nearly zero hysteresis. The minimum SS is 24 mV/dec (41 mV/dec) for device on 22 nm (11 nm) HZO. Next, we benchmark the performance of our MoS_2 NCFETs against current literatures using on/off ratio as a function of $\Delta V_{GS} = V_{GS} - V_{on}$ (Fig. 14, V_{on} is defined as the start of subthreshold region log-scale transfer curve).[8-14] Under V_{DD}= 0.5 V, our MoS_2 NCFET can realize on/off ratio over 10^7, which is over 40 times larger than reported 2D NCFETs [10-12]. It is also 70 times larger than 10 nm Si Fin-FET and 3.8 times larger than 14 nm Si NC Fin-FET [8,9].

High frequency operation is also critical for the application of NCFETs. Fig. 3b shows the transfer characteristics using different sweeping speed and pulse measurements down to 1μs pulse width. Under pulse mode, the detection limit of our setup increases due to short integration time. Even so, the transfer curves above the detection limit all overlap with DC sweep up to 1MHz, without any sign of degradation. At 10 kHz, our NCFET shows hysteresis-free transfer with on/off ratio of 10^5 and minimum SS of 54mV/dec. These pulse measurements suggest that our MoS_2 NCFETs can potentially operate at higher frequency than 1 MHz. However, we are currently limited by the low I_{off} to probe the frequency limit.

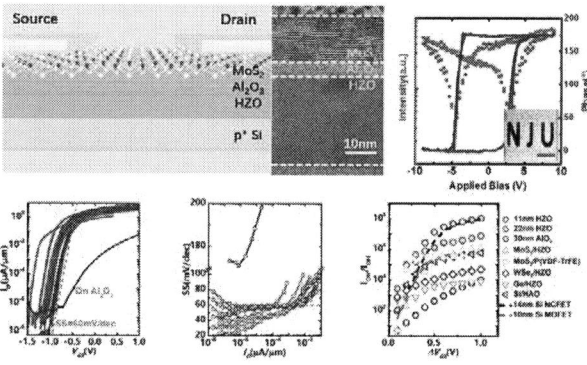

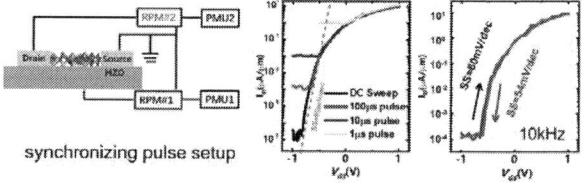

Figure 3. a) High speed synchronization pulse measurement setup using Keithley 4200SCS and HR-RPM. b) I_{DS}-V_G curves for t_{FE}=22nm NCFET under DC sweep and different pulse speed. All pulse measurements show excellent agreement with DC down to the detection limit. c) Double sweep I_{DS}-V_G curves under 10KHz for t_{FE}=22nm NCFET in Fig. 3b (black and red line: forward and reverse sweep respectively).

Figure 2. a) 3D schematic of the few-layer MoS_2 NCFET using HZO and AlOx as gate dielectric. b)High-resolution crosssection TEM image of MoS_2 NCFET after 450°C RTA with regular crystallographic orientation with regular crystallographic orientation. c) Single-point PFM characterization with 180° phase transition of HZO after 450°C RTA. d) Double sweep I_D-V_{GS} curves of a MoS_2 NCFET on 22nm HZO/2nm AlOx (red), 11nm HZO/2 nm AlOx(blue) and control device(black) at V_{DS} = 0.1V. e) Point SS of the MoS_2 NCFETs vs. I_D in Fig. 2d at RT. f) on/off ratio as a function of $\Delta V_{GS} = V_{GS} - V_{on}$ extracted from literatures compared with ours.

Evem with ultrathin high-k dielectric, MOSFETs cannot break SS of 60 mV/dec limited by Boltzmann tyranny. In order to achieve even lower power consumption and higher performance transistors, we recently focused on NCFETs using few-layer MoS_2 as channel and ferroelectric HZO/Al_2O_3 as gate dielectric (Fig. 2a and b). The HZO is deposited by alternating HfO_2 and ZrO_2 cycles in ALD followed by RTA. [7] The key to realize negative capacitance is the ferroelectric property of the HZO. To this end, we perform PFM and P-E measurements on 22nm thick HZO, which show stable ferroelectric and piezoelectric properties with single-point switching voltage about ±5V and maximum remanent polarization of ~26 μC/cm².

CONCLUSION

We show that interface engineering is important for improving the mobility and current density of 2D FETs. We develop a method to deposit ultra-thin EOT oxide of 1nm on various 2D materials. By combining HZO in the gate stack, our MoS_2 FETs show hysteresis-free steep switching characteristics up to 10 kHz, with sub-60mV/dec over 6 orders of I_D.

REFERENCES

[1] Z. Yu et al, Adv. Mater., 28, 547, 2016. [2] Y. Liu et al, Nano Lett., 16, 6337, 2016. [3] S. Li et al, Chem. Soc. Rev., 45, 118, 2016. [4] Z. Yu, et al., Adv. Func. Mater., 27, 1604093, 2017.[5] Z. Yu et al, Nature Commun., 5, 5290, 2014. [6]X. Chen et al, Nature Comm. 6, 6088, 2016. [7] Z. Yu et al, IEDM 2017, 23.6. [8] Z. Krivokapic et al., IEDM 2017, 15.1. [9] C. Auth et al. IEDM 2017, 29.1. [10] M. Si et al., Nature Nanotech., 18, 3682, 2017. [11] X. Wang et al., npj 2D Mater. & Applications. 1, 38, 2017. [12] M. Si et al., Nano Lett., 18, 3682, 2018. [13] J. Zhou et al., IEDM 2016, 12.2 [14] M. H. Lee IEDM 2017, 23.3

978-1-7281-0943-5/19 $31.00 © 2019 IEEE

TWO-DIMENSIONAL MATERIALS ELECTRON DEVICES: CONTACT AND DOPING

Yang Chai

Department of Applied Physics
The Hong Kong Polytechnic University, Hong Kong
E-mail: ychai@polyu.edu.hk

ABSTRACT

Owing to an ultrathin body, atomic scale smoothness, dangling bond-free surface, and sizable bandgap, transistors based on two-dimensional (2D) layered semiconductors show the potential of scalability down to the nanoscale, high-density three-dimensional integration, and superior performance in terms of better electrostatic control and smaller power consumption compared with conventional three-dimensional semiconductors (Si, Ge, and III-V compound materials). To apply 2D layered materials into complementary metal-oxide-semiconductor logic circuits, it is important to modulate the carrier type and density in a controllable manner, and engineer the contact (between metal electrode and 2D semiconductor) and the interface (between dielectrics and semiconducting channel) to get close to their intrinsic carrier mobility.

ELECTRICAL CONTACT

Electrical contact between metal and semiconductor is an indispensable part in very large-scale integrated circuit. The Schottky barrier (SB) at the contact region significantly affects the efficiency of charge injection and the majority carrier type of two-dimensional (2D) materials transistors. Here we construct van der Waals (vdWs) contact between metal electrode and 2D WSe$_2$ to improve the charge injection and control the carrier type in the field effect transistors (FETs). Compared with the evaporated metal electrode, the vdWs contact suppress the Fermi level pinning effect and reduce the hole Schottky barrier from 128.5 meV (evaporated contact) to 55.5 meV (vdWs contact) for WSe$_2$ transistors. The mobility is increased from 51.3 to 122.1 cm2 V-1 s-1. Moreover, thermal emission region is widened and steep SS is obtained within a large gate voltage range. The subthreshold slope of the WSe2 device on ZrO$_2$ is decreased from 226 mV dec^{-1} (evaporated contact) to 134 mV dec-1 (vdWs contact) subsequently. With HZO/Al$_2$O$_3$ negative capacitance gate stack WSe$_2$ FET with vdWs contact shows minimum subthreshold slope of 18.2 mV dec^{-1}, and the devices can be modulated by 5×10^4 within 220 mV as a result of the decreased Schottky barrier. This strategy can be also extended to other low-dimensional semiconductors for low-power electron devices.

Fig. 1a and 1d show the temperature dependent transfer curve of the WSe$_2$ FET with evaporated Pt and vdWs transferred electrode, respectively. The ON-state current of the WSe$_2$ FET with vdWs contact is 5 times larger than that with the evaporated Pt electrode. For the WSe$_2$ transistor with evaporated Pt contact, the mobility at 300 K and 100 K are 51.3 cm^{2}V^{-1} s^{-1} and 8.9 cm^{2}V^{-1} s^{-1}, respectively. For the WSe$_2$ transistor with vdWs contact, the field effect mobility at 300 K increase to 122.1 cm^{2}V^{-1} s^{-1} and the mobility at 100 K increases to 125.2 cm^{2}V^{-1}s^{-1}, which is much larger than the evaporated ones.

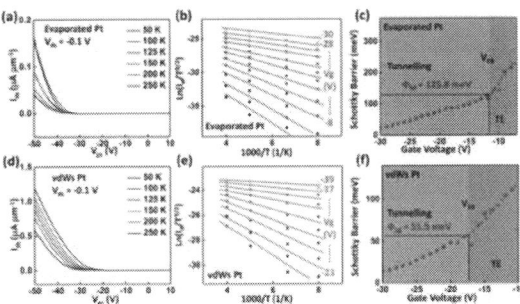

FIGURE 1. Temperature-dependent transfer curve of the FETs with (a) evaporated Pt and (d) vdWs Pt on 300 nm SiO$_2$. The Arrhenius plots for the FET with (b) evaporated Pt and (e) vdWs Pt. The slops of the lines are the Schottky barrier height for (c) evaporated Pt and (f) vdWs Pt devices with different back gate biases. The red part represents the thermal emission current and the blue part represents the thermal assisted tunneling current. The Schottky barrier of the evaporated Pt/WSe2 junction is 125.8 meV, while the vdWs Pt/WSe$_2$ junction is 55.5 meV.

DOPING

Control of the carrier type in two-dimensional (2D) materials is critical for realizing complementary logic computation. Here we present carrier type control in WSe2 field-effect transistors (FETs) via thickness engineering and solid-state oxide doping, which are compatible with state-of-the-art integrated circuit (IC) processing. It is found that the carrier type of WSe2 FETs evolves with its thickness, namely, p-type (< 4 nm), ambipolar (~ 6 nm), and n-type (> 15 nm). This layer-dependent carrier type can be understood as a result of drastic change of the band edge of WSe2 as a function of the thickness and their band offsets to the metal contacts. We also demonstrate the strong carrier type tuning by solid-state oxide doping, in which ambipolar characteristics of WSe2 FETs are converted into pure p-type, and the field-effect hole mobility is enhanced by two orders of magnitude. Our studies not only provide IC-compatible processing method to control the carrier type in 2D semiconductor, but also enable us to build functional devices, such as, a tunable diode formed with an asymmetrical-thick WSe2 flake for fast photodetectors.

978-1-7281-0943-5/19 $31.00 © 2019 IEEE

REFERENCES

[1] Yuda Zhao, Kang Xu, Feng Pan, Changjian Zhou, Feichi Zhou, and Yang Chai, "Doping, Contact and Interface Engineering of Two-Dimensional Layered Transition Metal Dichalcogenides Transistors", Advanced Functional Materials, 2017, 27, 1603484.

[2] Jingli Wang, Xuyun Guo, Zhihao Yu, Zichao Ma, Yanghui Liu, Mansun Chan, Ye Zhu, Xinran Wang, Yang Chai*,"Steep Slope P-Type 2D WSe2 Field-Effect Transistors with Van Der Waals Contact and Negative Capacitance", 2018 International Electron Devices Meeting (IEDM).

[3] Changjian Zhou, Yuda Zhao, Salahuddin Raju, Yi Wang, Ziyuan Lin, Mansun Chan, and Yang Chai*, "Carrier Type Control of WSe_2 Field-Effect Transistors by Thickness Modulation and MoO_3 Layer Doping", Advanced Functional Materials, 2016, 26, 4223-4230.

Synthesis and Electronic Devices of Atom-thin Transition Metal Dichalcogenides

Jiadong Zhou, Govindan Kutty R., Lixing Kang, Xiaowei Wang, Zheng Liu*

School of Materials Science and Engineering, Nanyang Technological University, Singapore 639798, Singapore
*E-mail: z.liu@ntu.edu.sg

SYNTHESIS OF ATOM-THIN TRANSITION METAL DICHALCOGENIDES (TMDs)

We demonstrated a universal bottom-up method to produce more than 40 different species of atom-thin, large-scale and high-quality two-dimensional (2D) crystals (Figure 1) [1]. In the last decade of 2D material research, only a few species such as graphene, h-BN and MoS2 can be directly produced from bottom-up techniques. The difficulty in producing a wide range of 2D materials that are theoretically predicted stem from the precise control of the mass flux and growth rate.

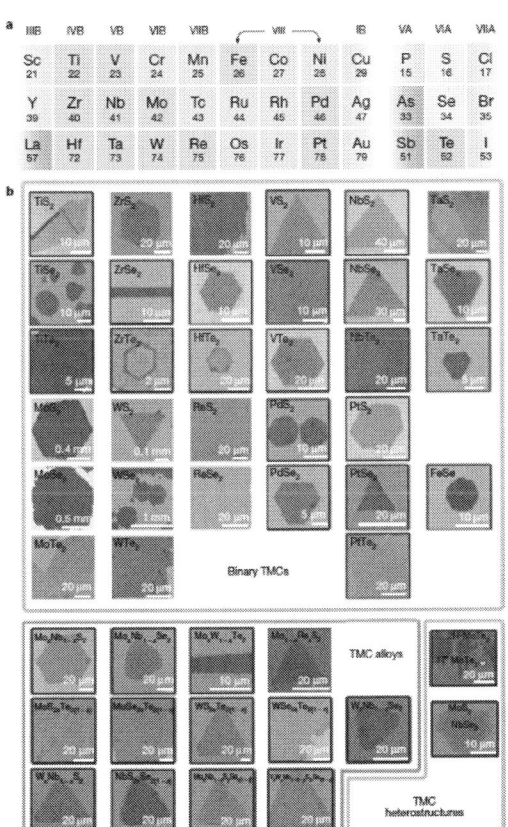

FIGURE 1. THE TRANSITION METALS AND CHALCOGENS USED, AND OPTICAL IMAGES OF THE RESULTING 47 DIFFERENT ATOMICALLY THIN TMCS AND HETEROSTRUCTURES.

TMD BASED SEMI-FLOATING AND NEGATIVE CAPACITANCE FETS

Being an interface between two types of semiconductor material, p-n junction is a ubiquitous building block in semiconducting electronics and optoelectronics. However, the performance of p-n junction considerably degrades when its thickness approaches to a few nanometers, where traditional technologies like doping and implantation become invalid and the surface effect dominates the behaviors of materials. In order to meet the increasing requirements of minimization of micro-electronic devices, high-performance ultra-thin p-n junctions are arousing wide attention by the geometrical engineering of 2D materials.

Based on our 2D TMDs, we report a two-dimensional nonvolatile programmable p-n junction device fabricated from the vertically stacked all 2D semiconductor/insulator/metal layers (WSe2/h-BN/Gr) with a semi-floating gate field-effect transistor configuration (Figure 2) [2]. The junction exhibits remarkable rectifying behavior with a rectifier ratio of 104 and photovoltaic properties with a power conversion efficiency up to 4.1% at incident light of 6.8 nW..

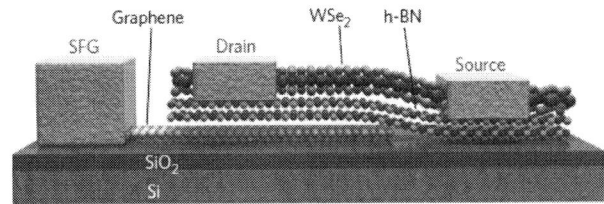

FIGURE 2. SCHEMATIC WSE2/H-BN/GRAPHENE SFG-FET DEVICE CONFIGURATION IN WHICH WSE2 SERVES AS THE TRANSPORT CHANNEL, GRAPHENE IS THE SFG, SI IS THE CONTROL GATE, AND H-BN AND SIO2 ARE THE DIELECTRICS. ONLY PART OF THE WSE2 IS ALIGNED OVER THE GRAPHENE TO FORM AN SFG-FET DEVICE ARCHITECTURE.

By trapping and erasing charges in the semi-floating gate, high-performance 2D p-n junctions are easily achieved and tuned by voltage pulses. No dopants or durative voltage are needed for the maintenance. The p-n junctions are extremely sensitive to voltage pulses and switched well between different states with the life time over 10 years and rectifier ratio of 10^4, and Memory, photovoltaic energy converter, photodetector, photovoltaic memory, logic rectifier and logic optoelectronic device are demonstrated in one device (Figure 3).

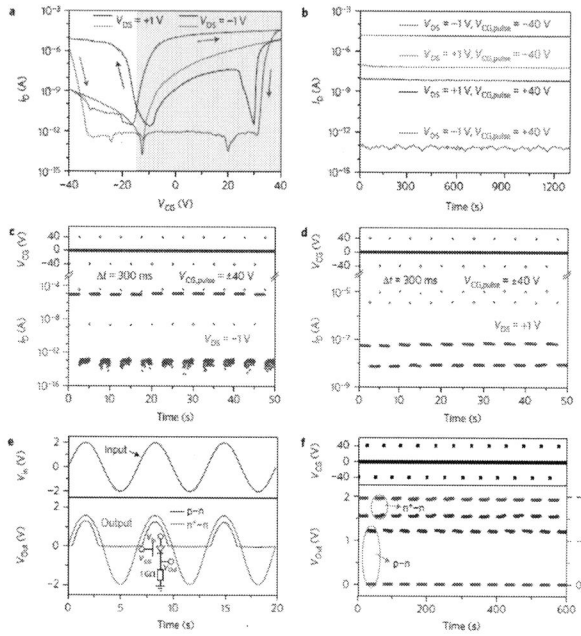

FIGURE 3. WSE2/H-BN/GRAPHENE DEVICE FOR NON-VOLATILE P–N JUNCTION MEMORIES AND LOGIC RECTIFIERS. A, I_D–V_{CG} CURVES OF THE DEVICE WITH $V_{DS}=\pm1$V. B, EVOLUTION OF THE I_D ACROSS THE WSE2 WITH DIFFERENT $V_{CG, PULSE}$ AND V_{DS} VALUES. C,D, SWITCHING BEHAVIOURS BETWEEN THE ERASE ('ON') AND PROGRAM ('OFF') STATES

REFERENCES

[1] Dong Li, Mingyuan Chen, Zhengzong Sun, PengYu, Zheng Liu, Pulickel M. Ajayan and Zengxing Zhang Two-dimensional non-volatile programmable p–n junctions, Nature Nanotechnology, 12, 901, 2017

[2] Jiadong Zhou Junhao Lin, Xiangwei Huang, Yao Zhou, Yu Chen, Juan Xia, Hong Wang, Yu Xie, Huimei Yu, Jincheng Lei, Di Wu, Fucai Liu, Qundong Fu, Qingsheng Zeng, Chuang-Han Hsu, Changli Yang, Li Lu, Ting Yu, Zexiang Shen, Hsin Lin, Boris I. Yakobson, Qian Liu, Kazu Suenaga, Guangtong Liu, Zheng Liu A library of atomically thin metal chalcogenides, Nature 556, 355, 2018

Interface engineering for 2D layered semiconductors

Kosuke Nagashio

The University of Tokyo, 7-3-1 Hongo, Bunkyo-ku, Tokyo, Japan
E-mail: nagashio@material.t.u-tokyo.ac.jp

INTRODUCTION

Although there are many building blocks to realize the high performance of Si-MISFET, the high-k/Si interface properties are the most critical. This is because dangling bonds of Si at the interface act as trap sites, which degrade the transistor performance. Therefore, many dedicated researchers have studied the interface properties in great detail until they became reliable and widely accepted for thorough interface characterization. In case of typical optical devices, GaN can be grown heteroepitaxially on Si substrate by properly selecting the buffer layer of AlN in order to reduce the lattice mismatch between GaN and Si. For both electron and optical devices, our goal is always to achieve the electrically inert interface so far for long time and will never been changed even for future.

Here, 2D layered heterostructure is focused, where there is no dangling bond for the basal plane because all the bondings are closed within the layer. Therefore, various layer stacking is possible without considering the lattice mismatch due to the van der Waals interface. 2D layered heterostructure provides new scheme for device application, because electrically inert interface is "ideally" expected. However, the detailed study on 2D heterostructure interface properties by the frequency dispersion in C-V is quite limited due to the large parasitic capacitance of electrode pad and high-dope Si substrate. It was realized that these parasitic capacitances were completely removed by using the quartz substrate. Based on the recent results obtained by current, capacitance and charge pumping methods, interface properties for bilayer graphene, MoS$_2$ and 2D-2D tunneling transistor are presented and the perspective on 2D electronics will be discussed.

ELECTRICALLY INERT h-BN/BLG INTERFACE[1]

BLG field-effect transistors (BLG-FETs), unlike conventional semiconductors, are greatly sensitive to potential fluctuations due to the charged impurities in high-k gate stacks since the potential difference between two layers induced by the external displacement electrical filed is the physical origin behind the band gap opening [2]. Therefore, the utilization of high-k oxide did not provide the high current on/off ratio even though the maximum band gap of ~250 meV is possible, as shown in **Fig. 2(a)**. On the other hand, the assembly of BLG with layered h-BN insulators into van der Waals heterostructure (**Fig. 1**) results in the drastic suppression of the off-current to the measurement limit even at a smaller band gap of ~100 meV, as shown in **Fig. 2(b)**. It should be noted that top gate metal electrode should be replaced by graphite since tiny polycrystalline metallic grains additionally introduce the potential fluctuation because different orientation has different workfunction. Moreover, by applying the conductance method in C-V measurement, we demonstrate that the electron trap/detrap response at such heterointerface is suppressed to undetectable level in the measurement frequency range. Interestingly, as shown in **Fig. 2(c)**, E_G exists for Y$_2$O$_3$/BLG/SiO$_2$ case even at zero displacement field, suggesting that E_G is formed due to the potential fluctuation of the surroundings. Therefore, the potential fluctuation energy is only ~1 meV with small E_G of ~100 meV for all-2D BLG heterostructure. **Fig. 2(d)** suggests that the potential fluctuation energy of ~10-100

meV are included in the conventional semiconductor systems as far as amorphous oxides are used as gate insulators. The electrically inert van der Waals heterointerface paves the way for the realization of future BLG electronics applications.

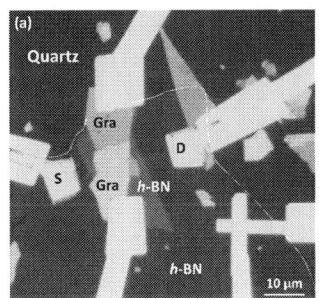
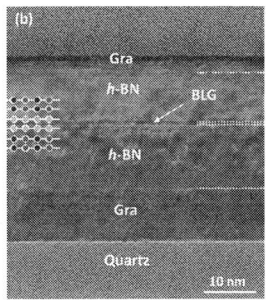

FIG. 1 (b) Optical image and (b) Cross sectional TEM image of all 2D BLG heterostructure FET.

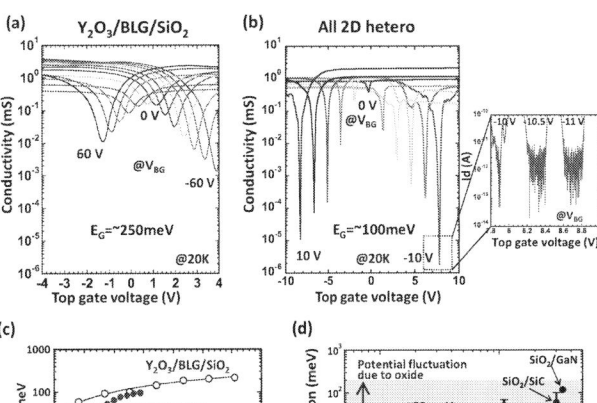

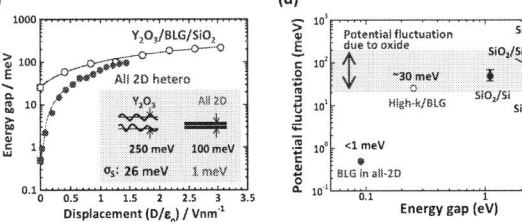

FIG. 2 Two-terminal conductivity as a function of top gate voltage for (a) high-k Y$_2$O$_3$/BLG/SiO$_2$ gate stack and (b) for all-2D heterostructure BLG gate stack. Magnified image shows that the current level reaches the measurement limitation. (c) E_g as a function of displacement field. (d) Comparison of potential fluctuations in various semiconductors in different gate stack structures.

ENERGY DISTRIBUTION OF D$_{IT}$ IN MOS$_2$ FET[3]

MoS$_2$ FET has been attracted as the atomically thin transistor to overcome severe short channel effects, because an effective channel length of ~3.9 nm has already been demonstrated [4]. Contrary to BLG where the all 2D heterostructure is required, high-k/MoS$_2$ FET has attracted attentions because large optical energy gap of ~1.9 eV is not largely affected by the potential fluctuation due to high-k oxide. In this study, monolayer MoS$_2$ FET with ALD-Al$_2$O$_3$ top gate (~10

978-1-7281-0943-5/19 $31.00 © 2019 IEEE

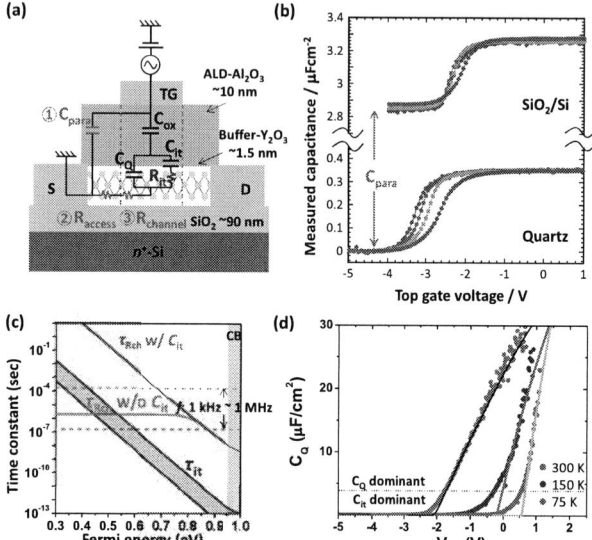

FIG. 3 (a) Equivalent circuit for MoS₂ FET. (b) C-V for MoS₂ FET on Si and quartz substrates. (c) Time constants vs E_F. (d) C_Q as a function of V_{TG}.

nm) was fabricated by the mechanical exfoliation of bulk MoS₂. As shown in the equivalent circuit in **Fig. 3(a)**, three issues should be considered before extracting the interface trap density (D_{it}). C_{para} by substrate can be removed in the device on the quartz substrate (**Fig. 3(b)**). Access resistance can be neglected due to normally on-state for typical MoS₂. We should be careful to the last one, that is, channel charging effect due to the high channel resistance of MoS₂. The simple calculations in **Fig. 3(c)** reveal that the time constant for interface states is shorter than that for channel resistance. Therefore, typically observed frequency dispersion in C-V measurement for 1L MoS₂ results from the channel charging effect, not from interface trapping. Although D_{it} for BLG with high mobility can be extracted, it is not possible for 1L MoS₂ because of low mobility. However, at the high frequency limit of 1 MHz ($1/C_{total} = 1/C_{ox} + 1/(C_Q + C_{it}(f)) \rightarrow 1/C_{ox} + 1/C_Q$, where C_Q is quantum capacitance), experimentally extracted C_Q agreed well with theoretical model, as shown in **Fig. 3(d)**. Therefore, C_{it} was extracted as a fitting parameter in the transfer curve (**Fig. 4(a)**). **Fig. 4(c)** shows the

energy distribution of D_{it} for 1L MoS₂. D_{it} extracted by charge pumping method exhibit the good agreement, suggesting the validity of D_{it} extraction from the transfer curve. The band tail distribution and mid gap states are observed. Here, the energy level for sulfur vacancy (Vs) of ~10^{13} cm⁻² is relatively deep [4]. This means this band-tail D_{it} is not directly related with Vs, but explained by the bond bending of Mo d-orbital because the conduction band of MoS₂ is composed of Mo d-orbital (**Fig. 4(b)**). The origin might be the extrinsic, that is, the strain at the high-k /MoS₂ interface or the surface roughness of SiO₂. In order to confirm this proposal, Dit obtained from 1L MoS2 heterostructure with h-BN is also plotted in Fig. 4(c). The band tail distribution as well as midgap states is reduced, suggesting that the origin for band tail distribution is the bond bending of Mo d-orbital. The top gate deposition method should be improved to obtain the high quality high-k/MoS2 interface.

2D-2D TUNNEL FETS[5]

2D materials are highly promising for tunnel FETs with low subthreshold swing and high drive current because the shorter tunnel distance and strong gate controllability can be expected from the van der Waals gap distance and the atomically sharp heterointerface formed independently of lattice matching. However, the common problem for 2D-2D TFETs is the lack of highly doped 2D materials with the high process stability as the sources. In this study, we have found that p^+-WSe₂ doped by charge transfer from a WO$_x$ surface oxide layer can be stabilized by transferring it onto a h-BN substrate. Using this p^+-WSe₂ as a source, we fabricate all-solid-state 2D-2D heterostructure TFETs with ALD-Al₂O₃ top gate insulator, i.e., type-II p^+-WSe₂/MoS₂ and type-III p^+-WSe₂/WSe₂. The band-to-band tunneling and negative differential resistance trends are clearly demonstrated at low temperatures. This work suggests that high doped 2D crystal of the charge transfer type is an excellent choice as sources for TFETs.

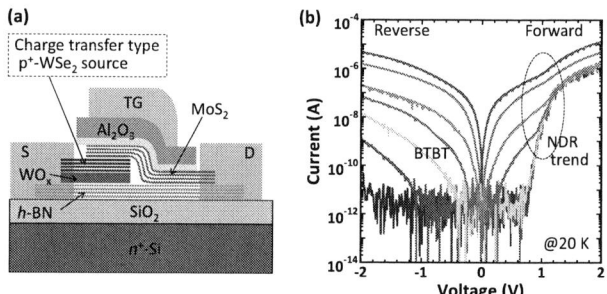

Fig. 5 (left) Schematic of all solid state 2D-2D TFET. (right) Diode properties showing NDR and BTBT.

ACKNOWLEDGEMENTS

This research was supported by the JSPS A3 Foresight program, Core-to-Core Program, A. Advanced Research Networks.

REFERENCES

[1] K. Kanayama, and K. Nagashio, Sic. Rep. 2015,5, 15789. & T. Uwanno, Y. Hattori, T. Taniguchi, K. Watanabe and K. Nagashio, ACS appl. mater. interfaces, 2018, 10, 28780.
[2] N. Fang, K. Nagashio, J. Phys. D, 2018, 51, 065110., N. Fang, K. Nagashio, ACS appl. mater. interfaces, 2018, 10, 32355. & K. Taniguchi, N. Fang, and K. Nagashio, Appl. Phys. Lett. 2018, 113, 133505
[3] S. B. Desai, et al., Science, 2016, 354, 99.
[4] P. Vancso, et al., Sci. Rep. 2016, 6, 29726.
[5] J. He, N. Fang, K. Nakamura, K. Ueno, T. Taniguchi, K. Watanabe, and K. Nagashio, Adv. Electronic Mater. 2018, 4, 1800207.

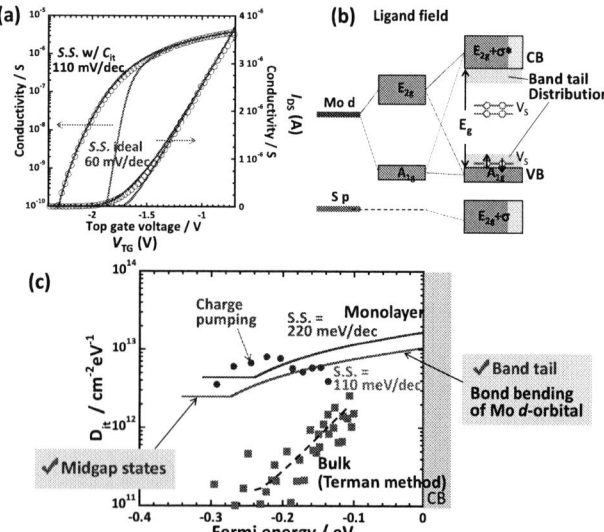

FIG. 4 (a) Transfer curve for 1L MoS₂ FET with fitting. (b) Electron dispersion originated from ligand field. (c) Energy spectrum of D_{it}.

978-1-7281-0943-5/19 $31.00 © 2019 IEEE

2D Layered Semiconductors beyond MoS$_2$

Wen-Hao Chang
National Chiao Tung University

Two-dimensional (2D) layered materials, particularly semiconducting transition metal dichalcogenides (TMDs), such as MoS$_2$ and WSe$_2$, have already attracted tremendous research efforts, due not only to their atomically thin thickness, but also to many exciting new phenomena associated with new materials. Recent advances in heterostructures formed by vertically stacked or laterally stitched 2D semiconductors further show their promise for novel device applications. Here, I will present our recent endeavors on the material synthesis and fundamental study of 2D TMDs grown by chemical vapor deposition (CVD). In particular, I will present CVD grown noble metal dichalcogenides, such as PtSe$_2$ and PdSe$_2$, with strongly layer number dependent properties. This new-type 2D material is promising for realizing metal-semiconductor junctions by local engineering the semiconducting and metallic phases via layer control, providing a feasible scheme to solve the issue of high metal contact resistance commonly occurred in 2D semiconductors.

Electronic Devices within Single Atomic Layer – Development of 2D Lateral Junctions

Jr-Hau He

King Abdullah University of Science & Technology

With the demanding requirement of nanotechnology in the semiconducting industry, the challenge will become unprecedented as the fabrication approaches the scaling limit in the next few years. The rise of 2D materials seems to be a probable solution for developing the next-generation semiconducting devices. As 2D lateral junctions bring a revolutionary breakthrough in the past few years, nanoscale sized devices are no longer limited to the vertical direction. Doping and structural design strategies that are totally different from conventional Si based devices can bring about more ideal and ultra-efficient electronic and optoelectronic devices. This perspective summarizes and compares different methods of 2D lateral junction designs (including electrostatic tunable p-n homojunction and direct growth of in-plane p-n heterojunction) and various material combinations (including metallic-insulating, semiconducting p-n, and ohmic junctions). In addition, examples of design strategies and what can be achieved by adopting these 2D lateral junctions have been provided to show the promising potential for the future development. It can be expected that over the next few years, 2D materials will dominate the semiconducting industry and holds the promise for keeping the Moore's law alive.

Can Magnetic Memory (MRAM) Displace DRAM?

Denny Tang

Western Digital Corporation

Magnetic Memory has been commercialized for more than a decade. It is offered as embedded memory of SoC and as discrete memory. The emb-MRAM has found a sweet spot in the market, replacing emb-FLASH at 28-nm node and below; and it is poised to replace SRAM of the last level cache at sub-10-nm node for its advantage over SRAM in density and in standby power. Can discrete MRAM displace a portion of the DRAM $70B (2017) market? That has been the question in people's mind. This presentation attempts to address this question from two viewpoints: (1) MTJ intrinsic device limit, thus its product implication, (2) extrinsic factors, such as the matching CMOS and memory architecture. CMOS is crucial to the memory density and DDR compliance is essential to data throughput performance.

Today, MRAM technology is maturing to the point of industrial production. At this stage, the technology has not reached its full performance potential. A continuous improvement will take place. This presentation gives a prediction how the scaled STT-MRAM may fare in the near future and how MRAM may evolve in the long term.

Interfacial engineering of SOT-MRAM to modulate atomic diffusion and enable PMA stability >400 °C

Chong Bi[1], Shy-Jay Lin[1,3], Xiang Li[1], Telem Simsek[1], M. Song[3], Wilman Tsai[3], and Shan X. Wang[1,2]

[1]Stanford University, Department of Electrical Engineering, Stanford, 94305, United States
[2]Stanford University, Department of Material Science and Engineering, Stanford, 94305, United States
[3]TSMC (Taiwan Semi-conductor Manufacturing Company)
E-mail: sxwang@stanford.edu

ABSTRACT

We report our work on the optimization of W/CoFeB/MgO structures to fulfill perpendicular magnetic anisotropy (PMA) requirements in the production of SOT-MRAM. By optimizing the natural oxidization process of deposited Mg layer and introducing different dust layers at W/CoFeB and CoFeB/MgO interfaces, PMA of W/CoFeB/MgO structures can be enhanced by about 100%, which is much higher than that in Ta-based structures. The origin of this PMA enhancement was further confirmed by transmission electron microscopy investigations. The corresponding SOT switching efficiency and current-induced effective fields were also investigated.

INTRODUCTION

Ta/CoFeB/MgO structures have been widely adopted in magnetic tunnel junctions (MTJs) because both PMA and tunneling magnetoresistance (TMR) can be achieved easily in this structure. However, due to pronounced Ta diffusion at 400 °C, the high performances of this Ta-based structure cannot be preserved after CMOS BEOL process, and therefore, it cannot be used to produce a practical spin-orbit torque magnetic random-access memory (SOT-MRAM). Replacement of Ta with W leads to much weaker diffusion and larger spin Hall angles and thus has attracted much attention in industry. However, establishment of PMA in W-based structures is not as easy as that in Ta-based structures, not only because of different crystalline structures between them, but also because Ta is a good boron sink during annealing processes post-deposition.

In this work, we present our work on the optimization of W/CoFeB/MgO structures by controlling both W/CoFeB and CoFeB/MgO interfaces to fulfill PMA requirements by industrial production. First, by optimizing the natural oxidization process of the Mg layer, the PMA of W-based structures can also be established in an as-deposited state like that in Ta-based structure. The natural oxidization process was then further optimized to improve PMA after 400 °C annealing. Second, the W/CoFeB interface was improved by introducing different dust layers, such as FeB, Mg, and MgO. As a result, the perpendicular anisotropic field can be enhanced by about 100%, which is much higher than that in Ta-based structures. Transmission electron microscopy (TEM) images further confirmed the roles of crystalline structure and atom diffusion on the establishment of PMA. Third, the corresponding SOT switching efficiency and the current-induced effective fields were also optimized by modifying both interfaces.

DEVICE FABRICATION AND CHARACTERIZATION

W/CoFeB/MgO structures were deposited by an 8-gun magnetron sputtering system with a base pressure below 1.6×10^{-8} Torr. The deposited samples were then annealed in All-Win 610 Rapid Thermal Process Systems at 400 °C for 30 mins. Finally, the samples were patterned into Hall bar structures with the width of 5 μm to measure PMA and SOT switching (Fig. 1) The SOT efficiency was estimated by measuring a current-induced additional perpendicular magnetic field (H_{zeff}, as shown in Fig. 2 and Fig. 3). To measure SOT switching, a large switching current pulse (1 ms) was applied first and then a small sense dc current (50 μA) was applied to detect magnetization switching direction through anomalous Hall effect.

RESULTS AND DISCUSSION

Fig. 4-6 show the sample structure and corresponding PMA measurements in as-deposited states and after annealing. The best oxidization condition for as-deposited samples is when $t_{Mg} = 0.7$ nm, which results in the appropriate oxygen concentration at the Mg/CoFeB interface to establish PMA. However, after annealing, $t_{Mg} = 0.7$ nm, 0.4 nm thicker Mg layer, becomes the optimized oxidization condition. Fig. 7 and Fig. 8 show the illustration of dust layers and the measured PMA with various dust layers, respectively. It shows that FeB and Mg insertion layers at the W/CoFeB interface can enhance PMA while MgO and Co insertion layers quench PMA. Fig. 9 shows the typical perpendicular effective field (H_k) extraction by Stoner model.

Fig. 10 shows typical current-induced SOT switching curves and the definition of critical SOT witching current (I_c). Fig. 11 shows the comparison of SOT efficiency and switching current under different dust layers. As we expected, one can see that a higher SOT efficiency corresponds to a smaller switching current. For the MgO insertion layer, the highest SOT switching efficiency is achieved.

As shown in Fig. 8, one surprising result is that the adoption of an ultra-thin FeB interface in W/CoFeB/MgO SOT stack was shown to contribute to its thermal stability up to 400 °C, such effects was never observed and reported in the literature [1-3]. Cross section TEM and EDX line profile analysis further indicate that the novel FeB interfacial layer blocks the interdiffusion of Co, Fe into W channel after 400 C annealing, which is postulated to be the key mechanism to enable the observed better PMA, as shown in Fig. 12 and Fig. 13.

CONCLUSIONS

Optimization of the performances of W/CoFeB/MgO structures is demonstrated in this work. We show both PMA and SOT switching efficiency can be highly improved by interfacial engineering.

ACKNOWLEDGEMENT

This work was supported in part by TSMC and Stanford SystemX Alliance.

REFERENCES

[1] S. Y. Jang et al., *J. Appl. Phys.*, vol. 107, p. 09C707, 2010.
[2] S. Peng et al., *Sci. Rep.*, vol. 5, p. 18173, 2015.
[3] K. U. Demasius et al., *Nat. Comm.*, vol. 7, p. 10644, 2016.

978-1-7281-0943-5/19 $31.00 © 2019 IEEE

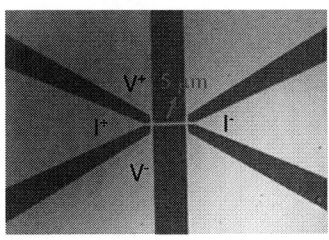

Fig. 1: The patterned Hall bar device for PMA and SOT switching measurement.

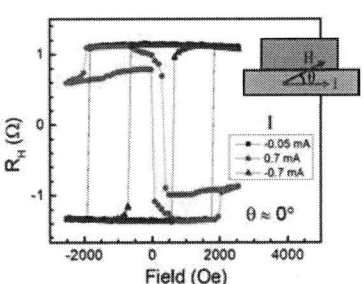

Fig. 2: Current dependent field-driven magnetic switching. The applied current can strongly change magnetic switching field like an additional perpendicular magnetic field. Inset shows the configuration of applied field and current.

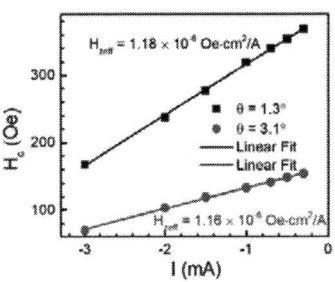

Fig. 3: The extracted current induced perpendicular field (H_{zeff}) under different θ, which shows that H_{zeff} does not depend on θ.

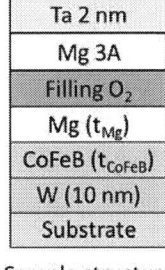

Fig. 4: The schematic of W/CoFeB/MgO stack.

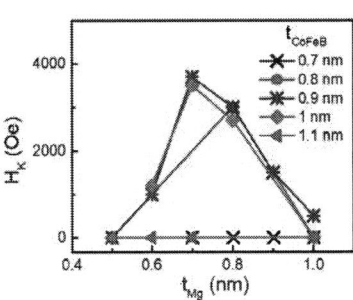

Fig. 5: Mg and CoFeB thickness dependence of PMA in as-deposited states.

Fig. 6: Annealing effects on PMA of W/CoFeB/MgO when t_{CoFeB} = 0.9 nm.

Fig. 7: Illustration of dust layer in W/CoFeB/MgO stack.

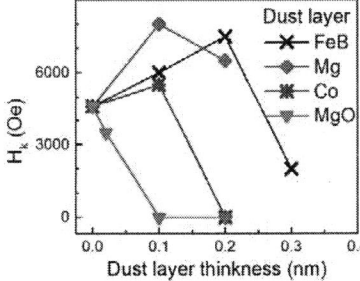

Fig. 8: PMA modified by different dust layers.

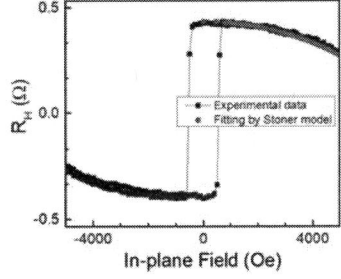

Fig. 9: Typical in-plane field dependent Hall resistance (R_H) and its fitting by the Stoner model with H_k as a fitting parameter.

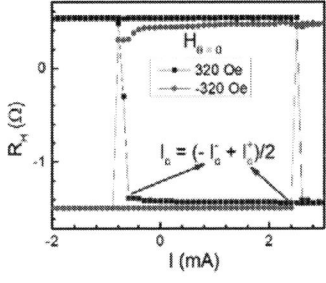

Fig. 10: Typical current induced SOT switching and the definition of H_c.

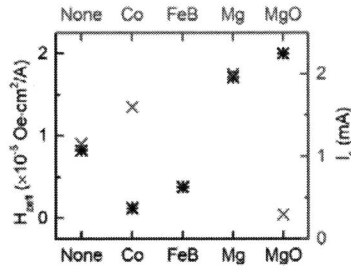

Fig. 11: Comparison of SOT switching efficiency and switching current with different dust layers.

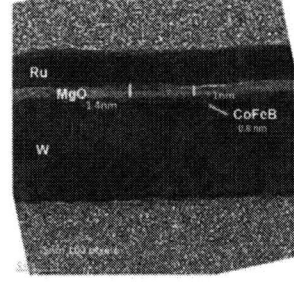

Fig. 12: Cross section TEM of SOT devices.

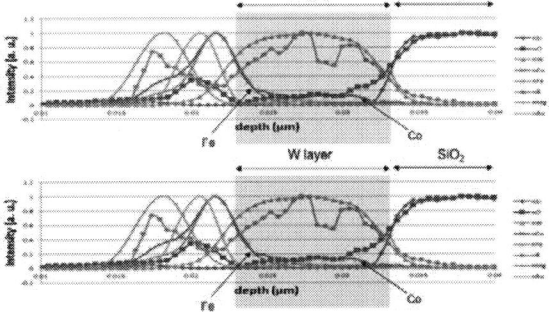

Fig. 13: EDX line profiles of SOT stacks with FeB dust layer (top) and without FeB insertion (bottom).

978-1-7281-0943-5/19 $31.00 © 2019 IEEE

Computational Random Access Memory (CRAM) and Applications

Jian-Ping Wang
University of Minnesota

To enable local or edge AI like family-based robots and self-driving cars without using cloud computing, the power consumption for any available computing system is still four to six orders of magnitude higher than needed. Among all proposed spintronic devices for computing, magnetic tunnel junctions (MTJs) [1] hold the brightest future with large practical impacts considering the maturity of manufacturing MTJs array. Magnetic tunnel junctions (MTJs) were proved to provide high endurance and low-cost solutions for non-volatile, stochastic and nonlinear activation functions and the combinations [2]. Those enable both the stochastic computing and probabilistic computing. This talk will report a new computing architecture based on MTJs: computational random-access memory (CRAM) [3,4], which will be a key build block for different AI accelerators.

[1] H. Meng, et al, "Programmable Spintronics Logic Device Based on a Magnetic Tunnel Junction Element", J. Appl. Phys. 97, 10D509. 2005

[2] J. P. Wang, et al, "A Pathway to Enable Exponential Scaling for the Beyond-CMOS Era", Proceedings of the 54th Annual Design Automation Conference 2017, Article No. 16;

[3] Z. Chowdhury, et al, "Efficient in-memory processing using spintronics", IEEE Computer Architecture Letters, 17, 1, 2018;

[4] M. Zabihi, et al, "In-memory processing on the spintronic CRAM: From hardware design to application mapping", IEEE Transactions on Computers, 2018

Comprehensive Reliability Study of STT-MRAM Devices and Chips for Last Level Cache Applications (LLC) at 0x Nodes

Jian Zhu, Yuan-Jen Lee, Huanlong Liu, Son Le, Jodi Iwata-Harms, Sahil Patel, Ru-Ying Tong, Vignesh Sundar, Santiago Serrano-Guisan, Dongna Shen, Renren He, Jesmin Haq, Jeffrey Teng, Vinh Lam, Yi Yang, Yu-Jen Wang, Tom Zhong, Luc Thomas, Hideaki Fukuzawa, Guenole Jan and Po-Kang Wang

TDK-Headway Technologies, Inc., Milpitas, CA, USA, E-mail: jian.zhu@headway.com

ABSTRACT

To consider STT-MRAM as an SRAM replacement, the reliability of the MTJ devices has to be demonstrated. A comprehensive study of degradation of STT-MRAM magnetic tunnel junction barrier under stress is presented in this paper. It is found that the breakdown mechanism of such devices follows a consistent path of soft breakdown (SBD) followed by hard breakdown (HBD). We discuss the strategy to improve the write margin by reducing the resistance area product (RA) of the tunnel barrier. Finally, we link our single device reliability studies to similar endurance studies performed on our fully functional chips.

BREAKDOWN METROLOGY

We explored two common methods of characterization of time dependent dielectric breakdown (TDDB), namely, the ramp voltage stress (RVS) and the constant voltage stress (CVS) methods as described in Fig 1. While the CVS method is closer to the typical stress the devices will be experiencing in the final product, it lacks in speed and ease of data collection. The RVS method on the other hand is fairly simple to implement and does not require prior knowledge of the breakdown voltage of the devices. A comparison of CVS and RVS (Fig. 2) has been performed on our devices, yielding very similar extracted parameters and lifetime extrapolation. As shown before [1], the breakdown voltage of the MTJs is fairly asymmetric with the write P state consistently having lower breakdown characteristics than the write AP state. This behavior is also present in our measurement as shown in Fig 3. For these reasons, only RVS measurements of the weaker breakdown on the P state will be considered for the remaining single device data in this paper.

BREAKDOWN MECHANISMS

A slight degradation of the MTJ properties has been observed before the typical HBD of the devices [2] and is attributed to SBD mechanism. This phenomenon is of a different nature from the low voltage gradual degradation observed in barriers with defects already present during fabrication [3]. While mostly ignored by previous studies, this degradation of transport properties will impact the MTJ performance in terms of reduced read margin and increased write current. To catch reliably the onset of this SBD, we developed a new metrology probing the MTJ after each applied voltage stress using a pulsed I-V (PIV) measurement where the resistance of the device is recorded under bias (Fig. 4). The PIV curves are limited excursion in voltage to minimize additional stress to the RVS experiment. Fig. 5(a) and (b) show the typical resistance drift of a device in RVS measurement. Reduction of P and AP state resistance could be observed. The TMR vs. resistance at AP state is plotted in Fig. 5(d). One clearly sees that the gradual degradation of the TMR and resistance follows a well-defined path, which could be explained by shunting of the barrier with a path of lower resistance. Computing the deviation from the original resistance at 3σ gives a reliable estimation of the onset of the SBD, which is illustrated in Fig. 5(c). Extracted SBD vs. HBD distributions show that the SBD reduces the write margin by about 100 mV compared to analysis considering HBD only (Fig. 6).

IMPROVE WRITE MARGIN BY REDUCING RA

Improving the write margin for LLC applications may be achieved by reducing RA of the barrier. As the switching current density remains constant for a given magnetic structure, switching voltage will be proportional to the RA. Thanks to the fact that RA is exponential to the thickness of the barrier while breakdown voltage is proportional to it, the breakdown voltage follows a logarithmic function of RA. As a result, the window between switching and breakdown voltages will be improved if RA is reduced.

Fig. 7 shows switching and breakdown voltage distributions for stacks using the same magnetic structure but with RA range from 3 to 17.5 ohm.um^2. The improvement of the write margin by reducing RA is clearly seen in Fig. 8 by normalizing the write margin to V_{sw}. Fig. 9 illustrates that the projected endurance for a 1ppm error rate using 10ns x 10^{12} and 10^{15} cycles shows a very good write margin when using a high performance stack design with writing voltage of 250mV and RA of 3.5 ohm.um^2 as reported in [4].

ENDURANCE TEST AT CHIP LEVEL

Total 420 fully functional 8Mb STT-MRAM chips described in [5] have been used for extensive write endurance test at chip level. To accelerate the test, 42 custom designed FPGA boards shown in Fig. 10(a) are used. Each board is capable of simultaneously testing 10 chips at a given stress condition. For each of the stress conditions, a subset of 5K bits distributed over the entire 8Mb chip array are tested. The applied bias across the MTJ for a given bit line and word line voltage will vary as the resistance of the MTJ changes as well as when the biasing polarity of the cell changes as shown in Fig. 10 (b-c). In Fig. 11, we performed both unipolar and bipolar tests on chips and the results show that the bipolar test on the chip has worse breakdown than the unipolar W0 and W1 tests, most likely due to the increased stress voltage on the W0 polarity before the device switching from the AP to P state. Meanwhile, the asymmetry between unipolar W0 and W1 is observed, and is consistent with TDDB results at device level.

Finally we compared the breakdown voltages extracted from single device with chip measurements using different bit line voltages with number of pulses up to 10^{11}, and we observe a discrepancy ~5% on the extrapolated breakdown voltage (Fig. 12). We conclude that the single device measurements are suitable as an assessment of the reliability of the STT-MRAM stack at chip level.

CONCLUSIONS

We have measured and characterized the degradation of the MTJ properties under stress. A SBD mechanism before HBD has been observed and characterized. A reduction of write margin by ~100mV is estimated to take into account this SBD. We also demonstrated that the write margin can be significantly improved by reducing the RA of the MgO tunnel barrier, allowing almost infinite endurance for STT-MRAM at lower RA using stacks with excellent switching characteristics. We also performed similar endurance studies at chip level and found very good agreement with the properties extracted from single device studies. This study confirms that reliability of STT-MRAM can be achieved using reported stacks with low RA and low switching voltages for LLC applications.

REFERENCES

[1] D. V. Dimitrov et al., *Appl. Phys. Lett. 94, 123110 (2009)*.
[2] G. Jan et al., VLSI Tech. Symp., T164-T165, 2015.
[3] B. Oliver et al., J. Appl. Phys. 95, 1315 (2004).
[4] G. Jan et al., VLSI Tech Symp., 65-66, 2018.
[5] G. Jan et al., VLSI Tech Symp., 42-43, 2014.

978-1-7281-0943-5/19 $31.00 © 2019 IEEE

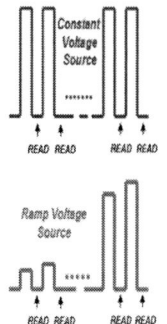

Fig 1. CVS vs. RVS voltage stress pulse train.

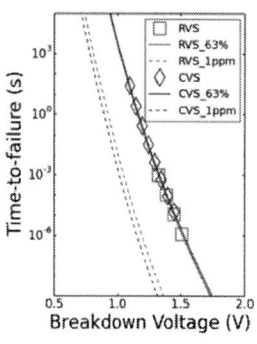

Fig 2. Lifetime extrapolation using RVS and CVS methods.

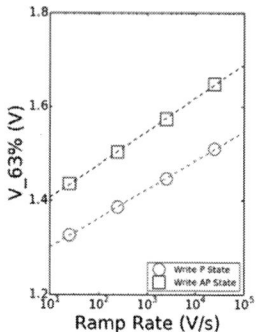

Fig 3. RVS breakdown probability distributions for different write polarities.

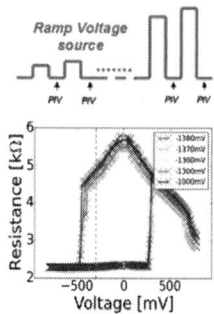

Fig 4. Pulsed-IV after different stress conditions during RVS measurement.

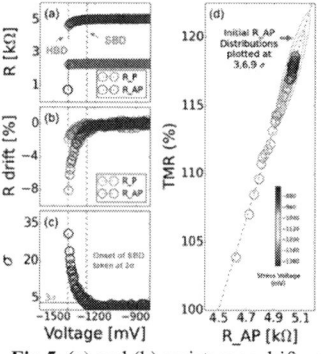

Fig 5. (a) and (b) resistances drift vs. stress voltages. (c) and (d) definition of the onset of SBD.

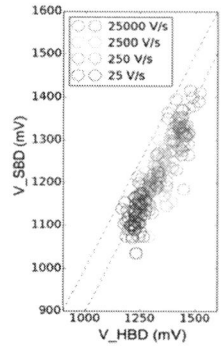

Fig 6. Extracted SBD and HBD voltages. SBD voltage is about 100mV smaller.

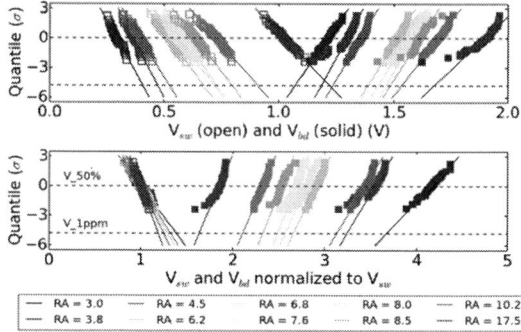

Fig 7. Top: distribution of switching (V_{sw}) and hard breakdown voltages (V_{bd}) vs. RA. Bottom: same plot normalized to V_{sw}.

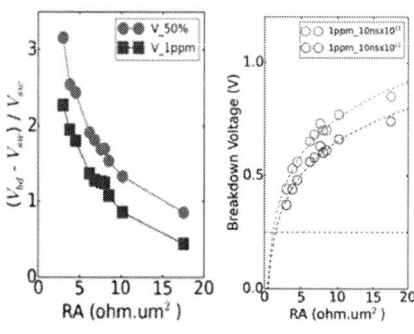

Fig 8. Normalized write margin vs. RA.

Fig 9. Lifetime extrapolation vs. RA.

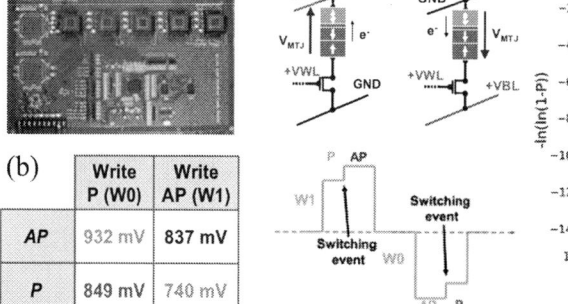

	Write P (W0)	Write AP (W1)
AP	932 mV	837 mV
P	849 mV	740 mV

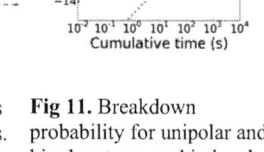

Fig 10. (a) FPGA boards used for chip testing. (b) Voltages on MTJ as a function of resistance state and bias conditions. (c) Voltages present on MTJ during a bipolar pulse train.

Fig 11. Breakdown probability for unipolar and bipolar stress at chip level.

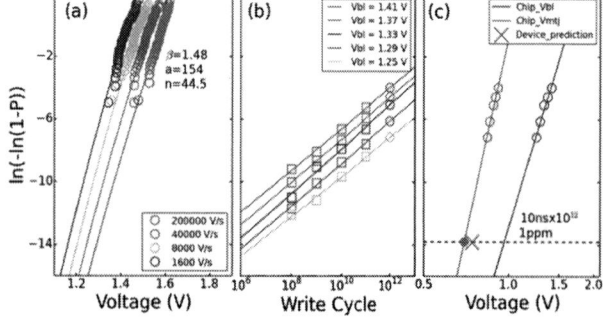

Fig 12. (a) Device level RVS measurement of breakdown voltage. (b) Chip level measurement of breakdown voltage using bipolar pulses with 10ns pulse width and different bit line voltages. The number of write cycles is up to 10^{11} (squares) with extrapolation to 10^{12} (circles). (c) Estimation of reliability for 10^{12} pulses of 10ns and an error rate of 1ppm using the chip (solid red circle) and single device (green cross) data.

978-1-7281-0943-5/19 $31.00 © 2019 IEEE

Spin-orbit torque driven one-bit magnetic racetrack devices – memory and neuromorphic applications

See-Hun Yang, Chirag Garg, Timothy Phung, Charles Rettner, and Brian Hughes

IBM Research – Almaden, San Jose, CA 95120, USA
E-mail: seeyang@us.ibm.com

ABSTRACT

Recent breakthroughs on spin-orbit torque have opened door to more versatile magnetic racetrack memory. Especially, spin-orbit torque driven one-bit three terminal racetrack memory is promising to develop field-free memory and neuromorphic devices that have many figure of merits. However, there are quite a few challenges to overcome to have working racetrack devices such as decent tunneling magnetoresistance, low threshold current density, and homogeneous films. We present here successfully fabricated one-bit three terminal magnetic racetrack devices in which the position of domain wall is readout by magnetic tunnel junctions.

INTRODUCTION

Current driven magnetic domain wall (DW) motion using spin transfer torque has attracted a great deal of attention due to its potential application to development of new type of storage class memory, e.g., magnetic racetrack memory[1]. Over the past decade significant progresses have been made to increase the chip density by incorporating perpendicular magnetic anisotropy[2]. In particular, discovery of a new torque, spin-orbit torque, has opened door to the development of even more powerful and versatile racetrack based devices[3, 4]. The spin orbit torque is generated by combined mechanism of spin Hall effect[5, 6] with Dzyaholshinsky-Moriya interaction[7, 8] at the interface between ultrathin magnetic layer and heavy nonmagnetic layer. Most recently, we discover more powerful torque, exchange coupling torque, from synthetic antiferromagnets that consist of two magnetic sublayers that are separated by ultrathin Ruthenium spacer layer[9]. Remarkably, the exchange coupling torque was found to move the DWs at more than 750 m/s as the net magnetization becomes almost compensated, thus allowing (1) even higher density due to minization of stray fields, (2) faster access time, and (3) more robust shift-register operation. Our model shows that the velocity exceeds 1 km/s when net magnetization becomes zero.

One-bit three-terminal (1B3T) racetrack device can be technologically more easily accessible compared to the original proposal based on multi-bits[10]. Especially, 1B3T racetrack memory is promising because (1) it does not require external in-plane field to switch compared to spin-orbit-torque based three terminal MRAM, (2) it can be significantly faster to write compared to spin transfer torque MRAM, (3) it requires much smaller power to write, and (4) neuromorphic operation is possible.

Here we show the spin-orbit torque and exchange coupling torque driven DW motion in nanowires formed from ultrathin perpendicularly magnetized single layer and synthetic antiferromagnetic layer, respectively. These devices clearly exhibit threshold behavior and symmetric DW displacement with respect to current pulse polarity, which is critical for neuromorphic operations. Next we present the device fabrications of 3T racetrack devices that can be applied not only to high speed devices but to neuromorphic chips[11].

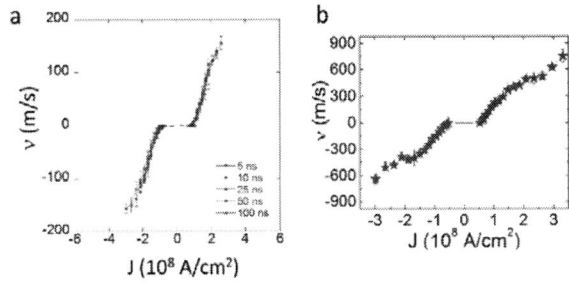

FIGURE 1.

RESULTS AND DISCUSSION

Fig. 1a,b show the current driven domain wall velocity versus current density in the wire with dimension of 50 μm × 2 μm formed from 5 Ta | 5 Ru | 1.5 Pt | 0.3 Co | 0.7 Ni | 0.15 Co | 5 TaN (single layer)[12] and 2 TaN | 1.5 Pt | 0.3 Co | 0.7 Ni | 0.3 Co | 0.8 Ru | 0.3 Co | 0.7 Ni | 0.3 Ni | 5 TaN (synthetic antiferromagnets)[9] where each film thickness is in nm. Note here that Kerr microscope is used to measure the DW velocity by locating DW position displacement after the application of sequence of 5 ns current pulses. Threshold current densities are $\sim 5 \times 10^7$ A/cm^2 and $\sim 3 \times 10^7$ A/cm^2 for single layer and synthetic aniferromagnets, respectively. This threshold behavior forms a critical component for neuromorphic devices. Note that the DW moves along the current flow direction in both cases, which is in contrast with conventional spin transfer torque driven DW motion where the DW moves along the electron flow direction in 3d transition metal based wires [1]. The main driving mechanisms for single layer and synthetic antiferromagnets are spin-orbit torque and exchange coupling torque, respectively, and especially, exchange coupling torque becomes very powerful as the exchange coupling strength increases, net magnetization decreases and current density increases, thus allowing ultrafast switching with low power consumption in 1B3T high speed memory.

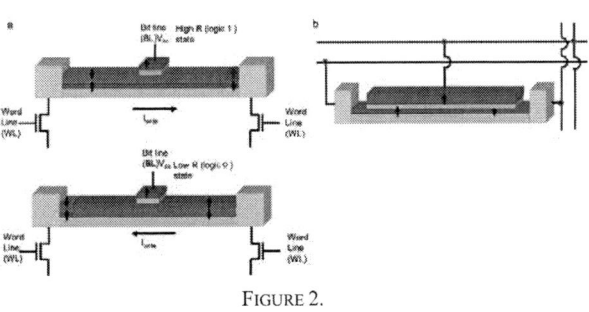

FIGURE 2.

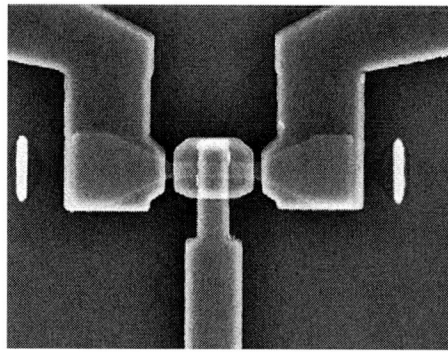

FIGURE 3.

The concepts of 1B3T high speed memory and neuromorphic devices are described in Fig. 2a,b, respectively. Let us discuss how these devices work. First one DW is created at the end of racetrack. Note that no additional DW is generated anymore afterwards. Application of sequence of current pulses move the DW back and forth depending the polarity of current pulses. High speed memory requires high velocity of DW motion such that each set of current pulses move the DW from one side end to the other end of magnetic tunnel junction. On the other hand, a slow but reliable symmetric current driven motion of DW is needed for neuromorphic device since the synaptic weight is determined by the position of DW such that the readout resistance across magnetic tunnel junction forms a feedback to the neural network for deep machine learning.

Fig. 3 shows the SEM image of 1B3T racetrack device we fabricated using novel approaches. To have successfully working devices, quite a few challenges should be overcome. First, smooth edge of wire and smooth and homogeneous film are required to reduce the threshold current density. Second, large tunneling magnetoresistance is needed to have high on-off ratio to speed up reading and reduce signal-to-noise.

CONCLUSION

Our discovery of new powerful torques, spin-orbit torque and exchange coupling torque, allows versatile one-bit three-terminal magnetic racetrack devices that can be applied to develop memory and neuromorphic chips. We successfully fabricated these devices by employing novel patterning approaches that would form a building block in the full integrated chips in the near future.

REFERENCES

[1] S. S. P. Parkin, M. Hayashi, and L. Thomas, "Magnetic Domain-Wall Racetrack Memory," *Science,* vol. 320, pp. 190-194, 2008.

[2] S. Parkin and S.-H. Yang, "Memory on the racetrack," *Nat. Nano.,* vol. 10, pp. 195-198, 2015.

[3] K.-S. Ryu, L. Thomas, S.-H. Yang, and S. Parkin, "Chiral spin torque at magnetic domain walls," *Nat. Nano.,* vol. 8, pp. 527-533, 2013.

[4] S. Emori, U. Bauer, S.-M. Ahn, E. Martinez, and G. S. D. Beach, "Current-driven dynamics of chiral ferromagnetic domain walls," *Nat. Mater.,* vol. 12, pp. 611-616, 2013.

[5] A. Hoffmann, "Spin Hall Effects in Metals," *IEEE Trans. Magn.,* vol. 49, pp. 5172-5193, 2013.

[6] J. Sinova, S. O. Valenzuela, J. Wunderlich, C. H. Back, and T. Jungwirth, "Spin Hall effects," *Rev. Mod. Phys.,* vol. 87, pp. 1213-1260, 2015.

[7] I. E. Dzyaloshinskii, "Thermodynamic theory of weak ferromagnetism in 350 antiferromagnetic substances," *Sov. Phys. JETP,* vol. 5, pp. 1259-1272, 1957.

[8] T. Moriya, "Anisotropic Superexchange Interaction and Weak Ferromagnetism," *Phys. Rev.,* vol. 120, pp. 91-98, 1960.

[9] S.-H. Yang, K.-S. Ryu, and S. Parkin, "Domain-wall velocities of up to 750 m s−1 driven by exchange-coupling torque in synthetic antiferromagnets," *Nat. Nano.,* vol. 10, pp. 221-226, 2015.

[10] S. Lee and K. Lee, "Emerging Three-Terminal Magnetic Memory Devices," *Proceedings of the IEEE,* vol. 104, pp. 1831-1843, 2016.

[11] T. Gokmen and Y. Vlasov, "Acceleration of Deep Neural Network Training with Resistive Cross-Point Devices: Design Considerations," *Frontiers in Neuroscience,* vol. 10, 2016-July-21 2016.

[12] K.-S. Ryu, S.-H. Yang, L. Thomas, and S. S. P. Parkin, "Chiral spin torque arising from proximity-induced magnetization," *Nature Communications,* vol. 5, p. 3910, 2014.

Device Structural Effects, SPICE Modeling and Circuit Evaluation for Negative-Capacitance FETs

Pin Su and Wei-Xiang You

Department of Electronics Engineering & Institute of Electronics, National Chiao Tung University, Taiwan

E-mail: pinsu@faculty.nctu.edu.tw

Introduction

Employing a hafnium-oxide based ferroelectric compatible with present CMOS gate stack, the negative-capacitance FET (NCFET) [1] has garnered substantial interest as it may enable the supply-voltage/power scaling of logic transistors [2]. Several NCFETs, with 2D (Fig. 1) or short-channel FinFET structures, have been experimentally demonstrated with negligible hysteresis (see, e.g., [3][4]). With its steep slope and similar current transport mechanism to the MOSFET, the NCFET has become a promising beyond-CMOS device candidate.

In this talk, we will discuss several device structural effects on NCFETs with emphasis on the wide-range steep slope [5]. We will explain the intrinsic difference between SOI and double-gate 2D NCFETs. The impact of drain coupling on the NC effect and its implication on the design of short-channel NC-FinFETs will be addressed. In addition, with the aid of our developed short-channel NC-FinFET SPICE model, we will discuss the functionality and performance gain of large logic circuits employing NC-FinFETs [6].

Intrinsic Difference between SOI and DG 2D NCFETs

We have analytically derived a general model [7] for the SOI and double-gate (DG) 2D NCFETs (Fig. 1), by which the design spaces for the two structures can be constructed and explored. Fig. 2 shows that, for the long-channel DG structure, it is difficult to achieve an average subthreshold swing (SS$_{ave}$) below 60 mV/decade at room temperature (Fig. 2(a)). While for the SOI with thin BOX structure the sub-$2.3kT/q$ SS$_{ave}$ can be achieved (Fig. 2(b)). This can be explained by the internal polarization charge (Q_{int}) at the ferroelectric interface (Fig. 3). Different from the DG device with merely a constant depletion charge in subthreshold (Fig. 3(b)), a bias-dependent internal charge mirrored from the back-gate ($-Q_b$ in Fig. 3(a)) is present in the SOI structure. This bias-dependent internal charge can polarize the ferroelectric and enable the NC effect in subthreshold for the SOI device, as indicated by the internal voltage gain (Av) shown in Fig. 3(c).

In addition, an adequate back-gate bias can be used to optimize the NCFET with the SOI with thin BOX structure [8].

Short-Channel Effects & NC-FinFET Model

Different from the long-channel DG device, the subthreshold Av of a short-channel NC-FinFET can be larger than one and sub-$2.3kT/q$ SS$_{ave}$ can be achieved (Fig. 4). This is because, in short-channel NC-FinFETs, the internal polarization charge (Q_{int}) in subthreshold is no longer a constant due to the drain capacitive-coupling effect [9]. Our NC-FinFET model captures the impact of drain coupling on the NC effect through an accurate Q_{int} model with good scalability in gate length (Fig. 5). The subthreshold Av and SS$_{ave}$ can be further improved by adequate utilization of the fringe fields through the spacers (Fig. 6).

Our NC-FinFET model is compatible with BSIM-CMG, and the Landau-Khalatnikov equation has been integrated into a unified surface-potential framework (which can also be used for the FinFET as shown in Fig. 4) without the need of sub-circuits or local charge integration.

Evaluation of Logic Circuits

With the aid of the NC-FinFET model, we have examined the functionality and performance gain of large logic circuits (e.g., dynamic 4-bit Manchester carry-chain adder) employing 14nm ultra-low-power NC-FinFETs with strong inverse V_{ds}-dependency of threshold voltage (which can be observed in Fig. 4) [6]. Our study indicates that, in addition to steep slope, the unique inverse V_{ds}-dependency of V_T is beneficial to the performance of logic circuits, especially for the pass-transistor logic style.

Acknowledgements

The authors acknowledge the support from the Ministry of Science and Technology, Taiwan, under contracts MOST-106-2221-E-009-148-MY2, MOST-107-2221-E-009-090-MY2, MOST-107-3017-F-009-002 and MOST-107-2633-E-009-003, and "Center for Semiconductor Technology Research" from the Featured Areas Research Center Program by the Ministry of Education in Taiwan. They are also grateful to Prof. Chenming Hu for his help during the work.

References

[1] S. Salahuddin and S. Datta, *Nano Lett.*, vol.8, no.2, p.405, 2008.

[2] *IRDS* (https://irds.ieee.org/)

[3] Z. Krivokapic *et al.*, *IEDM*, Dec. 2017, 15.1.1-15.1.4.

[4] M. Si *et al.*, *Nature Nanotechnol.*, vol.13, p.24, Dec. 2017.

[5] P. Su and W.-X. You, *IEEE S3S*, San Francisco, Oct. 2018.

[6] W.-X. You, P. Su and C. Hu, *IEEE S3S*, San Francisco, Oct. 2018.

[7] W.-X. You and P. Su, *IEEE TED*, vol.65, no.10, p.4196, 2018.

[8] W.-X. You and P. Su, *IEEE TED*, vol.64, no.8, p.3476, 2017.

[9] W.-X. You, C.-P. Tsai, P. Su, *IEEE TED*, vol.65, no.4, p.1604, 2018.

978-1-7281-0943-5/19 $31.00 © 2019 IEEE

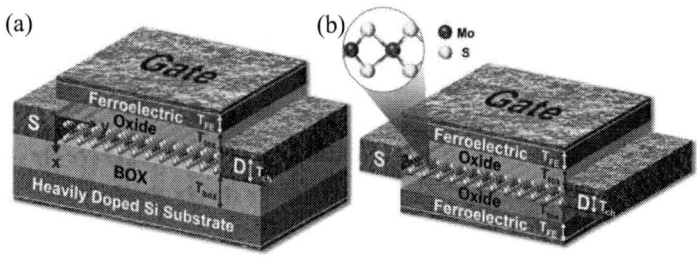

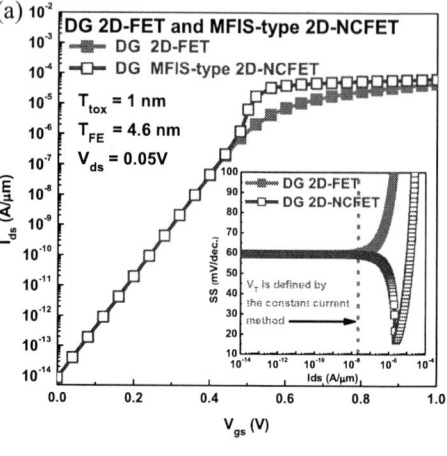

Fig. 1. Schematics of the (a) SOI and (b) symmetrical-DG metal-ferroelectric-insulator-semiconductor (MFIS) NCFETs with a 2D MoS$_2$ channel (T_{ch} = 0.65 nm, N_{ch} = 6.5E7 cm^{-2}). T_{FE} is the ferroelectric thickness. $T_{tox/box}$ are interlayer oxide and BOX thicknesses, respectively.

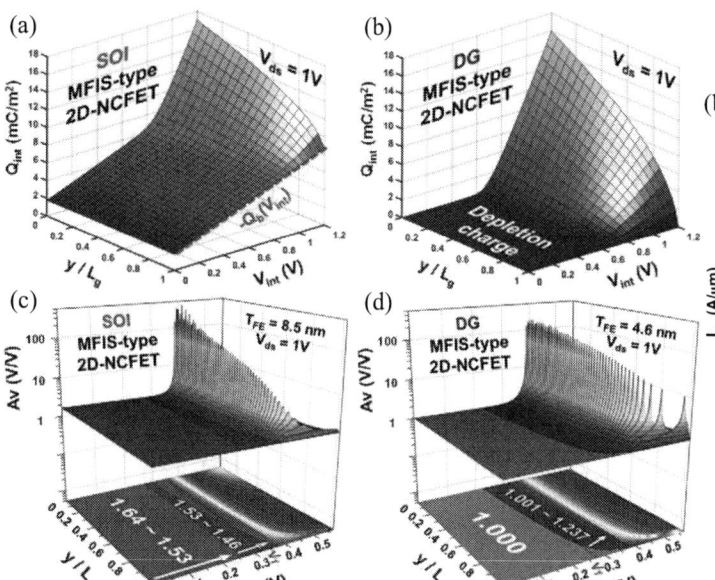

Fig. 3. Distributions of the internal-voltage (V_{int}) dependent internal charge (Q_{int}) and the gate-bias dependent voltage gain (A_v) along the channel [5][7]. The depletion charge in (b) is not enough to polarize the ferroelectric, so the A_v in (d) cannot be larger than one and steep slope cannot occur until strong inversion for the long-channel DG device.

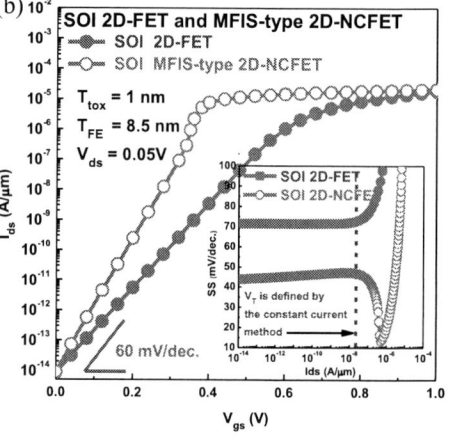

Fig. 2. Comparison of long-channel (a) DG and (b) SOI 2D-FET and 2D-NCFET. While the improvement in SS cannot be seen until strong inversion for the DG case, significant improvement in SS can be achieved over the whole subthreshold region for the SOI with thin BOX structure (T$_{box}$ = 5 nm). Note that the T_{FE} is determined based on the hysteresis-free capacitance matching [7].

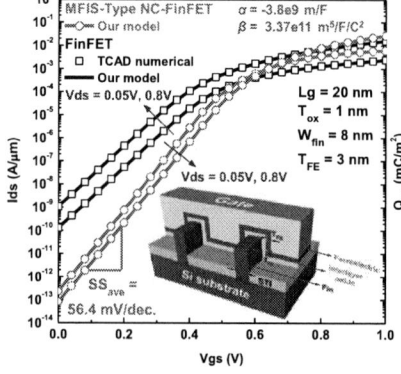

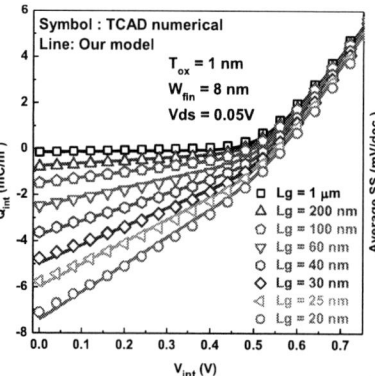

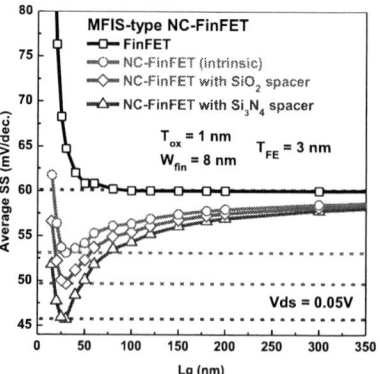

Fig. 4. Sub-60 mV/decade SS$_{ave}$ can be achieved for the short-channel NC-FinFET. Note that, different from the FinFET counterpart, the threshold voltage at high V$_{ds}$ can be larger than that at low V$_{ds}$ for the NC-FinFET [9].

Fig. 5. Our NC-FinFET model captures the impact of drain coupling on the NC effect through accurate gate-length dependence of Q_{int} [5]. This scalable Q_{int} model is derived through an analytical short-channel surface-potential equation [9].

Fig. 6. The SS$_{ave}$ can be further improved by adequate design of the spacers for short-channel NC-FinFETs.

978-1-7281-0943-5/19 $31.00 © 2019 IEEE

Ferroelectric Si-doped HfO$_2$ Capacitors for Next-Generation Memories

Ava J. Tan[1,2], Zhongwei Zhu[2], Hwan Sung Choe[2], Chenming Hu[1], Sayeef Salahuddin[1], Alex Yoon[2]

[1]Department of Electrical Engineering & Computer Sciences, University of California Berkeley, Berkeley, CA, USA
[2]Lam Research Corporation, Fremont, CA, USA
E-mail: avatan@berkeley.edu

ABSTRACT

In this work, we assess the viability of a CMOS-compatible ferroelectric material, silicon-doped hafnium oxide (Si-HfO$_2$), for future capacitor-based ferroelectric memories by investigating process development, structural/electrical characterization, and memory device benchmarking. Ferroelectric capacitors with iridium electrodes and Si-HfO$_2$ thicknesses from 8 nm down to 4 nm are fabricated. The Si-HfO$_2$ layer is grown by atomic layer deposition (ALD) and subjected to a variety of post-processing annealing conditions. Polarization-voltage (PV) loops are taken to extract ferroelectric parameters such as remanent polarization (P_R) and coercive field (E_C), and standard memory tests for fatigue/endurance, retention, and imprint are conducted.

INTRODUCTION

The discovery of ferroelectricity in HfO$_2$, a gate dielectric used in virtually all high performance logic transistors today, opens a new avenue to realize novel CMOS-compatible ferroelectric devices. It has been shown that dopants such as Zr [1], Si [2], Al [3], and others can stabilize the non-centrosymmetric, orthorhombic crystalline phase of HfO$_2$ which has been linked to ferroelectricity. Indeed, there have been several demonstrations of negative capacitance transistors (NCFETs) and ferroelectric memories (FeRAM) [4-6] in the past several years which leverage the ferroelectric properties of HfO$_2$. Most notably, ferroelectric HfO$_2$ has obviated the need for exotic materials such as PZT and SBT in ferroelectric memories, thereby making FeRAM a true contender among other emerging memory technologies. Moving forward, it is therefore imperative to develop and understand processing techniques for ferroelectric HfO$_2$ grown by ALD and to conduct feasibility tests to predict the performance of doped hafnium oxides in fabricated memory devices.

EXPERIMENTAL PROCEDURE

The devices characterized in this work were fabricated by a simple 4-step process as outlined in Fig. 1. Silicon substrates were first sputtered with a blanket layer of 10 nm iridium for the back electrode. The ferroelectric HfO$_2$ layer was grown by ALD using alternating growth cycles of HfO$_2$ and SiO$_2$, appropriately chosen to achieve specific thicknesses and doping concentrations. Film thicknesses from between 4 nm to 8 nm were grown, targeting a silicon doping concentration in the range of 5-6 mol% Si to maximize the ferroelectric properties of the resulting film [7]. Next, 10 nm iridium was sputtered on top of the stack through a shadow mask with 200 μm diameter holes to form an array of capacitors. The samples were then diced and annealed under varying conditions.

RESULTS & DISCUSSION

To properly design ferroelectric devices for various applications, the first and foremost step should be to choose the material parameters appropriately. All of the samples with various film thicknesses were separately annealed at temperatures ranging from 600 – 1000 °C, and afterwards were measured with a Radiant Precision LC II Ferroelectric Tester. From the measured PV loops, the P_R (retained polarization at zero applied voltage) and E_C (field at which ferroelectric switching initiates) were extracted. Both of these

ferroelectric parameters are plotted as a function of anneal temperature and film thickness in 2D contour plots shown in Fig. 2.

Standard memory tests such as fatigue, retention, and imprint were conducted on the samples. Fatigue testing was done by cycling the film with a 10 kHz bipolar voltage pulse train of +/- 2.5 MV/cm until failure or 10^9 cycles was reached. Retention tests were conducted using the methods as described in [6] and subjecting the programmed samples to an elevated temperature of 100 °C. Imprint of each programmed device was assessed by comparing the PV characteristics of the device after 3 weeks at 100 °C to the fresh device prior to thermal testing.

As seen in Fig. 3a, the write endurance appears to be inversely proportional to film thickness, with the 4 nm film witnessing a gradual decrease in remanent polarization over its lifetime and the 8 nm film failing abruptly after 10^6 cycles. This behavior is correlated with the magnitude of the leakage current through the film (Fig. 3b). These observations, coupled with the fact that the magnitude of the peak switching current for the 8 nm film is 343 mA/cm^2 vs. 213 mA/cm^2 for the 4 nm film, suggests that higher switching currents cause greater damage over time and thereby trigger the onset of early device failure.

Retention testing on a representative 8 nm ferroelectric film is shown in Fig. 3c. After 10 days at 100 °C, the film retains most of its polarization, and is projected to retain more than 50% of the initial opposite state switched polarization (which appears to be the limiting state due to imprint – see Fig. 4a and 4b to observe the imprinting behavior) after 10 years.

Film crystallinity is verified through X-ray diffraction (XRD) as shown in Fig. 5a. The representative 4 nm film shows a strong peak around 30.5°, which corresponds to the orthorhombic crystalline phase. A TEM image of a memory device with a ferroelectric layer thickness of roughly 8 nm is shown in Fig. 5b.

CONCLUSION

We have investigated the processing space for ferroelectric HfO$_2$, verified its ferroelectric properties via various electrical and material characterization techniques, and assessed the performance of capacitor memories based on FE-HFO$_2$. These critical feasibility studies demonstrate the many promising characteristics of CMOS-compatible ferroelectric memories for future memory technologies.

ACKNOWLEDGEMENTS

The authors acknowledge Evans Analytical Group for XRD, XPS and TEM characterization. A. J. Tan acknowledges fellowship support by the National Defense Science and Engineering Graduate Fellowship.

REFERENCES

[1] J. Müller et al., Nano Lett., 2012, pp. 4318-4323. [2] U. Schroeder et al., ECS J. Solid State Sci. Technol., 2013, p. 69-72. [3] S. Mueller et al., Adv. Funct. Mater., 2012, 2412-2417. [4] K. Chatterjee et al., IEEE Electron Device Lett., 2017, 1379-1382. [5] D. Kwon et al., IEEE Electron Device Lett., 2018, 300-303. [6] S. Mueller et al., IEEE Trans. Device Mater. Rel., 2012, pp. 93-97. [7] P. Lomenzo et al., J. Vac. Sci. Technol. B, 2014, 03D123.

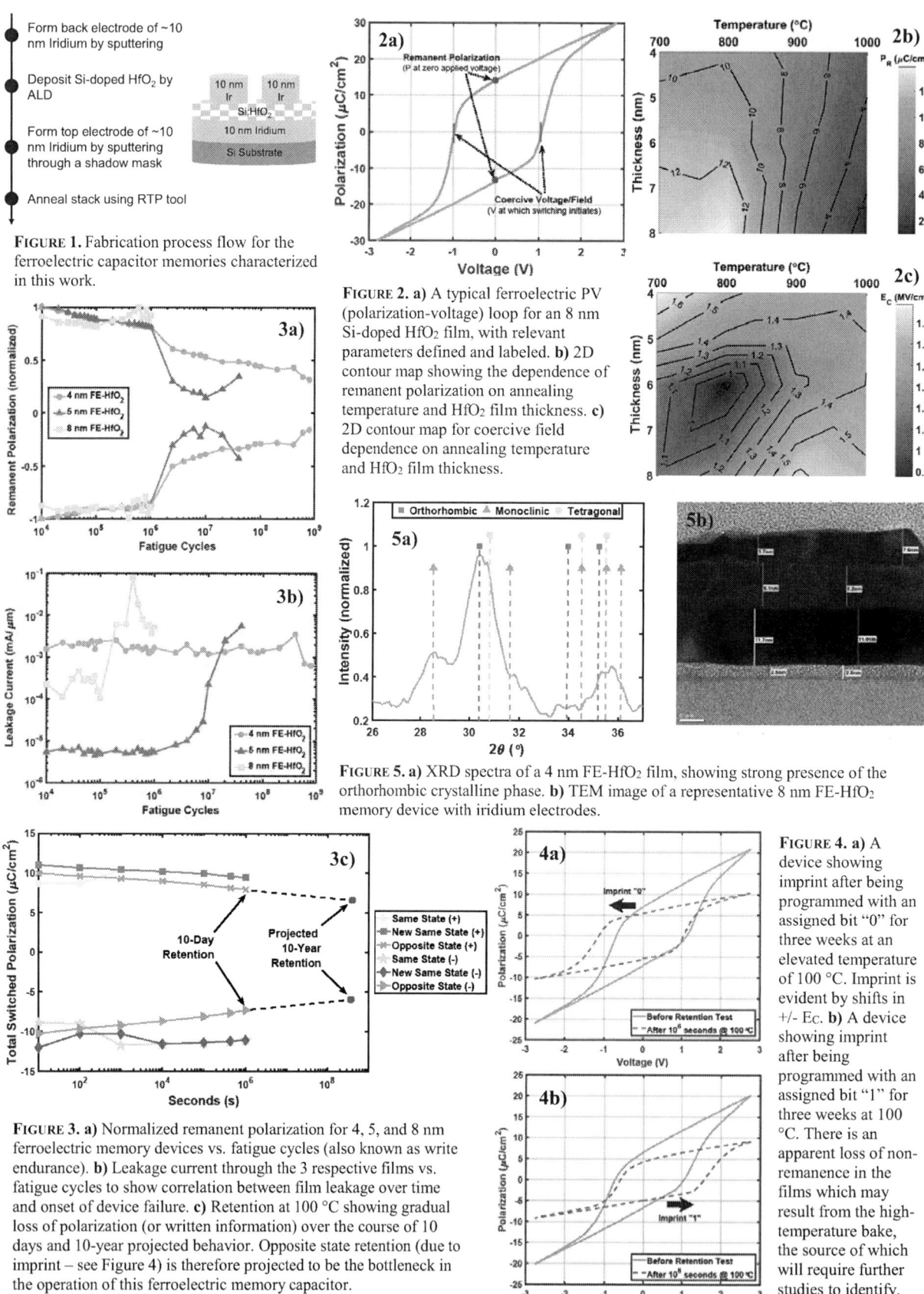

- Form back electrode of ~10 nm Iridium by sputtering
- Deposit Si-doped HfO₂ by ALD
- Form top electrode of ~10 nm Iridium by sputtering through a shadow mask
- Anneal stack using RTP tool

FIGURE 1. Fabrication process flow for the ferroelectric capacitor memories characterized in this work.

FIGURE 2. a) A typical ferroelectric PV (polarization-voltage) loop for an 8 nm Si-doped HfO₂ film, with relevant parameters defined and labeled. **b)** 2D contour map showing the dependence of remanent polarization on annealing temperature and HfO₂ film thickness. **c)** 2D contour map for coercive field dependence on annealing temperature and HfO₂ film thickness.

FIGURE 5. a) XRD spectra of a 4 nm FE-HfO₂ film, showing strong presence of the orthorhombic crystalline phase. **b)** TEM image of a representative 8 nm FE-HfO₂ memory device with iridium electrodes.

FIGURE 3. a) Normalized remanent polarization for 4, 5, and 8 nm ferroelectric memory devices vs. fatigue cycles (also known as write endurance). **b)** Leakage current through the 3 respective films vs. fatigue cycles to show correlation between film leakage over time and onset of device failure. **c)** Retention at 100 °C showing gradual loss of polarization (or written information) over the course of 10 days and 10-year projected behavior. Opposite state retention (due to imprint – see Figure 4) is therefore projected to be the bottleneck in the operation of this ferroelectric memory capacitor.

FIGURE 4. a) A device showing imprint after being programmed with an assigned bit "0" for three weeks at an elevated temperature of 100 °C. Imprint is evident by shifts in +/- E_C. **b)** A device showing imprint after being programmed with an assigned bit "1" for three weeks at 100 °C. There is an apparent loss of non-remanence in the films which may result from the high-temperature bake, the source of which will require further studies to identify.

The Guideline on Designing a High Performance NC MOSFET by Matching the Gate Capacitance and Mobility Enhancement

Y. C. Luo[1], F. L. Li[2], E. R. Hsieh[1], C. H. Liu[2], Steve S. Chung[1], T. P. Chen[3], S. A. Huang[3], T. J. Chen[3], and Osbert Cheng[3]

[1]*Department of Electronics Engineering, National Chiao Tung University, Taiwan*   [2]*National Taiwan Normal University, Taiwan*
[3]*United Microelectronics Corporation (UMC), Taiwan*

Abstract- In this paper, to provide a guideline of designing a high-performance NCFET, we explored not only the capacitance matching between ferroelectric HZO MIM and MOSFET but also how effective mobility is affected by HZO dipoles. For capacitance matching, we observe a 50x enhancement of overall gate capacitance triggered by NC effect, while, however, it generated an adverse degradation of the mobility. This mobility degradation is induced by the remote scattering from the ferroelectric HZO dipoles. Fortunately, if suitable polarization can be formed to align the HZO dipoles, the effects of remote scattering can be mitigated. From a trade-off between gate capacitance and the mobility, an NCFET with desirable characteristics can be achieved. Besides, we showed that improved SS is possible when the derivative of the voltage across the ferroelectric MIM is negative and is corresponding to the release of energy from the ferroelectric MIM which boosts SS.

1. Introduction

Negative-capacitance field effect transistor (NCFET) has drawn a lot of attention recently because of their potential for providing steep subthreshold swing (SS) (< 60mV/dec) along with large on current (I_{on}) [1-4]. To achieve both desirable characteristics in NCFET, capacitance matching between ferroelectric MIM and MOSFET has been intensively studied [5-9]. While efforts have been made to match capacitances, how NC effect influences the mobility, which is also an important factor in designing high I_{on} in NCFET, has seldom been mentioned. In this work, it was observed that while NC effects do enhance the gate capacitance of NCFETs by up to 50x, these effects have also severely degraded their mobility by few orders due to enhanced remote scattering triggered by ferroelectric HZO dipoles. Therefore, we provide a guideline to designing an NCFET so that high I_{on} can be achieved by combining less degraded mobility and large gate capacitances. Also, it is shown that one can achieve higher-performance NCFET if voltage across the HZO MIM (V_{MIM}) is more negative when positive gate voltage is applied. Moreover, our experimental data showed that negative derivatives of V_{MIM} leads to SS improvement.

2. Device Preparation

TiN/HZO(5nm)/TiN samples whose area= 100×100 μm^2 were prepared in post metal annealing temperature of 550^{0}C. MIM is electrically connected in series to the gate of 28 nm-nMOSFET w/ Width = 10, 5, 3, 2, 1, 0.75, 0.5, and 0.25, and Length = 0.04 (μm) (Fig. 1).

3. Results and Discussions

A. Enhanced Ion and SS in NCFET

Figs. 1 and 2 are the schematic and equivalent circuit model of our experimental setup respectively, in which an HZO MIM is electrically connected to the gate of nMOSFET at V_{ds}= 0.1V. The internal voltage, $V_{interal}$, in Fig. 2 is measured. In Fig. 3, I_dV_{gs} of the NCFET with width = 1 μm shows enhanced I_{on}, steeper SS, and smaller off-current. In Fig. 4, the same HZO MIM is connected to another MOSFET with different widths. However, in this case, not only SS and off-current show zero improvement, the I_{on} value is further degraded from the original MOSFET. In fact, when connected to HZO MIM with a fixed area, the HZO-gated MOSFET with W/L = 1/0.04 μm has superior I_{on} level and SS in comparison to the others (Fig. 5). In Fig. 6, similarly, 22x enhancement of on/off ratio of the HZO-gated MOSFET with width = 1 μm is observed while the on/off ratio of the others show less than 5x of improvement. In Fig. 7, we compare all the I_{on} enhancement of NCFETs with widths from 10 to 0.5 μm and observed that only the one with width = 1 μm shows enhanced I_{on}. Hence, it is of interest to find out what affect the I_{on} level. We begin to examine the current formula of MOSFETs (equation (1) in Fig. 7), and narrow down to two key factors. One is the gate capacitances of the NCFET (C_{total}) and MOSFET (C_{MOS}). The other is effective mobility (μ_{eff}). It should be noted that W/L does not matter because W has been normalized and L is fixed for all samples. Also, Threshold voltage (V_{th}) is shifted to the same value.

B. Enhanced Gate Oxide Capacitance in NCFET

C_{MOS} and C_{total} for both MOSFET and NCFET are measured by applying DC voltage with 1MHz small signal on their top gates with grounded sources, floating drains and floating body. In Fig. 8, due to NC effect, C_{total} is greatly enhanced compared to C_{MOS}. Furthermore, the capacitance contributed by the HZO MIM (C_{MIM}) is calculated, for C_{total} can be characterized as C_{MIM} in series with C_{MOS}. The negative extracted C_{MIM} in Fig. 8 is reasonable because positive C_{MIM} and C_{MOS} would have only resulted in a non-enhanced C_{total}. It can be observed from Fig. 9 that, in fact, all the C_{total} is enhanced and there exists a global maximum at width = 1μm. In Fig. 10, it is shown that the NCFET with width = 1μm, whose current is greatly enhanced in Fig. 7, having 50 times of C_{total} enhancement. However, I_{on} of the NCFETs which should be proportional to the gate oxide capacitances but does not increase as much as expected (Fig. 7). This implies that there is another factor which caused the decrease of I_{on} current.

C. Mobility Degradation in NCFET

To calculate μ_{eff} in NCFETs, we first measure the drain transconductance (g_d). Fig. 11 shows that g_d of the NCFET with width = 1μm is approximately the same as the MOSFET while another NCFET with W/L = 5/0.04 μm shows smaller g_d than the MOSFET due to weaker NC effects (Fig. 12). For μ_{eff}, it is calculated as $g_d \times L/W/Q_{inv}$. Q_{inv}, inversion charge, is computed by integrating C_{ox}. In Fig. 13 and Fig. 14, μ_{eff} of the NCFET is degraded by almost 2 orders compared to that of the original MOSFET. Fig. 15 and Fig. 16 further show that μ_{eff} degrades in all NCFETs. However, the NCFET with width = 1μm exhibits the smallest μ_{eff} degradation which is accounted for its enhanced current. The universal degraded μ_{eff} could be explained as remote scattering, HZO dipoles fluctuation causing phonon scattering along the channel. For the NCFET with less degraded μ_{eff}, it is the maximum polarization (Fig. 9) that aligns HZO dipoles and less dipole fluctuation, thus, mitigates the effect of remote scattering (Fig. 17).

D. Correlation Between V_{MIM} and Improved SS

From Fig.18, V_{MIM} is calculated from measured internal voltage (Fig. 2). V_{MIM} with width = 1 μm has the most negative V_{MIM} when a positive gate voltage is applied. The most negative V_{MIM} implies that it stores the most amount of energy against the applied voltage. The energy later contributes to the enhancement of I_{on}. Furthermore, stronger disturbance of V_{MIM} in the forward branch (Fig.18) implies more existing negative derivatives of V_{MIM}, which is explained as electrons transferring from one energy minimum to another. The energy release results in SS improvement (Fig. 19). While ideal W-shape free-energy plots only has two minimum, multiple local minimum of SS shows that multiple domains exist in the HZO MIM (Fig. 19) [10-11]. The other evidence of negative derivatives of V_{MIM} improving SS is that, in Fig. 20, the forward SS is much smaller than that of the backward one which is consistent with the fact that more negative derivatives of V_{MIM} found in the forward path shown in Fig. 17.

The purpose of negative capacitance from HZO MIM is to increase the I_{on} current. However, mobility is degraded by orders in NCFET due to remote scattering, canceling out the effects of larger gate capacitances and results in smaller I_{on}. Effects of remote scattering can be reduced by properly polarizing HZO dipoles. Therefore, we may conclude that to realize high performance NCFET, both a less degraded mobility and a strongly enhanced C_{total} from capacitance matching are needed. Besides, we also show that, when a positive top voltage is applied, the more negative the V_{MIM} is, the larger on-current the NCFET becomes. It was also demonstrated that negative derivatives of V_{MIM} represent the energy release from the ferroelectric dipoles, which eventually contributes to the improvement of the subthreshold swing.

Acknowledgments *This work was supported in part by the Ministry of Science and Technology, Taiwan, under contract MOST107-2633-E009-003.*

References: [1] S. Salahuddin et.al., *Nano Lett.*, p. 405, 2008. [2] Z. Krivokapic et. al., *IEDM*, p. 357, 2017. [3] H. Ota et. al., *IEDM*, p. 318, 2016. [4] K. S. Li, et al., *IEEE IEDM*, p. 620, 2015. [5] H. Agarwal, et al., *IEEE EDL*, p. 1254, 2018. [6] A. I. Khan, et al., *APL*, p. 113501, 2011. [7] H. Agarwal, et al., *IEEE TED*, p. 1211, 2018. [8] P. Bidenko, et al., *IEEE J-EDS*, p. 910, 2018. [9] J. Jo, et al., *IEEE EDL*, p. 245, 2016. [9] A. I. Khan, et al., *Nat. Mat.*, p. 182, 2015. [10] J. A. Kittl, et al., *APL*, 042904, 2018.

978-1-7281-0943-5/19 $31.00 © 2019 IEEE

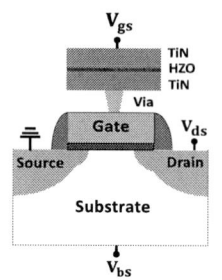

Fig. 1 Schematic of HZO-gated nMOSFET. An HZO film sandwiched by two TiN layers is electrically connected to the gate of an nMOSFET.

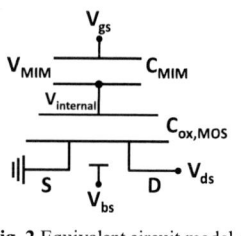

Fig. 2 Equivalent circuit model of **Fig. 1**. The capacitance is modeled by C_{MIM}(capacitance of HZO MIM) in series with C_{MOS} (gate capacitance of MOSFET). V_{MIM} is the voltage across the HZO MIM.

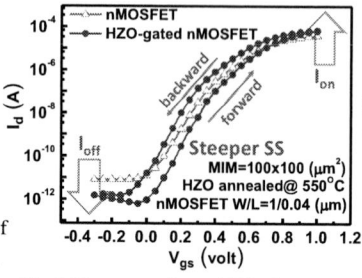

Fig. 3 The comparison of I_dV_{gs} between an NCFET and a MOSFET. The NCFET shows enhanced I_{on}, SS, and lower I_{off} compared to the original MOSFET.

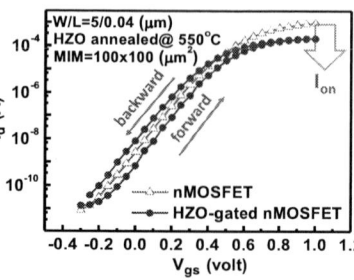

Fig. 4 I_dV_{gs} of an NCFET and a MOSFET with W/L = 5/0.04 (μm) connected to HZO MIM with an area of 100×100 μm^2 annealed at 550 ℃. The NCFET shows degraded I_{on}.

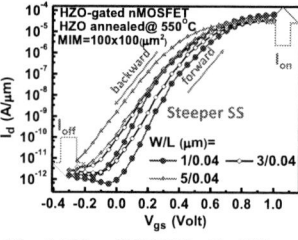

Fig. 5 I_dV_{gs} of NCFET with different width connected with HZO MIM with an area of 100×100 μm^2 annealed at 550 ℃. The NCFET with W/L = 1/0.04 μm shows better I_{on}, SS and I_{off} than the others do.

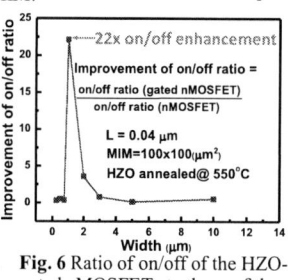

Fig. 6 Ratio of on/off of the HZO-gated nMOSFETs to those of the nMOSFETs. Greatly enhanced on/off ratio happens at width = 1 μm.

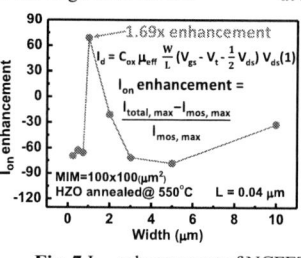

Fig. 7 I_{on} enhancement of NCFET compared with that of MOSFET. $I_{total,\ max}$ is the maximum Ion values of NCFET, $I_{mos,\ max}$ is for MOSFET.

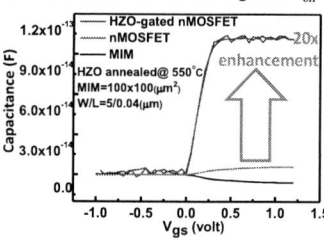

Fig. 8 Gate oxide capacitance of HZO-gated nMOSFET (C_{total}) is greatly enhanced by NC effect.

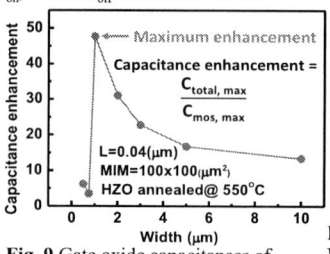

Fig. 9 Gate oxide capacitances of HZO-gated nMOSFETs (C_{total}) are all greatly enhanced by NC effects. Maximum occurs at width = 1 μm.

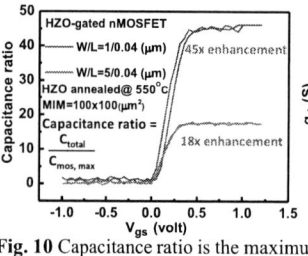

Fig. 10 Capacitance ratio is the maximum NCFET gate capacitance divided by maximum MOSFET gate capacitance. NCFET with width = 1 μm shows the largest enhancement of the gate capacitance.

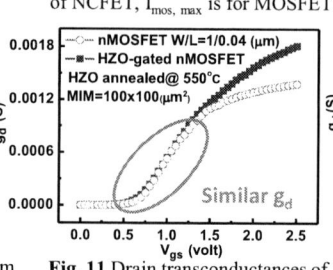

Fig. 11 Drain transconductances of a HZO-gated MOFET and a MOSFET with W= 1μm are measured. They show similar values below 1 volt.

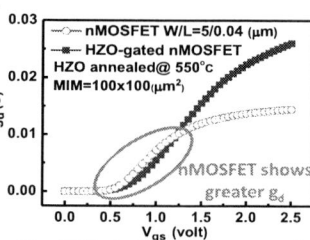

Fig. 12 Drain transconductances of a HZO-gated MOFET and a MOSFET with W= 5um are measured. The values of nMOSFET are larger than those of the HZO-gated nMOSFET below 1 volt.

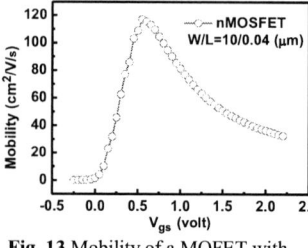

Fig. 13 Mobility of a MOFET with W/L = 10/0.04 μm is measured. The peak value is close to 120 cm^2/Vs.

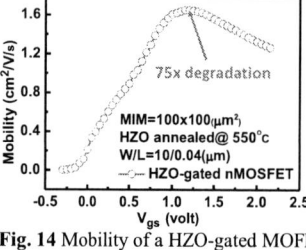

Fig. 14 Mobility of a HZO-gated MOFET with W/L = 10/0.04 μm is measured whose mobility has severely degraded compared with that of the MOSFET in Fig. 13.

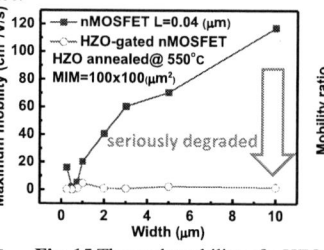

Fig. 15 The peak mobility of a HZO-gated MOFET and a MOSFET with different widths is measured. Mobility is seriously degraded with width.

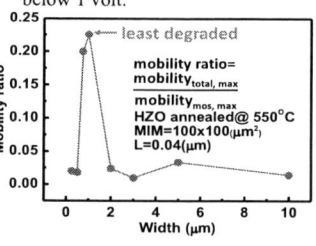

Fig. 16 The ratio of maximum effective mobility of HZO-gated MOFETs to that of MOSFETs. The least degraded mobility happens at width = 1 μm

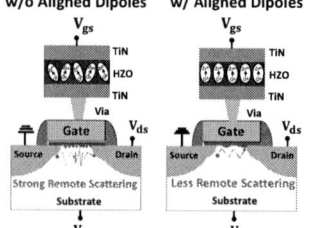

Fig. 17 (left) Remote scattering from HZO dipoles affects the effective mobility. (right) suitable polarization with aligned HZO dipoles results in less remote scattering and higher effective mobility.

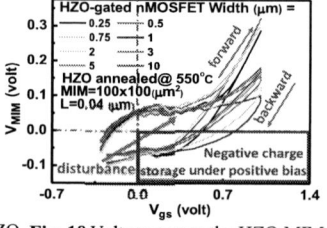

Fig. 18 Voltage across the HZO MIM (V_{MIM}) is measured. The most negative value of V_{MIM} happens at width = 1 μm. Besides, more disturbance of V_{MIM} happens in the forward path which is highly related to SS improvement.

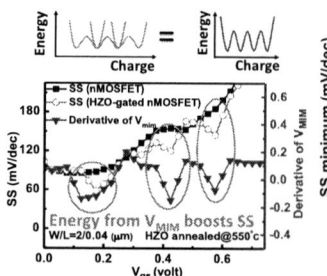

Fig. 19 The local minimum of SS happens when the derivative of V_{MIM} goes negative, which implies that electrons jump from one energy minimum to another.

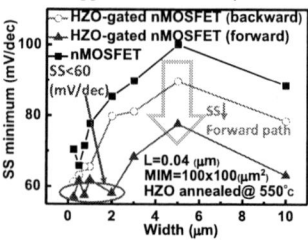

Fig. 20 Minimum SS in the forward path is smaller than that of the backward path due to more negative derivatives of V_{MIM} in the forward branch as shown in Fig. 18.

978-1-7281-0943-5/19 $31.00 © 2019 IEEE

Impact of Gate Stack Design on Improving Subthreshold Swing Behaviors in Ferroelectric-Gate Field-Effect Transistors

Shinji Migita, Hiroyuki Ota, Akira Toriumi*, and Takashi Matsukawa

National Institute of Advanced Industrial Science and Technology (AIST), Japan
*The University of Tokyo, Japan
E-mail: s-migita@aist.go.jp

ABSTRACT

Ferroelectric-gate field-effect transistor (FE-FET) is an attractive device for both memory and logic applications. The metal-ferroelectric-metal-insulator-semiconductor (MFMIS) gate stack structure is useful to solve the issue of charge unbalance between ferroelectric and metal-insulator-semiconductor (MIS) capacitors in FE-FETs. In this study, the function of the MFMIS gate stack for the improvement of subthreshold behaviors in FE-FETs is analyzed. The results indicate that the design of capacitor-area-ratio in an MFMIS structure enhances the ferroelectric poling component relative to the paraelectric component and increases the charge boost efficiency, thereby leading to improved subthreshold swing.

INTRODUCTION

Ferroelectric-gate field-effect transistors (FE-FETs) in which ferroelectric capacitors are integrated at the gate of the FETs are expected to constitute the ultimate memory device. Since the first report in 1963 [1], several experimental studies have focused on using perovskite-type ferroelectric materials. Recently, the discovery of ferroelectric HfO_2 accelerated the developments of FE-FETs. Additionally, HfO_2 ferroelectrics were used to fabricate vertical type FE-FETs [2] and the operation of FE-FETs prepared via the 22-nm technology-node [3] are demonstrated. Another aspect corresponds to the application of FE-FETs as logic FETs. Salahuddin and Datta proposed an idea in 2008 that a hysteresis-free and a steep subthreshold swing behavior might be attainable in FE-FETs when a ferroelectric capacitor and paraelectric capacitors are coupled in series under an optimized capacitances design [4]. The operation mode of this FE-FET is termed as the negative-capacitance FETs (NC-FETs), and NC-FETs are currently extensively studied around the world. The potential of the ferroelectric capacitor is key for FE-FET applications in memory and logic devices.

A major issue in FE-FETs corresponds to the charge unbalance between the capacitors of the ferroelectric (MFM) and the metal-insulator-semiconductor (MIS). As an example, we discuss our experimental results for the MFM capacitor and MIS-FET in **Fig. 1**. The MFM capacitor consists of $TaN/Hf_{0.5}Zr_{0.5}O_2$ (10 nm)/TaN structure deposited via sputtering and is further annealed at 700 °C [5]. The yield of the ferroelectric $Hf_{0.5}Zr_{0.5}O_2$ film reaches 20 $\mu C/cm^2$ remnant polarization and exceeds 30 $\mu C/cm^2$ at the maximum applied voltage. The MIS-FET prepared with 3.8-nm thick SiO_2 exhibits a successful operation, and the charge density generated by the gate electrode is less than 2 $\mu C/cm^2$. It is observed that the breakdown of the SiO_2 occurs when the charge density reaches 2.7 $\mu C/cm^2$. The results clearly indicate the existence of charge imbalance between the MFM and MIS capacitors. A saturated hysteresis in the MFM capacitor yields a charge exceeding 20 $\mu C/cm^2$ while the operation of MIS-FET accepts a charge of less than 2.7 $\mu C/cm^2$. Thus, the ferroelectric capacitor must be operated in a minor loop mode when the ferroelectric film is directly deposited on the insulator and forms the metal-ferroelectric-insulator-semiconductor (MFIS) gate stack in FE-FETs.

A solution to the issue was proposed by Tokumitsu *et al.* in 1999 [6]. They applied an internal metal gate and designed an MFMIS gate stack structure in which the area of MFM capacitor is lower than the area of MIS capacitor. The mismatch in the charge density was solved by adjusting the charge between the capacitors. They demonstrated a high memory window and improved retention properties using $SrBi_2Ta_2O_9$ ferroelectric.

We apply the MFMIS structure using $Hf_{0.5}Zr_{0.5}O_2$ ferroelectric and successfully demonstrate FE-FETs with a large memory window and low operation voltage [7]. Furthermore, the results indicate that the MFMIS structure is also effective in improving the subthreshold swing behaviors in FE-FETs. In the study, we analyze the mechanism of the MFMIS gate stack in the FE-FET.

CHARGE BOOSTS VIA GATE STACK DESIGNS

Two types of FE-FETs, MFIS and MFMIS gate stack structures, and their operation modes are compared in **Fig. 2**. They are differentiated by the capacitor-area-ratio, which is defined by A ($=S_I/S_F$, where S_I denotes the area of MIS and S_F denotes the area of MFM). In both cases, the operation of FE-FETs should be considered under the limitation of breakdown in SiO_2 (Fig. 2(b)). The charge density must be less than 2.7 $\mu C/cm^2$. In case of the MFIS structure ($A = 1$), the criterion is satisfied via the minor loop operation of ferroelectric capacitor. In case of the MFMIS structure (for example $A = 11$), a saturated hysteresis loop is available because the ferroelectric charge itself decreases. As shown in the P–V curves in Fig. 1(a), the A-value should exceed 3 and be less than 11.

The response of the ferroelectric capacitor consists of the ferroelectric poling component and paraelectric component. We use the P–V curves shown in Fig. 2(b), and the components are extracted in Fig. 2(c). In the MFIS case, the paraelectric component exceeds

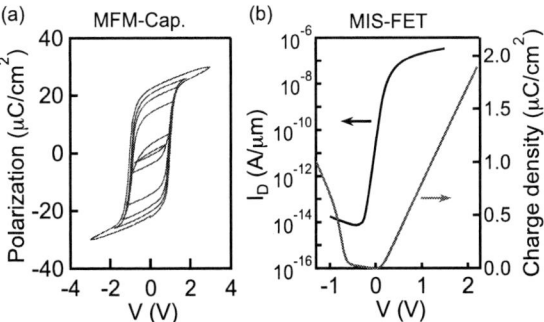

Fig. 1. (a) P–V characteristics of the 10-nm thick $Hf_{0.5}Zr_{0.5}O_2$ capacitor measured with different sweep ranges. (b) I_D-V_G characteristic ($V_D = 0.10$ V) and charge density of MIS-FET (L/W = 50 μm/50 μm). Insulator corresponds to 3.8-nm thick SiO_2.

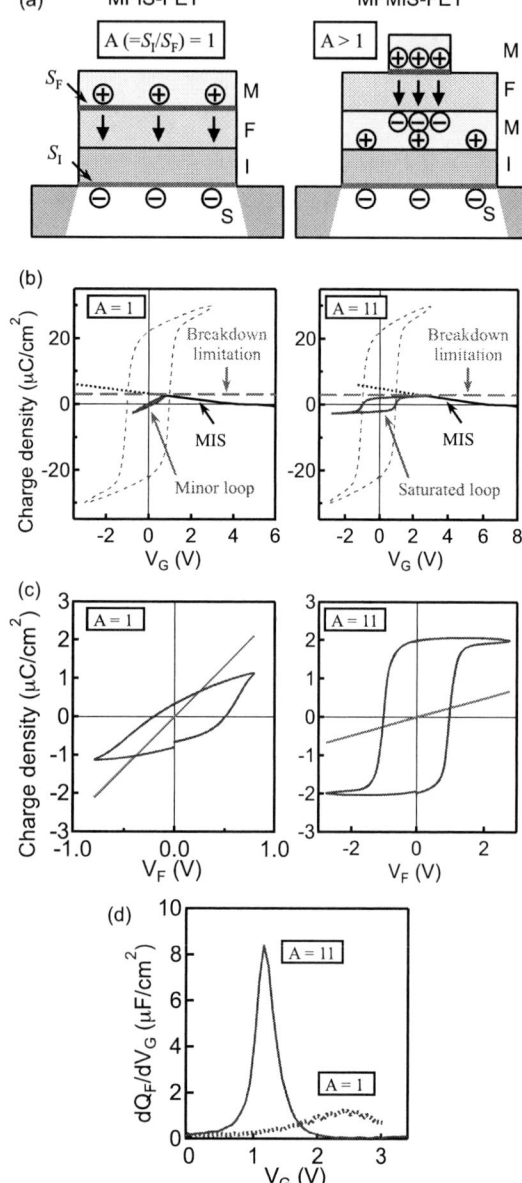

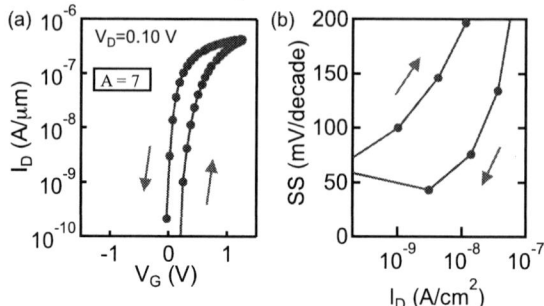

Fig. 3. (a) I_D-V_G characteristic of the MFMIS-FET (L/W = 50 μm/50 μm) designed with A (S_I/S_F) = 7, measured by V_G swing between -1.25 and 1.25 V. (b) Relationship between the subthreshold swing and I_D.

2(d). The efficiency of the MFIS gate stack is not extremely high because the charge is mostly yielded by the paraelectric response. In contrast, the MFMIS gate stack exhibits a high charge boost efficiency. Thus, ferroelectric poling contributes to the emergence of high charge within a low gate bias range.

The effect of MFMIS gate stack design is examined via the fabrication of FE-FET in which the MFM capacitor and MIS-FET in Fig. 1 are combined with an area-ratio $A = 7$. The results are shown in **Fig. 3** and demonstrate an improved subthreshold swing in conjunction with the memory operation.

CONCLUSION

Charge matching between MFM and MIS capacitors is essential to maximize the use of ferroelectric performance in FE-FETs. The MFMIS gate stack is an important method to realize an optimized capacitor-area-ratio. It contributes to memory applications as well as logic applications. The potential of FE-FETs is further enhanced via the design of materials and devices.

ACKNOWLEDGEMENT

This work was partly supported by JST CREST Grant Number JPMJCR14F2, Japan.

REFERENCES

[1] J. L. Moll and Y. Tarui, IEEE Trans. Electron Devices **ED-10**, 333 (1963).
[2] K. Florent, S. Laczzari, L. Di. Piazza, M. Popovici, E. Vecchio, G. Potoms, G. Groeseneken, and J. Van Houdt, 2017 Sympo. VLSI Technol. Dig., T158.
[3] S. Dünkel, M. Trentzsch, R. Richter, P. Moll, C. Fuchs, O. Gehring, M. Majer, S. Wittek, B. Müller, T. Melde, H. Mulaosmanovic, S. Slesazeck, S. Müller, J. Ocker, M. Noack, D.-A. Löhr, P. Polakowski, J. Müller, T. Mikolajick, J. Höntschel, B. Rice, J. Pellerin, and S. Beyer, Tech. Dig. IEDM 2017, p. 485.
[4] S. Salahuddin and S. Datta, Nano Lett. **8**, 405 (2008).
[5] S. Migita, H. Ota, H. Yamada, K. Shibuya, A. Sawa, and A. Toriumi, Jpn. J. Appl. Phys. **57**, 04FB01 (2018).
[6] E. Tokumitsu, G. Fujii, and H. Ishiwara, Appl. Phys. Lett. **75**, 575 (1999).
[7] S. Migita, H. Ota, H. Yamada, K. Shibuya, A. Sawa, T. Matsukawa, and A. Toriumi, 2018 IEEE Silicon Nanoele. Workshop, 3.1.

Fig. 2. (a) Comparison of MFIS-FET and MFMIS-FET structures and definition of parameter A $(=S_I/S_F)$. (b) Relationship between the P-V characteristic and load line of the MIS capacitor under the limitation of dielectric breakdown in the MIS capacitor. The case corresponding to A = 11 is shown for the MFMIS-FET. (c) Separation of paraelectric and ferroelectric components in the ferroelectric capacitors for MFIS and MFMIS gate stacks. (d) Efficiencies of charge boost for the MFIS (A = 1) and MFMIS (A = 11) gate stacks.

the ferroelectric component. This implies that the FE-FET with MFIS gate stack is mainly operated by the paraelectric charge and considered as the high-k gate FET with a low ferroelectric poling. Conversely, the ferroelectric poling component is extremely high in the MFMIS case. More than 70% of the charge in the FE-FET is controlled by the ferroelectric property.

From the logic device application viewpoint, the charge boost efficiency by the gate stack is of interest, which is compared in Fig.

Fabrication of Ω-gated Negative Capacitance FinFETs and SRAM

P.-J. Sung[1,3], C.-J. Su[1], D. D. Lu[2], S.-X. Luo[2], K.-H. Kao[2], J.-Y. Ciou[2], C.-Y. Jao[4], H.-S. Hsu[4], C.-J. Wang[1], T.-C. Hong[3], T.-H. Liao[4], C.-C. Fang[4], Y.-S. Wang[2], H.-F. Huang[2], J.-H. Li[2], Y.-C. Huang[3], F.-K.Hsueh[1,3], C.-T.Wu[1], Y.-C. Huang[3] W. C.-Y. Ma[4], K.-P. Huang[5], Y.-J. Lee[1,*]T.-S. Chao[3], J. -Y. Li[6], W.-F. Wu[1], W.-K. Yeh[1], Y.-H. Wang[7]

[1]National Nano Device Laboratories, Hsinchu, Taiwan ;[2]National Cheng Kung University, Tainan, Taiwan;; [3]Dept. of Electrophysics, National Chiao Tung University, Hsinchu, Taiwan; [4]Dept. of Electrical Engineering, National Sun Yat-Sen University, Kaohsiung, Taiwan; [5]Mechanical and Systems Research Laboratories, Industrial Technology Research Institute, Hsinchu, Taiwan; [6]Dept. of Electrical Engineering and Graduate Institute of Electronics Engineering, National Taiwan University, Taipei, Taiwan. [7]National Applied Research Laboratories, Taipei, Taiwan
Tel: +886,-3-5726100-7793, Fax: +886-3-5722715, *Email: yjlee@narlabs.org.tw

Abstract

Ω-gated negative capacitance (NC) FinFETs, CMOS inverters and SRAM are fabricated and analyzed. Forming gas annealing (FGA) is performed and found to not only enhance ferroelectricity (FE) but also the NCFET electrostatics, in terms of higher I_{ON}, smaller hysteresis and subthreshold slop (SS). The SS is less than 60 mV/dec for both N-FinFET and P-FinFET in this work. Moreover, the CMOS inverter shows more symmetric and larger voltage gain after FGA.

Introduction

Low power devices that operate at a very low supply voltage (V_{dd}) are attractive for their energy-saving benefits. To extend the state-of-the-art CMOS technology, NCFETs have been proposed and demonstrated to achieve a SS less than 60 mV/dec [1]. The availability of HfO$_2$-based FE materials [2] enables CMOS-compatible NCFET demonstrations in both planar [3-4] and multiple-gate FETs [5]. Typical FGA has been shown to effectively passivate defects in the gate stacks [6]. Nevertheless, FGA effects on ferroelectricity and NCFETs are not discussed extensively yet. In this work, Ω-gated NCFETs, CMOS inverters and SRAM on silicon-on-insulator (SOI) substrates are fabricated for thorough and systematical investigation.

Device Fabrication

Fig. 1 shows the NCFET fabrication process flow. Ω-gated FinFETs are fabricated on a thinned-down SOI wafer by designing fin widths and the following etching process. After patterning the active and fin regions, the gate stack consisting of a 0.6 nm-thick SiO$_2$ interfacial layer (IL) and 6 nm-thick HfZrO$_2$ (HZO) layer with equal amounts of Hf and Zr, was deposited sequentially. Subsequently, the devices were capped with TiN metal gates by ALD and PVD. Rapid thermal annealing (RTA) at 700 °C for 30s [3] was carried out to transform the HZO film into the crystalline state. After the gate patterning, ion implantation was carried out in source and drain by using As$^+$ for NFETs and BF$_2^+$ for PFETs, respectively. Microwave annealing (MWA) was performed for the dopant activation. After the conventional contact metallization, FGA was executed at 400°C to reduce potential traps at material interfaces and FE film grains. Fig. 2 depicts the schematics of the CMOS inverter fabrication. Fig. 3 shows the cross-sectional TEM image of an Ω-gated NC-FinFET with 30 nm fin height and 20 nm fin width. Fig. 4 shows the polarization-voltage (P-V) characteristics of a planar HZO capacitor measured at different stages during the device fabrication process. RTA enables FE-phase transformation and thus the polarization hysteresis loop. The subsequent MWA preserves the P-V characteristics, suggesting that MWA as the dopant activation does not influence the polarization loops since the HZO crystallization. After FGA, the polarization is enhanced and ascribed to the defect passivation in the gate stacks. Verilog-A model based on Landau's theory is calibrated to the final P-V curve (after FGA) to extract α, β, γ, and ρ parameters.

Device Characterization

Fig. 5 shows the J_g-V_g characteristics of the HZO MOSCAPs with and without FGA. Note that the leakage currents at V_g around 0 to -2 V is lower for the MOSCAP with FGA due to D_{it} passivation. We believe FGA could enhance the ferroelectricity by removing imperfections in the HZO film as shown in Fig. 4. Fig. 6 shows the Id-Vg transfer characteristics of an Ω-gated NFET with and without FGA. Without FGA, N-FinFET shows the lowest SS is 66 mV/dec. and P-FinFET shows 62 mV/dec., respectively. With FGA, N-FinFET shows steeper SS of 49 mV/dec and tighter hysteresis, while the P-FinFET with FGA also shows the low SS of 50 mV/dec in the Fig. 7. Therefore, the scaled NCFETs with W/L=20/50 nm are demonstrated with SS below 60 mV/dec after FGA. The I_{on}/I_{off} more than 10^7 at V_d = 0.1 V for both N and P-FinFET can be achieved thanks to the Ω-gated FinFET structure.

To explore NCFET circuit behavior, N-FinFET and P-FinFET devices are fabricated with a common gate stack. Fig. 8 illustrates typical voltage transfer characteristics (VTCs) of a NCFET inverter which becomes more symmetric due to a FGA process. The inset shows that the voltage gain can be improved from 35 to 54 V/V after FGA. Fig. 9 shows the I_d-V_g transfer characteristics of the CMOS inverter in Fig. 8. The threshold voltage and SS become smaller and steeper after FGA. After FGA, the average SS at I_d between 10^{-10}~10^{-8} A/μm of the P-FinFETs is decreased from 71 to 63 mV/dec, and the N-FinFET is from 196 to130 mV/dec. Fig. 10 shows the characteristics of 6T static random access memory (SRAM) fabricated with the same process in this work. The measured typical butterfly curves show SNM at different operation voltage. A SNM is 250 mV at V_d=0.8V and 600 mV at V_d=2V.

Conclusion

The fabricated NC Ω-gated FinFETs on SOI show significant SS improvement and enhanced ferroelectric characteristics with FGA. N-FinFET with SS of 49 mV/dec and P-FinFET with SS of 50 mV/dec are achieved. Moreover, the CMOS inverter shows more symmetric voltage transfer curves and lager voltage gain with FGA. We also demonstrate a 6T SRAM with negative capacitance with a SNM of 250 mV at V_d =0.8V.

Acknowledgment

This work was performed by the National Nano Device Laboratories facilities and supported by the Ministry of Science and Technology under the grant numbers MOST -105-2221-E-492-029-MY2, 105-2628-E-492-002-MY3, 106-2221-E-492-034,107-2628-E-492-001-MY3,107-2633-E-009-003,107-2634-F-006-008, 107-2636-E-006-004.

References

[1] S. Salahuddin et al., Nano Lett., 2008, p. 405. [2] T. S. Böscke et al., Appl. Phys. Lett., 2011, p. 102903. [3] M. H. Lee et al., IEDM, 2016, p. 306. [4] M. H. Lee et al., IEDM, 2017, p. 565. [5] K. S. Li et al., IEDM, 2015, p. 620. [6] L. Do Thanh et al., J. Electronchem. Soc., 1988, p.1797.

978-1-7281-0943-5/19 $31.00 © 2019 IEEE

- SOI Thinning Down to 30 nm.
- Active Region Defined by E-beam.
- Fin Etch (Dry Etch).
- Ω–gate Formed by HF Dip.
- Gate Oxide Formation:
  - □ 1st Interfacial layer formation by H_2O_2
  - □ 2nd HZO(Hf:Zr=1:1) Deposition by ALD.
- 50 nm TiN for Metal Gate Formation by ALD and PVD.
- HZO crystallization by RTA 700 °C for 30 s
- Gate Patterning.
- CMOS: Source/Drain Lithography & Implantation :
  - □ P-FinFETs Region:
    1st Imp: BF_2, 1×15 cm^{-2}, 10KeV.
  - □ N-FinFETs Region:
    2nd Imp: As, 1×15 cm^{-2}, 10KeV.
- Microwave Annealing.
- Passivation and Contact Holes.
- Metallization: Al-Cu.
- FGA by RTA 400 °C for 5 mins

Fig. 1. Process flow for the fabrication of Ω-gated and planar NCFETs on SOI.

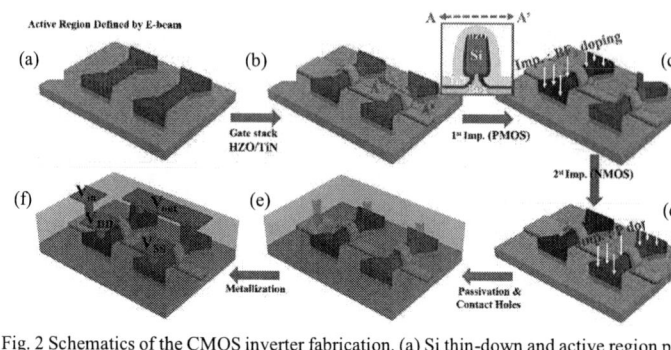

Fig. 2 Schematics of the CMOS inverter fabrication. (a) Si thin-down and active region pattering. (b) Formation of HZO gate stacks followed by RTA 700 °C for 30 s, and then gate patterning, and (c) & (d) source/drain implantation for (c) P-FinFETs (BF_2/10 keV/10^{15} cm^{-2}) and (d) N-FinFETs (As/10 keV/10^{15} cm^{-2}), followed by MWA activation. (e) & (f) oxide passivation, contact and metallization for CMOS inverter.

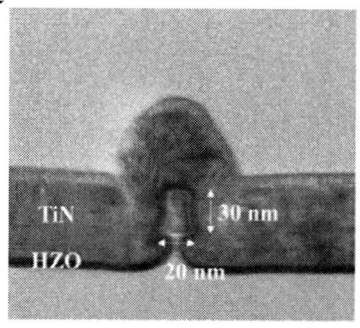

Fig. 3. Cross-sectional TEM image of an Ω-gated NCFET. The fin width and height are about 20 nm and 30 nm, respectively.

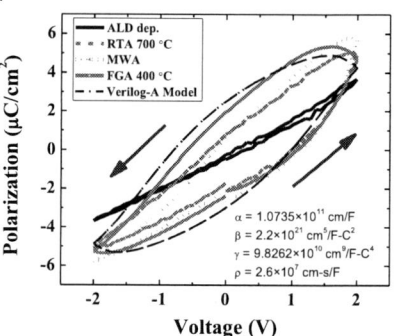

Fig. 4. P-V characteristics of a 6 nm HZO ferroelectric capacitor on an N^+ substrate ($N_d \sim 10^{20}/cm^3$). RTA facilitates FE crystallization; MWA preserves it; FGA further enhances FE polarization. P-V calibrated to Verilog-A model.

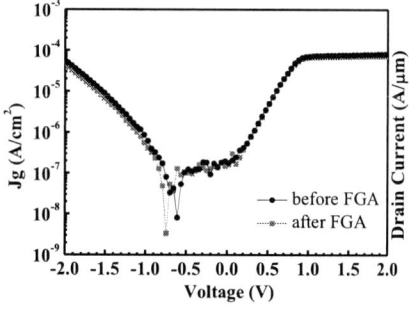

Fig. 5. I-V characteristics before and after FGA. Lower leakage currents around -2 to 0V with FGA is due to Dit passivation.

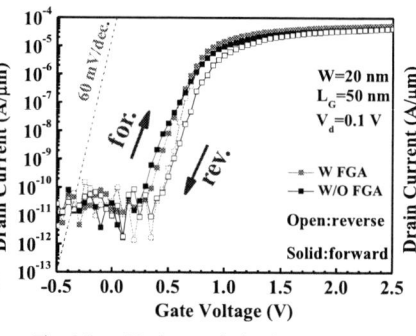

Fig. 6. I_d vs. V_g characteristic of a steep-swing Ω-gated N-FinFET with W/L_G=20 nm/50 nm. The lowest SS is 49 mV/dec with FGA.

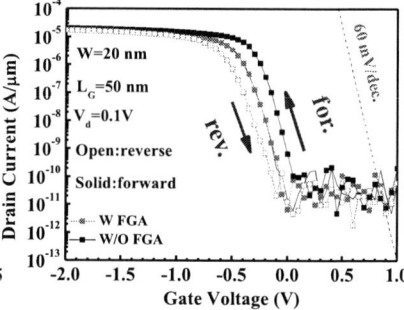

Fig. 7. Id vs. Vg characteristic of a steep-swing Ω-gated P-FinFET with W/L_G=20 nm/50nm. The lowest SS is 50 mV/dec. with FGA.

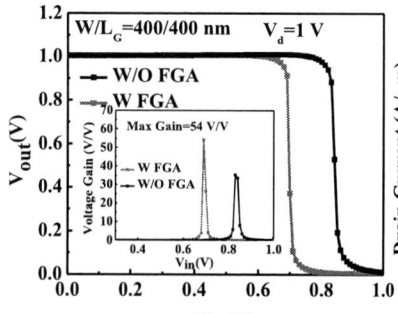

Fig. 8 Voltage transfer characteristics of an NCFET CMOS inverter with and without FGA. The inset shows CMOS voltage gain is enhanced by FGA.

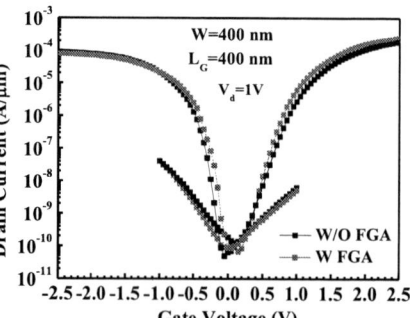

Fig. 9 I_d-V_g curves of N-FinFET and P-FinFFET in a CMOS inverter shown in Fig. 8.

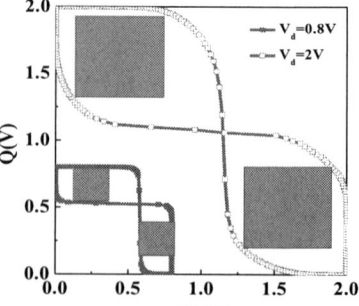

Fig. 10 6T-SRAM and measured butterfly curves (W/L_G= 30 nm/70 nm). A SNM is 250 mV at V_d=0.8V.

Electron Enhanced Atomic Layer Deposition (EE-ALD)

Steven M. George

University of Colorado Boulder

Electron enhanced atomic layer deposition (EE-ALD) can dramatically reduce the temperatures required for film growth. Temperature reduction is possible because of electron stimulated desorption (ESD) of surface species. The desorption process creates highly reactive "dangling bond" surface sites. Precursors can then adsorb efficiently on the dangling bonds. Our work has demonstrated the EE-ALD of GaN, BN, Si and Co at room temperature. Film growth was performed using alternating exposures of chemical precursors and low energy electrons. In situ ellipsometry measurements have monitored linear film growth versus number of reaction cycles. Additionally, we have observed the dependence of the EE-ALD growth rates on electron energy. Maximum growth rates have varied from 0.3 Å/cycle for Si films at 100-150 eV to 3.2 Å/cycle for BN films at 80-160 eV. Recent measurements have also obtained maximum growth rates of 0.5 Å/cycle for Co films at 125 eV. EE-ALD is also topographically selective. This area selectivity is derived from the directionality of the electron flux. Surfaces normal to the incident electrons receive full electron flux, whereas surfaces parallel to the incident electrons receive no electron flux. EE-ALD should be useful for the bottom-up fill of high aspect ratio structures.

978-1-7281-0943-5/19 $31.00 © 2019 IEEE

Analytical Estimation of LER-like Variability in GAA Nano-Sheet Transistors

Amita, Ajinkya Gorad and Udayan Ganguly *member* IEEE
Department of Electrical Engineering, IIT Bombay, India. E-mail: udayan@ee.iitb.ac.in

ABSTRACT

FinFET and gate-all-around (GAA) devices such as Nanowire FET (NWFET) and Nano-sheet FET (NSFET) are prone to process induced variability. Line edge roughness (LER), Metal grain granularity (MGG) and Random dopant fluctuation (RDF) are conventional sources of variability. Due to the complex fabrication process of NSFET, a new source of variability has emerged, termed as sheet thickness variation (STV). The STV causes variation in quantum confinement (QC), hence results in performance variability. In this paper, we present an analytical model to estimate key performance parameters e.g. $\sigma[V_T]$, $\sigma[I_{on}]$, $\sigma[SS]$ etc., for FinFET and NWFET due to LER and for NSFET due to STV. The model, implemented on MATLAB, is 300× efficient in comparison to TCAD. Such models, when incorporated in existing SPICE model files, will help in predicting the circuit and system level performance and impact of scaling.

INTRODUCTION

A major challenge in advanced technology nodes (sub-20nm) is process induced variability [1]. A comparative study of advanced switching device architectures, w.r.t process induced variability, is presented in Table I. The LER remains one of the dominant sources of variability for FinFET and NWFET. New sources of variability may come into existence for NSFET viz. metal thickness variation (MTV) due to the inter-sheet spacing (T_{sp}) variation, and sheet thickness (T_{sh}) variation (STV) due to roughness caused by deposition and etching process. The MTV affects the work-function of the gate metal and hence the overall V_T-distribution. The STV impacts NSFET performance in a similar fashion as LER does to FinFET and NWFET. As the NSFET width (W_{sh}) is large (~20nm [2]), random variations in QC can be neglected. However, in the vertical direction, a slight variation in the T_{sh} (STV) will result into a significant change in performance parameters due to the QC. A significant amount of work has been done to model LER for FinFET [3] and NWFET [4-6]. The STV dependence in NSFET is not yet explored. In this paper for the first time, we have presented a universal model to capture the impact of LER on FinFET and NWFET, and STV on NSFET. The model is capable of determining V_T, I_{on}, SS and DIBL distributions for all the three device architectures in a computationally efficient way. Such a model can be used to predict the system and circuit level performance at current and advanced technology nodes.

METHODOLOGY

A. STV data extraction using NSFET TEM image

The key dimensions of NSFET like T_{sh} and T_{sp} are extracted from the TEM image of 8 NSFET devices (24 sheets) presented in [2, 7]. The NSFET TEM image is converted into spatial co-ordinates and an intensity (RGB scheme) map associated with each of the (x,y) co-ordinates using MATLAB™. Further, the generated co-ordinates are scaled as per the dimensions given in [7]. Next, using the scaled 2D intensity profile, a horizontal cut (perpendicular to sheets) is taken to extract respective thicknesses from the intensity profile with the help of an edge detection technique. Fig.1 shows T_{sh} and T_{sp} extraction from the intensity profile (Fig. 1(b)). A finite T_{sh} variation of 0.43nm is observed as opposed to ideal uniform nano-sheet, despite using epitaxial growth for Si and SiGe layers.

B. A Generalized Analytical Model for LER and STV

To model the LER and STV impact on V_T, I_{on}, SS and DIBL of FinFET, NWFET and NSFET, we have used the analytical model presented in [3] and modified it as per the respective device architectures. The model requires a look-up-table (LUT) of sensitivity check (SC) data. For FinFET, the W_{fin} SC data (calibrated against Intel's 25nm Tech. node) is presented in [3]. The diameter (D) SC data for NWFET is taken from [5]. The SC data for NSFET is generated from TCAD. The NSFET TCAD deck is calibrated against results reported in [2]. The sub-threshold slope and DIBL of $I_d - V_G$ characteristics of nominal NSFET agrees well with the reference as shown in Fig. 3. The W_{fin}, D and T_{sh} SC data is converted into an area (A) SC data. The methodology flow and associated mathematical expressions are presented in Fig. 4.

RESULTS AND DISCUSSION

The $\sigma[V_T]$, estimated for $\sigma_{LER} = 0.66nm$, $\rho = 0.7$, and $W_{fin} = 5nm$ in case of FinFET [3], shows a good match with TCAD in Fig. 5(a). In case of NWFET, the $\sigma[V_T]$ estimated for input parameters taken from [6], agrees well with the reference data as shown in Fig. 5(b). Using the $\sigma[T_{sh}]$ and $\mu[T_{sh}]$ extracted, and assuming autocorrelation length (Λ) of 50nm and correlation coefficient of 0.7 between the top and the bottom surface of a single sheet, the $\sigma[V_T]$ is extracted (Fig. 5(c)) for NSFET. For the nominal $T_{sh} = 4.5nm$, the variability is low. However, it can be significantly high for high QC scenario when low effective mass materials like SiGe, Ge, InAs etc. are used as channel material. A comparison of maximum $\sigma[V_T]$ induced due to variability in each of the devices is plotted in Fig. 5(d), which shows NSFET is a better choice in sub-7nm tech. node in comparison to NWFET. Further, $\sigma[I_{on}]$, $\sigma[SS]$, and $\sigma[DIBL]$ with respect to L_{sh} for NSFET is plotted in Fig. 6(a-c). The stack architecture in NSFET helps reducing the overall variability. However, a rapid increase in σ of each key parameter with L_{sh}, shows an increase in variability in the NSFET with the scaling. Due to the stacked architecture of the NSFET, the mean of each transistor parameter depends on variability in individual sheet transistor as shown in Fig. (d).

CONCLUSION

We present a universal analytical model to estimate performance parameters of FinFET, NWFET (due to LER) and NSFET (due to STV). The model shows a good match against well-calibrated TCAD data and against existing literature. The model is computationally efficient in comparison to TCAD and can be used in SPICE model files in the form of an LUT to predict circuit and system level performance.

REFERENCES

[1] X. Wang, et. al., *IEDM*, pp. 5.4.1-5.4.4 2011. [2] J. Zhang et al., *IEDM*, p.22.1.1-4, 2017. [3] Amita, et. al., *IEEE TED*, 2018, p 4772-4779. [4] Jinyoung Park, et. al., *IEEE TED*, 2016, p 5048-5054. [5] Dogyun Son, et. al. *IEEE SNW*, 2014. [6] Dokyun Son, et al, *JNN* 2017, 7179-7182. [7] N. Loubet et. al., *Symp. VLSI Tech.* T230, 2017. [8] Sushant Mittal, et. al., *IEEE TED*, 2017, p 3489-3493.

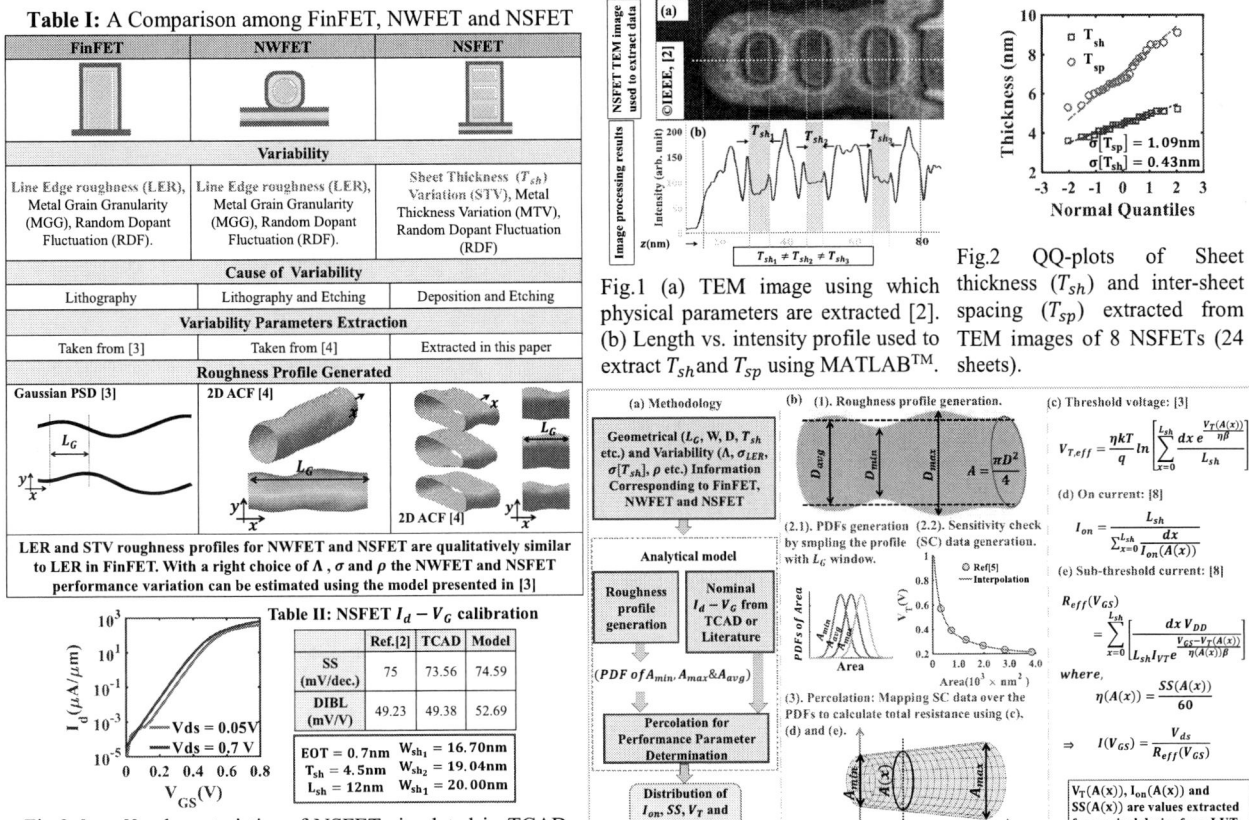

Table I: A Comparison among FinFET, NWFET and NSFET

FinFET	NWFET	NSFET
Variability		
Line Edge roughness (LER), Metal Grain Granularity (MGG), Random Dopant Fluctuation (RDF).	Line Edge roughness (LER), Metal Grain Granularity (MGG), Random Dopant Fluctuation (RDF).	Sheet Thickness (T_{sh}) Variation (STV), Metal Thickness Variation (MTV), Random Dopant Fluctuation (RDF)
Cause of Variability		
Lithography	Lithography and Etching	Deposition and Etching
Variability Parameters Extraction		
Taken from [3]	Taken from [4]	Extracted in this paper
Roughness Profile Generated		
Gaussian PSD [3]	2D ACF [4]	
LER and STV roughness profiles for NWFET and NSFET are qualitatively similar to LER in FinFET. With a right choice of Λ, σ and ρ the NWFET and NSFET performance variation can be estimated using the model presented in [3]		

Fig.1 (a) TEM image using which physical parameters are extracted [2]. (b) Length vs. intensity profile used to extract T_{sh} and T_{sp} using MATLAB™.

Fig.2 QQ-plots of Sheet thickness (T_{sh}) and inter-sheet spacing (T_{sp}) extracted from TEM images of 8 NSFETs (24 sheets).

Table II: NSFET $I_d - V_G$ calibration

	Ref.[2]	TCAD	Model
SS (mV/dec.)	75	73.56	74.59
DIBL (mV/V)	49.23	49.38	52.69

EOT = 0.7nm, W_{sh_1} = 16.70nm
T_{sh} = 4.5nm, W_{sh_2} = 19.04nm
L_{sh} = 12nm, W_{sh_3} = 20.00nm

Fig.3 $I_d - V_G$ characteristics of NSFET simulated in TCAD for the dimensions taken from [2] which is used to generate LUT for STV.

Fig. 4 A flow of the model used to extract the performance parameters.

Fig.5 V_T variation caused by LER and STV: (a) $\sigma[V_T]$ vs. Λ for FinFET, (b) $\sigma[V_T]$ vs. Λ_x for NWFET, (c) $\sigma[V_T]$ vs. L_{sh} for NSFET, and (d) Comparison of maximum $\sigma[V_T]$ present in each of the device architecture. The impact of LER/STV on $\sigma[V_T]$ increases when going from FinFET to NWFET whereas drops drastically in case of NSFET.

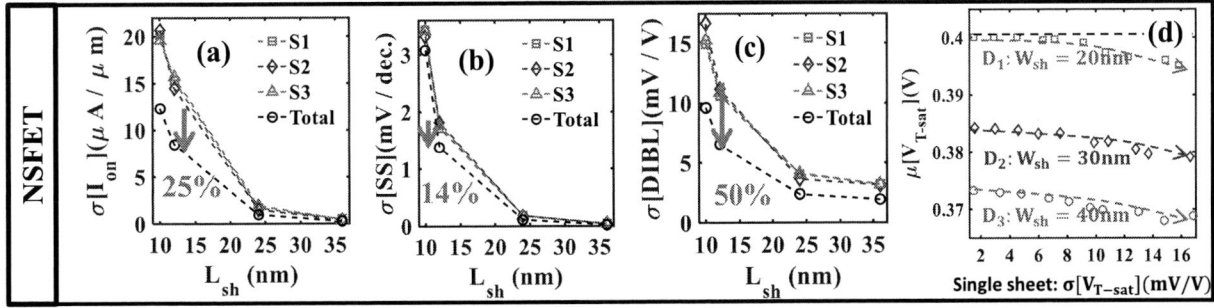

Fig.6 (a) $\sigma[I_{on}]$ vs. L_{sh}, (b) $\sigma[SS]$ vs. L_{sh}, $\sigma[DIBL]$ vs. L_{sh}, and (d) $\mu[V_{T-sat}]$ vs. $\sigma[V_{T-sat}, individual]$ of a single individual sheet due to STV. This remarkable dependence of mean on the std. deviation of single sheet transistor is due to the stacked architecture of the NSFET and independent variations of single sheet transistor parameters. Such stacking helps in reducing the variability of each of these parameters significantly, yet overall performance will still degrade with scaling.

978-1-7281-0943-5/19 $31.00 © 2019 IEEE

Low temperature junctionless device stacking enabled by leading edge sequential 3D integration

Guillaume Besnard, Gweltaz Gaudin, Walter Schwarzenbach, Ludovic Ecarnot, Ionut Radu, Bich-Yen Nguyen, Anne Vandooren* and Nadine Collaert*

SOITEC, Parc Technologique des Fontaines, 38190 Bernin, France
*IMEC, Kapeldreef 75, 3000 Leuven, Belgium
E-mail: guillaume.besnard@soitec.com

ABSTRACT

Sequential or monolithic 3D integration requires a combination of advanced manufacturing processes and highly reliable integration scheme at both device and substrate level. In order to achieve this, front-end layer transfer technology looks as key element of 3D integration flow. In this paper, we will bring an insight on technology challenges as well as solutions to enable sequential 3D device integration. We will also exhibit successful demonstration of low temperature device stacking using junctionless top devices.

INTRODUCTION

Sequential 3D (S3D) integration draws more and more attention since a decade now. Readiness of such disruptive technology is hardly predictable because of real application driver and maturity. For several reasons, industry is continuously evaluating opportunities of stacking multiple layers of devices in front-end manner. The S3D requires a new single-crystalline template atop an already processed semiconductor substrate (CMOS, DRAM, etc). Hence, a new silicon surface on which new devices can be processed is available for front end of line device integration. One of the first attempt introducing S3D technology in roadmap was for pure scaling purpose [1,2,3]. Then, it has been proposed as solution for post-Silicon era to facilitate co-integration of different materials such as Germanium and III-V materials [4,5]. Hence, breaking down front-end part of a chip over two tiers allows these materials to be used at their full potential. Obviously, such integration scheme implies significant adaptation of device processing and design work. Nevertheless, S3D integration scheme will bring benefits for both More than Moore and More Moore applications, and recent developments in this field will help technology moving forward bringing more options to device stacking.

More recently, S3D has shown up in other fields of the wide 3D landscape, especially for More than Moore applications. Contrary to More Moore that mostly relies on standard 2D scaling, S3D has slightly shifted to functional scaling [6]. Functional scaling, also called heterogenous integration, targets to co-integrate multiple functions of a system in the ultimately-scaled System-on-Chip (SoC). To enable this, S3D positioned itself at different point of 3D granularity. Rather than interconnecting transistors, functional scaling interconnects larger parts over several tiers, like cells or even blocks of a circuit. Here, S3D would bring much shorter interconnect delay to reduce transmission bottlenecks between critical components or helping in co-integration of very different technologies like Logic and Radio-Frequency (RF) building blocks. Despite promising higher interconnects density at lithography resolution, S3D will face off advanced packaging technologies (Fan-Out) that are also constantly improving. To be successful, S3D performance gain needs to largely overcome its manufacturing constraints and complexity.

KEY TECHNOLOGIES TO ENABLE HIGH DENSITY 3D INTEGRATION

Low Temperature CMOS processing

Lowering the processing temperature of CMOS integration flow, which is already heavily demanding in terms of process capability, is truly a masterpiece of S3D technology. A lot of work has been done on process side to get good performance at low thermal budget and integration optimizations have been used to keep high reliability level [6]. Junctionless devices featuring in-situ doped Si and SiGe Source/Drain Epi and Replacement Metal Gate have been fabricated with maximum temperature of 525°C on 10nm thick Si channel transferred on top of CMOS devices (Fig.1). For the channel implantation, dopant activation has been made prior to the layer transfer from SOI donor wafer which has in this case no thermal budget restriction.

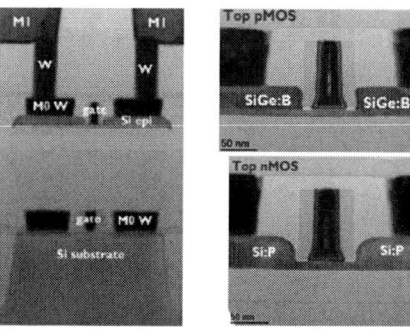

Fig.1 – TEM cross-section of transistors sequentially stacked on top of each other (left) and top NMOS and PMOS devices featuring low temperature processing (right).

Both NMOS and PMOS devices exhibit very good $I_D(V_G)$ curves, even for short gate length (~40nm). In addition to good performance, junctionless architecture enables devices with sufficient reliability due to reduced vertical electric field (Fig.2). This allows to remove some thermal treatment like gate stack drive-in anneals which ensure low temperature compatibility.

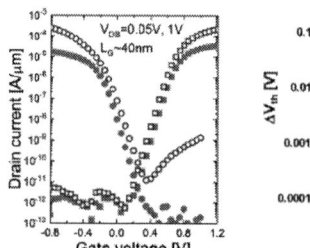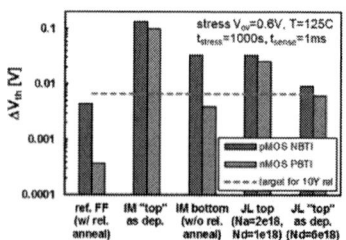

Fig.2 – Output characteristics of NMOS and PMOS devices processed at low temperature (left) and associated 10-years reliability (right).

978-1-7281-0943-5/19 $31.00 © 2019 IEEE

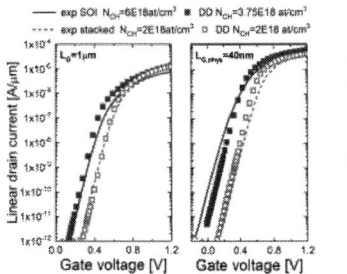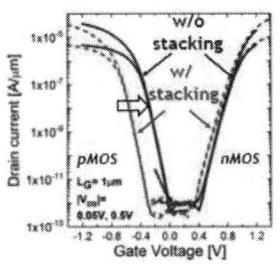

Fig.3 – $I_D(V_G)$ curves of SOI built-in and stacked (i.e. with layer transfer) devices (left). Output characteristics of bottom devices before then after low temperature top device processing (right).

Impact of layer transfer process has been evaluated by fabricating low temperature devices directly on SOI. On Fig.3, we show this step is seamless for top devices with no degradation. Bottom devices also obtained good results, but V_{th} shift is observed on PMOS side while keeping same current level. It can be explained by remaining thermal budget applied with top device manufacturing. In this case, bottom devices do not receive any reliability-specific optimization.

Smart Cut™

As a key element of high volume SOI manufacturing, Smart Cut™ is the long last demonstrated process for transferring a very thin single-crystalline layer from a donor substrate to a new handling substrate. This process can be used for S3D when the typical Si handle substrate is replaced by a CMOS processed handle wafers. Benefits of having integrated layer transfer process to CMOS flow are numerous: "Front-End" compatible solution, higher throughput and cleaner solution compared to standard backside grinding flow, reduced cost since donor wafer is reusable. All these points make Smart Cut™ process a reliable solution and technology of choice for highly integrated 3D applications.

In the frame of SOI industrialization, SOITEC developed unique Front-End bonding expertise of thin SiO_2 and Si layers. This expertise allows low defectivity, high yield and bonding energy at low temperature preventing degradation of past and subsequent CMOS flows. It allows bonding process to be seamlessly integrated to device manufacturing flow without requirement of nonstandard materials. Thanks to recent developments needed for fully-depleted (FD) SOI applications, Smart Cut™ process has been optimized to provide ultra-thin layers (SOI and BOX) with low roughness and excellent thickness uniformity [7] (Fig. 4). All process changes can be leveraged to layer transfer process of S3D flow.

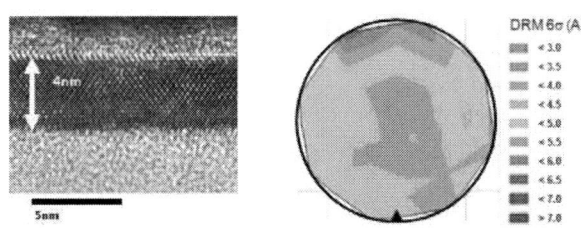

Fig.4 – Demonstration of 4nm thin SOI layer (left) and local thickness variability obtained through Differential Reflectance Microscopy (right) thanks to advanced Smart Cut™ processing.

RESULTS TOWARDS SUCCESSFUL S3D TECHNOLOGY

Both low temperature Smart Cut™ and low temperature CMOS processing are critical steps for successful S3D integration. Once bonding of donor wafer on handle wafer is done, high temperature steps must be removed from the process flow while keeping consistent performance level. Since high temperature smoothing is not an option for S3D finishing after layer transfer, the low temperature processing has been investigated. It is shown that this integration has comparable

results versus standard high temperature finishing with roughness <1Å RMS and thickness uniformity range below 5Å (6σ) (Fig.5).

In addition to 3D-compliant layer transfer process, we developed a specific low temperature layer reconstruction and re-activation using Solid-Phase Epitaxy Regrowth (SPER) technique [8,9]. SPER process can recover all post-Smart Cut™ defectivity and perform dopants re-activation at temperature below 500°C.

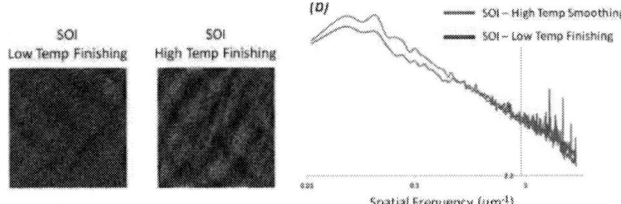

Fig.5 – Performance of low temperature versus high temperature finishing process.

CONCLUSION

In this paper, we reviewed opportunities for Sequential 3D and stated on readiness of key building blocks by showing latest results in this field. CMOS flow and Smart Cut™ process are driving down maximum process temperature around 500°C to be compliant with standard, high temperature CMOS devices. The feasibility is demonstrated using junctionless top devices stacked on a conventional CMOS. Future work will consist in merging both approaches into a fully integrated S3D flow.

ACKNOWLEDGEMENTS

Authors would thank IMEC research platform for devices fabrication and characterization.

REFERENCES

[1] P. Batude et al., "3DVLSI with CoolCube process: An alternative path to scaling", Proc. IEEE Symp. on VLSI Technol., pp. T48-T49, Jun. 2015.

[2] L. Brunet et al., "First demonstration of a CMOS over CMOS 3D VLSI CoolCube integration on 300 mm wafers", Proc. IEEE Symp. on VLSI Technol., pp. 1-2, Jun. 2016.

[3] A. Vandooren et al., "First Demonstration of 3D stacked Finfets at a 45nm fin pitch and 110nm gate pitch technology on 300mm wafers", IEDM Tech. Dig., Dec. 2018 – to be published.

[4] T. Irisawa et al., "Demonstration of InGaAs/Ge dual channel CMOS inverters with high electron and hole mobility using staked 3D integration", Proc. IEEE Symp. on VLSI Technol., pp. T56-T57, Jun. 2013.

[5] V. Deshpande et al., "Advanced 3D monolithic hybrid CMOS with sub-50 nm gate inverters featuring replacement metal gate (RMG)-InGaAs nFETs on SiGe-OI fin pFETs", IEDM Tech. Dig., pp. 8.8.1-8.8.4, Dec. 2015.

[6] A. Vandooren et al., "3D sequential stacked planar devices on 300mm wafers featuring replacement metal gate junction-less top devices processed at 525°C with improved reliability", Proc. IEEE Symp. on VLSI Technol., Jun. 2018.

[7] W. Schwarzenbach et al., "Beyond advanced FDSOI: LowTemp SmartCut for enabling High Density 3D SoC applications", S3S unified conf., Oct. 2018.

[8] I. Radu et al., "Novel Low Temperature 3D Wafer Stacking Technology for High Density Device Integration" in ESSDERC, Bucharest, 2013.

[9] G. Gaudin et al., "Physical and Electrical Properties of Thin Doped Silicon Films Obtained by Low Temperature Smart Cut and Solid Phase Epitaxy", ECS Journal of Solid State Science and Technology, 2 (12) P534-P538, 2013.

New Observation and Analysis of Layout Dependent Effects in Sub-40nm Multi-Ring and Multi-Finger nMOSFETs for High Frequency Applications

Zu-Cheng Li, Jyh-Chyurn Guo, Senior Member, IEEE, and Jinq-Min Lin

National Chiao Tung University, Hsinchu, Taiwan 30010, R.O.C.
E-mail: zc_li.ee04g@g2.nctu.edu.tw, jcguo@mail.nctu.edu.tw, jmlin2000@nctu.edu.tw

ABSTRACT

Multi-finger (MF) and multi-ring (MR) nMOSFETs were designed and fabricated in 40nm CMOS technology to explore the layout dependent effects in key device parameters and parasitic RC responsible for RF performance. For the first time, the experimental proves the advantages of MR nMOSFETs, such as the increase of effective mobility (μ_{eff}), transconductance (g_m), and channel current (I_{DS}), and smaller parasitic source resistance (R_S), all of which are in favor of higher speed and higher frequency. However, the undesired increase of 3-D fringing capacitances may bring a critical trade-off influencing high frequency performance. In this paper, new observation and in-depth analysis of the complicated layout dependent effects can facilitate the device layout optimization in the right direction for RF and mm-wave design and applications.

I. INTRODUCTION

MF devices have been widely used in RF and analog circuits design attributed to the effective reduction of gate resistance (R_g). However, the very small finger width (W_F) associated with very large finger number (N_F) aimed at sufficiently low R_g, may lead to the penalties, such as g_m degradation due to lower μ_{eff} caused by STI compressive transverse stress $\sigma_\perp$ and larger R_S from longer source line, and the increase of gate capacitance (C_{gg}) originated from the gate sidewall and finger-end fringing capacitances (C_{of} and $C_{f(poly-end)}$) [1]-[4]. All of the mentioned factors result in significant impact on the high frequency performance like f_T and f_{MAX}. In this paper, a new device layout, namely multi-ring (MR) MOSFET is proposed as a potential solution to reach higher μ_{eff} and improvement of g_m. The basic idea is that STI compressive $\sigma_\perp$ can be minimized in the MR layout due to miniaturized STI area around two ends of the gate finger. As for the ultimate goal of higher f_T and f_{MAX}, the parasitic RC like C_{of}, $C_{f(polyend)}$, and R_S appear as critical parameters deserving extensive investigation to explore a complete spectrum of the layout dependent effects and optimization guideline for RF performance enhancement.

II. EXPERIMENTAL RESULTS AND ANALYSIS

MF and MR nMOSFETs were fabricated in 40nm CMOS technology with 40nm drawn length and V_{DD}=0.9V. Fig.1 (a)~(b) illustrate the MF MOSFET layouts with various W_F and N_F at fixed $W_F \times N_F = W_{tot}$=32μm. Note that the smaller W_F and larger N_F for lower R_g may lead to lower μ_{eff} due to increased STI compressive $\sigma_\perp$. Thus, MR MOSFET as shown in Fig. 2(a)~(b) with various W_F and ring number (NR, N_F=4*NR) are proposed to effectively suppress the STI compressive $\sigma_\perp$ attributed to very small STI area near two ends of every gate finger. Fig. 3 (a) presents the source line routing with contacts to source/drain for MF MOSFET, which suggests significant increase of R_S in case of very large N_F. Fig. 3 (b) shows the R_S vs. N_F determined by our proprietary matrix method in which the MR nMOS can achieve much smaller R_S than MF nMOS with the same N_F due to wider metal width and more contacts in parallel allowed in MR layout. More interestingly, the threshold voltage in linear and saturation region, V_{Tlin} and V_{Tsat} vs. W_F shown in Fig. 4(a) and (b) indicate obvious inverse narrow width effect (INWE) for MF nMOS whereas some abnormal trend for MR nMOS, such as INWE in V_{Tlin} vs. W_F but narrow width effect (NWE) in V_{Tsat} vs. W_F . Moreover, MR nMOS reveal apparently lower V_{Tlin} and V_{Tsat} than MF nMOS and particularly large drop in case of W_F=2μm. It appears as a new observation and suggests different channel dopant concentrations in MF and MR nMOS, due to the difference in the compressive stress induced retardation of boron diffusion [5]. For MR nMOS with smaller compressive $\sigma_\perp$, thus less retardation, i.e. faster boron diffusion may lead to lower boron concentration, worse short channel effect (SCE), and then lower V_{Tlin}. As for the V_{Tsat} at V_{DS}=V_{DD}=0.9V and DIBL=V_{Tlin}-V_{Tsat}, two more factors like finger-end fringing field and effective V_{DS} considering IR drop through parasitic resistances R_S and R_D should play an important role. As shown in Fig. 5(a), both MF and MR nMOS indicate smaller DIBL associated with the narrower W_F, which suggests the finger-end fringing field a dominant factor. As for the comparison between MF and MR nMOS, the finger-end fringing field from Raphael simulation, shown in Fig. 5(b) reveals the larger one achieved by MR nMOS and suggests the smaller DIBL, which can match the case of W_F=0.5μm but is against that of W_F=2μm. It means that the worse SCE due to lower boron concentration and higher effective V_{DS} due to smaller R_S may become dominant factors resulting worse DIBL in MR nMOS with W_F=2μm. Fig. 6(a) and (b) demonstrate promising results that MR nMOS can realize significant increase of I_{DS} and g_m through V_{GT} in comparison with MF nMOS. This improvement is considered originated from two major factors, such as higher μ_{eff} and smaller R_S. Table I provides a summary of basic device parameters extracted from MF and MR nMOS by using our proprietary extraction method [6]. The results show sub-35nm gate length, L_g =32.45nm/32.63nm and $T_{ox(inv)}$=19.501Å/19.502Å for MF/MR nMOS, i.e. very minor difference between two types of layout, but some significant difference in ΔW due to STI top corner rounding, such as 32.16nm and 50.32nm for MF and MR nMOS. This 1.56 times larger ΔW in MR nMOS can lead to the increase of W_{eff}=(W_F+ΔW)N_F and becomes another key parameter responsible for the increase of I_{DS} and g_m. Further investigation has been done by μ_{eff} extraction based on the linear I-V given by (1) with I_{DS}(V_{GT}) from measurement and key device parameters like L_g, $C_{ox(inv)}$, and W_{eff} as determined and summarized in Table I. As shown in Fig. 7 (a) and (b), the μ_{eff} vs. V_{GT} extracted from MF and MR nMOS indicate similar layout dependence, such as the lower μ_{eff} associated with the smaller W_F, which accounts for the impact from the increase of STI compressive $\sigma_\perp$. The comparison of MF and MR nMOS given by Fig. 7(c) indicates μ_{eff} enhancement realized by MR nMOS attributed to the smaller STI compressive $\sigma_\perp$. The final and most important verification has been made on high frequency parameters, such as g_m=Re(Y_{21}), C_{gg}= Im(Y_{11})/ω, C_{gd} = -Im(Y_{12})/ω, and f_T@model =g_m/2π(C_{gg}-C_{gd})$^{1/2}$ in saturation region as shown in Fig. 8 (a) ~ (d). First, the MR nMOS demonstrates the advantage of higher g_m through V_{GT} than MF nMOS, primarily due to lower R_S and larger W_{eff}. However, the MR nMOS reveals the penalty of larger C_{gg} owing to the increase of $C_{of}W_{eff}$ and $C_{f(poly-end)}N_F$ (Table 1) originated from separate source/drain for every gate finger and longer poly extension over STI. As a result, the increase of C_{gg} overwhelms that of g_m and leads to f_T degradation in MR nMOS compared to MF nMOS. As shown in Fig. 8(d). the MF nMOS can reach peak f_T up to 303~308GHz whereas the MR nMOS shows peak f_T around 280~289GHz. Thus, how to effectively reduce C_{gg} and keep higher g_m becomes the major challenge worthy of further research effort for high frequency performance improvement.

$$\mu_{eff} = \frac{I_{DS}}{(V_{DS} - I_{DS}(R_S + R_D))} \frac{L_g}{C_{ox(inv)}(V_{GS} - V_T - \lambda V_{DS})W_{eff}}, \; 0 < \lambda < \frac{1}{2} \quad (1)$$

III. CONCLUSION

MR nMOS have been proven with the advantages of higher g_m due to μ_{eff} enhancement and R_S reduction. However, the undesired increase of C_{gg} originated from the 3-D fringing capacitances like C_{of} and $C_{f(poly\text{-}end)}$ overwhelms the g_m improvement and leads to the penalty of f_T degradation compared to the MF nMOS. This indepth analysis provides a useful guideline for layout optimization in RF mm-wave devices design. Some more innovative layout solutions, aimed at higher g_m and lower C_{gg} for the ultimate goal of f_T boost to well above 300GHz become a new challenge worthy of extensive research effort.

REFERENCES

[1] J.-C. Guo and Y.-Z. Lo, IEEE TED-62, no.9, pp.3004-3011, 2015.
[2] K.-L. Yeh and J.-C. Guo, *IEEE TED*-60, no. 1, pp.109 -116, 2013.
[3] K.-L. Yeh and J.-C. Guo, *IEEE TED*-58, pp.3140~3146, 2011.
[4] K.-L. Yeh and J.-C. Guo, *IEEE TED*-58, pp.2838~2846, 2011.
[5] P.R. Chidambaram *et al.*, *IEEE TED*-53, no.5, pp.944~964, 2006.
[6] J.-C. Guo and K.-L. Yeh, US patent 8,691,599 B2, 2011 ~ 2032.

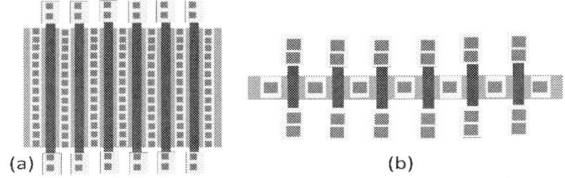

Fig. 1 Schematics of MF nMOS layouts with various W_F and N_F at fixed $W_F \times N_F$=32μm (a) W2N16 (b) W025N128.

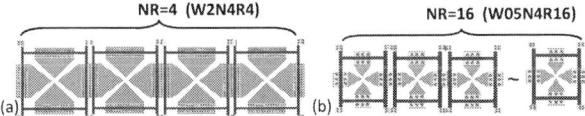

Fig. 2 Schematics of MR nMOS layouts with various W_F, N_F, and NR at fixed $W_F \times N_F \times NR$ =32μm (a) W2N4R4 (b) W05N4R16

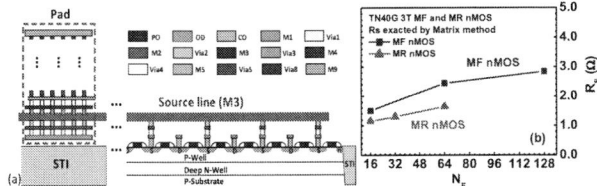

Fig. 3 (a) The cross section of MF device with the source line including contacts to S/D (b) R_S extracted by matrix method for MF and MR nMOS

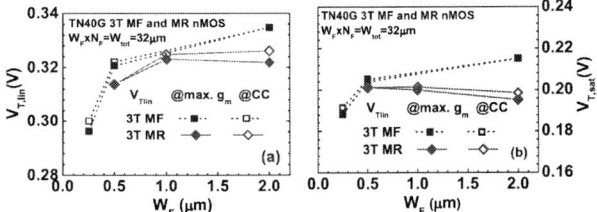

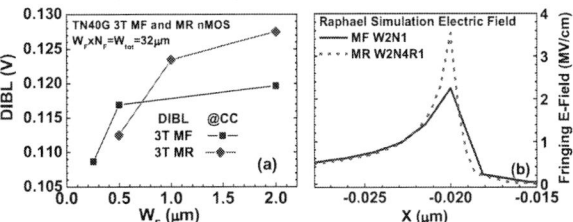

Fig. 4 Comparison of MF and MR nMOSFETs (a) linear region : V_{Tlin} vs. W_F at V_{DS} = 50 mV (b) saturation region : V_{Tsat} vs. W_F at V_{DS} = 0.9 V.

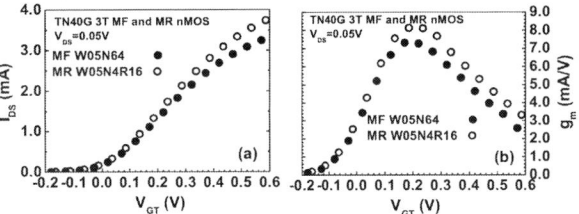

Fig. 5 Comparison of MF and MR nMOSFETs (a) DIBL versus W_F (b) finger-end fringing field along the channel direction and near the OD/STI corner region calculated by Raphael simulation

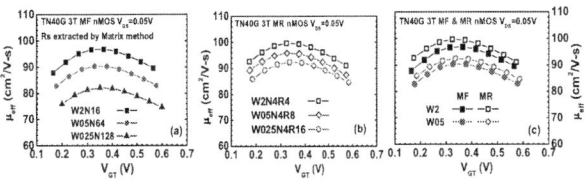

Fig. 6 Comparison of MF and MR nMOS : W05N64 and W05N4R16 (a) I_{DS} vs.V_{GT} (b) g_m vs. V_{GT} (V_{DS}=0.05V)

Table I Device parameters of the MF and MR nMOSFETs

TN40G Parameters	Unit	MF nMOS	MR nMOS
L_g	μm	0.03245	0.03263
$T_{ox(inv)}$ (target)	A	19.5	19.5
$C_{of,sim}$	fF/μm	0.30517	0.29134
$C_{f(poly\text{-}end),sim}$	fF	0.03344	0.1187
$C_{ox(inv)}$	fF/μm²	17.707	17.706
$T_{ox(inv)} = \varepsilon_0\varepsilon_{ox}/C_{ox(inv)}$	A	19.501	19.502
$\Delta W = (\alpha - C_{f(poly\text{-}end)})/(C_{ox(inv)}L_g)$	μm	0.03216	0.05032

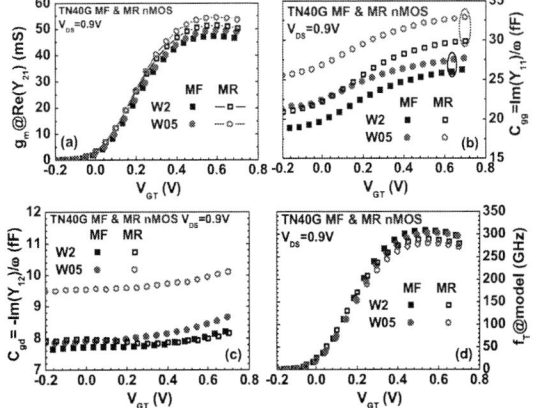

Fig. 7 μ_{eff} vs. V_{GT} extracted from linear I-V characteristics (a) MF nMOS (b) MR nMOS (c) comparison of MF and MR nMOS with various W_F and N_F

Fig. 8 Comparison of MF and MR nMOS (a) g_m =Re(Y_{21}) (b) C_{gg} = Im(Y_{11})/ω (c) C_{gd} = -Im(Y_{12})/ω and (d) f_T @model =g_m/2π(C_{gg}-C_{gd})$^{1/2}$ vs. V_{GT} (V_{DS}=0.9V)

978-1-7281-0943-5/19 $31.00 © 2019 IEEE

Backside Si passivation: leading to high performance UTB GeOI structures for monolithic 3D integrations

Wen Hsin Chang, Toshifumi Irisawa, Hiroyuki Ishii, Noriyuki Uchida and Tatsuro Maeda

National Institute of Advanced Industrial Science and Technology (AIST), Japan
E-mail: wh-chang@aist.go.jp

ABSTRACT

Backside Si passivation for ultrathin boy (UTB) GeOI has been verified to suppress Coulomb scattering from Ge/buried oxide (BOX) interface. With improvement of backside interfacial quality, primary carrier scattering factors in GeOI channel have been effectively reduced, resulting in significant enhancement of hole mobility. The hole mobility improvement has been proved through device characterizations of UTB GeOI pMOSFETs and Hall measurements. UTB GeOI platform formed by only low thermal budget processes is very promising for future Ge large scale integrated (LSI) circuits devices in monolithic 3D (M3D) integration scheme.

INTRODUCTION

Monolithic 3D (M3D) integration has gained a lot of attention recently, as it offers possible solutions for logic and memory LSI with significant benefits of increasing device packing density and reducing the length of interconnects. Ge-on-insulator (GeOI) structures fabricated by layer transfer technique have been considered as promising building blocks for M3D integration (Fig. 1) because only low thermal budget processes are needed to fabricate ultrathin boy (UTB) GeOI n/pMOSFETs. Owing to Ge low melting point, the S/D activation temperature for Ge devices can be lowered below 400 °C [1], which is favored for M3D integration not to damage the underlying devices and interconnect layers.

Although Ge possesses 4 times higher bulk hole mobility than that of Si, the hole mobility demonstrated in GeOI pMOSFETs with body thickness (T_{body}) less than 10 nm was still much lower than bulk Si pMOSFETs [2, 3]. Poor backside Ge/buried oxide (BOX) interface has been suggested to degrade the device performance of UTB GeOI pMOSFETs, which will definitely limit the GeOI applications. In this work, device performances of GeOI channels with and without backside Si passivation have been thoroughly investigated. Through the mobility characterizations of front/back GeOI channels and Hall measurements, physical origins for mobility enhancement have been discussed and verified.

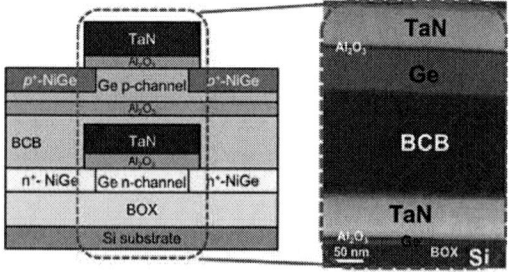

Fig. 1 Monolithic 3D Ge CMOS on the double GeOI structure. Top GeOI pMOSFET has been successfully transferred on the UTB-GeOI nMOSFET as shown in the TEM image.

EXPERIMENTAL PROCEDURES

The GeOI structures with and without backside Si passivation were prepared with ELO [4] and HELLO [5] technologies. Since Si passivation was proved to be effective in reducing interfacial trap density (D_{it}) between Ge and oxide [6], Si passivation with the thickness of 0.5 nm was performed on initial Ge surface of Ge donor wafer. The fabricated Ge/Si/BOX/Si hetero-structure (Fig. 2(a)) has been examined by high resolution scanned energy dispersive X-ray spectroscopy (EDX) analysis. The scanned EDX analysis of Si and Ge signals indicates the existence of Si passivation layer underneath GeOI layer (Fig. 2(b)).

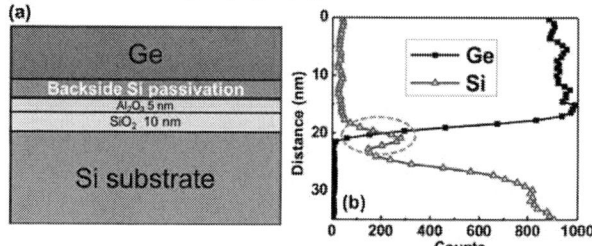

Fig. 2 (a)Schematic cross-sectional image of Ge/Si/BOX/Si hetero-structure (b)High resolution EDX analysis of Ge/Si/BOX/Si hetero-structure. Si passivation layer on top of BOX was confirmed.

After Ge mesa isolation and recessed channel process, 5-nm-thick Al_2O_3 was deposited through plasma enhanced ALD. 60-nm-thick TaN metal gate was defined by sputtering and reactive ion etching after photolithography. Subsequently, self-aligned NiGe metallic S/D regions were fabricated through rapid thermal annealing (RTA) and selective etching of unreacted Ni in HCl. For further reducing parasitic resistance, low thermal budget ion implantation after germanidation (IAG) technique was performed at 400 °C [1]. Finally, Ti/Al metal stacks were deposited by an e-beam evaporator as measurement pads. The processing temperature from GeOI structure to UTB GeOI pMOSFETs was kept below 400 °C, demonstrating M3D integration compatibility.

RESULTS AND DISCUSSION

To investigate the impact of backside Si passivation on GeOI quality, device performance of GeOI pMOSFETs was measured from front and back gates, respectively. Table 1 shows subthreshold slope (SS) values extracted from transfer curves at V_D of -50 mV of I_D-V_G characteristics of 15-nm-thick GeOI pMOSFETs with and without backside Si passivation, respectively. Using SS and CET values of front gate dielectric of 2.5 nm and back gate dielectric of 12.5 nm, D_{it} of GeOI front and back interfaces were also derived. For front channel, similar D_{it} values were obtained for both GeOI devices with and

978-1-7281-0943-5/19 $31.00 © 2019 IEEE

without backside Si passivation. For back channel without Si passivation, comparable D_{it} values was attained, indicating similar interfacial quality at the front and the back ALD Al$_2$O$_3$/Ge interfaces. On the other hand, much smaller D_{it} was confirmed by back gate mode operation in device with backside Si passivation, where SS was significantly improved through backside Si passivation. This reduced D_{it} owing to backside Si passivation should suppress Coulomb scattering from back interface and enhanced the hole mobility.

Table 1. SS and D_{it} of front and back channel operation of UTB GeOI pMOSFETs with and without backside Si passivation.

	SS (mV/decade)		D_{it} (cm^{-2}eV^{-1})	
	front	back	front	back
w/ Si passivation	113	109	5×10^{12}	8.6×10^{11}
w/o Si passivation	117	247	5.2×10^{12}	5.7×10^{12}

Thanks to better backside interfacial quality, V_{sub} can effectively control V_{th} in UTB GeOI pMOSFETs with backside Si passivation. The relationship between V_{sub} and V_{th} of GeOI devices with and without backside Si passivation is shown in Fig. 3(a) and 3(b), respectively. Comparing these two devices, the effect of V_{sub} was much weaker for those without backside Si passivation. On the other hand, the relationship between V_{sub} and V_{th} with backside Si passivation showed a clear linear trend and a high substrate bias factor ($\Delta V_{th}/\Delta V_{sub}$) of 0.3 was obtained, which is thanks to both thin BOX and high-quality Ge/BOX interface.

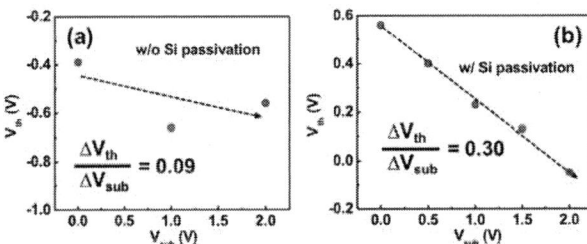

Fig. 3 Relationship between V_{sub} and V_{th} for GeOI devices (a) without and (b) with backside Si passivation.

From the peak G_m value at V_D of - 50 mV (Fig. 4 (a)), the peak field-effect hole mobility (μ_{FE}) of GeOI pMOSFETs with backside Si passivation was extracted through the following formula.

$$\mu_{FE} = L_G \cdot G_m / W \cdot C_{ox} \cdot |V_D|$$

where C_{ox} of both front and back gates were derived from C-V curves. The thickness dependence of peak μ_{FE} extracted from GeOI front and back channel was shown in Fig. 4(b). The closed and open symbols indicate the back and front GeOI channel mobility, respectively. Larger mobility of ~ 200 cm^2/Vs was obtained from GeOI back channel even with GeOI thickness scaling down to 4 nm. However, obvious mobility degradation with scaling T_{body} was also observed for back channel, indicating that the impact of poor front interfacial quality becomes stronger while T_{body} scaling down.

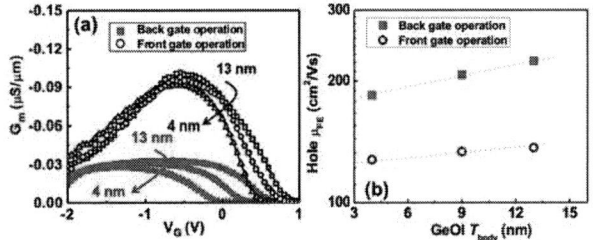

Fig. 4 T_{body} dependence of (a) transconductance and (b) peak filed-effect mobility extracted from front and back channel of GeOI pMOSFETs with backside Si passivation.

Besides μ_{FE}, we also investigate hole Hall mobility of our GeOI structures. The thickness dependence of hole Hall mobility for GeOI structures with and without backside Si passivation is plotted in Fig. 5. Hall mobility in UTB GeOI with backside Si passivation are higher than those without backside Si passivation. Obvious mobility degradation for GeOI structure without backside Si passivation can be attributed to severe carrier scattering at the relatively poor backside interfacial quality. On the other hand, higher mobility in GeOI structure with backside Si passivation maintains around 200 cm^2/V · s while T_{body} scaling down to 4 nm, indicating the successful suppression of carrier scattering due to backside Si passivation.

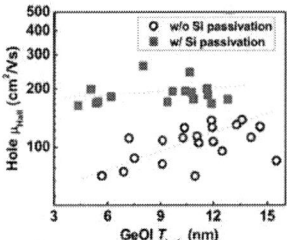

Fig. 5 GeOI thickness dependence of hole Hall mobility for GeOI structure with and without backside Si passivation.

CONCLUSION

Backside Si passivation has been proved to be effective for reducing interfacial trap density between Ge and BOX, which contributes to the suppression of Coulomb scattering from the back interface, thus achieving high hole mobility in UTB GeOI. The trend of mobility extracted from Hall measurement is consistent of what we obtained from device characterizations. UTB GeOI structures with backside Si passivation in conjunction with low thermal budget process are promising platform for future M3D integration.

ACKNOWLEDGMENT

The authors would like to thank the support from Sumitomo Chemical Co., Ltd. and Kokusai Electric Inc. for Ge layer transfer technology and AIST-NPF for device fabrication.

REFERENCES

[1] W. H. Chang et al., *EDL* **37**, 253 (2013). [2] C. H. Lee et al., *IEEE SOI Conf.* 1 (2011). [3] X. Yu et al., *IEDM* 20 (2015). [4] T. Maeda et al., *ME* **109**, 133 (2013). [5] T. Maeda et al., *APL* **109**, 262104 (2016). [6] J. Mitard et al., *IEDM* 873, (2008).

Hyper-Selective Co Metal ALD on Metals vs. SiO_2 Without Passivation

Steven Wolf*, Mike Breeden*, Scott Ueda*, Andrew Kummel**
*Materials Science and Engineering Program
**Department of Chemistry and Biochemistry
University of California San Diego, La Jolla, CA 92093
Email: sfwolf@eng.ucsd.edu

ABSTRACT

Integration of logic and memory into a single, monolithic 3D system-on-a-chip (SOC) allows for more efficient packing that can dramatically shorten (by 50 fold) computation times while using less power. For this 3D SOC vertical integration, high aspect ratio vias and interconnects need to be selectively deposited to induce the formation of larger grains and reduced resistivity. Here, low-temperature hyper-selective Co ALD was achieved utilizing an organometallic Co precursor and two different co-reactants that can allow for bottom-up via fill.

I. INTRODUCTION

It is desired to have selective metal deposition for bottom-up fill for monolithic 3D SOCs in order to improve performace by connecting logic and memory on a single chip. This would induce formation and growth of larger grains, which are expected to decrease via and interconnect resistance by reducing grain boundaries and decreasing surface roughness (Fig. 1). In addition, bottom-up growth may remove the need for nucleation layers on low-k dielectrics (SiCOH) since the nucleation will occur only on the bottom surface. The key metals for bottom-up growth include cobalt and ruthenium; cobalt is particularly important since it used as both a capping layer on Cu to protect it from oxidation [1], and in sub 10 nm vias, where Co is consider to be a better conductor than Cu due to Co having a smaller electron mean free path and problems with Cu electroplating in sub 10 nm vias [2].

II. METHODS

ALD cobalt metal was explored using a metal-organic cobalt precursor, Bis(1,4-di-tert-butyl-1,3-diazadienyl) cobalt ($Co(dad)_2$), and either a co-reactant of formic acid (HCOOH) or tert-butylamine (TBA) at 180°C on Cu, Pt, and SiO_2 substrates. The deposited Co films were studied using in-situ x-ray photoelectron spectroscopy (XPS) and atomic force microscopy (AFM). Cross-sectional scanning electron microscopy (SEM) and 4-point probe measurements were performed to check film thickness and resistivity, respectively.

III. RESULTS

A. Selective Co from HCOOH

It was found that there is nearly infinite selective deposition of Co on a conductor and not an SiO_2 for 180°C ALD with $Co(dad)_2$ and HCOOH. Fig. 2 shows the XPS of performing 100 ALD cycles followed by an additional 100 cycles on UHV annealed Pt vs SiO_2. On Pt, a thick (>10 nm)

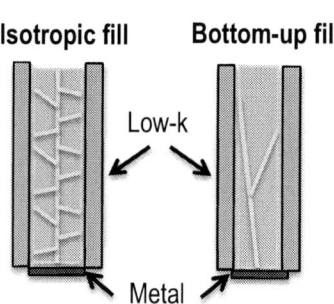

Fig. 1. Selective bottom-up Co. By depositing Co selectively on metal vs. low-k for via fill, growth can induce large grains that provide lower electrical resistance.

Co^{+0} film deposits while virtually no deposition results on SiO_2. AFM images show no change on SiO_2 before and after Co ALD cycles consistent with no nuclei formation, while the Co on Pt surface roughness remains below 1.8 nm (Fig. 3). To verify self-limiting precursor exposures consistent with ALD, a saturation study was performed and monitored with XPS.

Fig. 4 highlights the effect of individual additional half cycle amounts that result in self-limiting $Co(dad)_2$ and HCOOH exposures consistent with ALD. Additionally, this study revealed a novel mechanism about the reaction. Previously it was thought the HCOOH dissociatively chemisorbed to produce atomic H which removed the ligands from $Co(dad)_2$. Instead, XPS showed that HCOOH did not remove the ligands but instead induced a ligand-exchange process. Fig. 5 shows

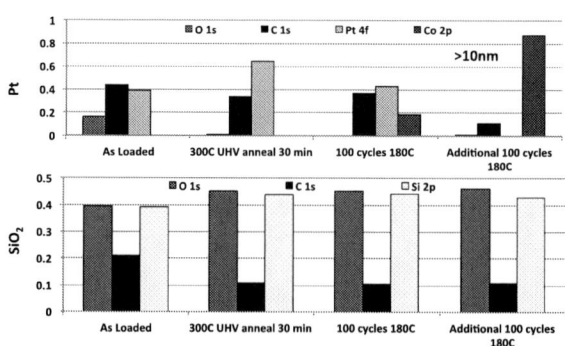

Fig. 2. XPS of UHV annealed Pt and SiO_2 Substrates that underwent 100 followed by an additional 100 ALD cycles of $Co(dad)_2$ + HCOOH at 180°C. On Pt, a completely buried Pt signal is consistent with a film > 10 nm thick, while on SiO_2, no deposition has occurred consistent with infinite selectivity.

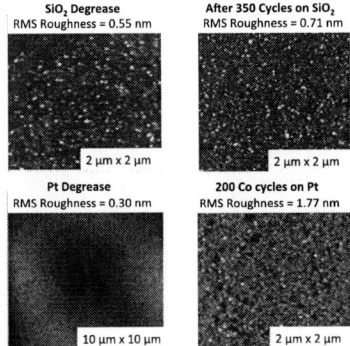

SiO₂ Degrease
RMS Roughness = 0.55 nm

After 350 Cycles on SiO₂
RMS Roughness = 0.71 nm

2 μm x 2 μm   2 μm x 2 μm

Pt Degrease
RMS Roughness = 0.30 nm

200 Co cycles on Pt
RMS Roughness = 1.77 nm

10 μm x 10 μm   2 μm x 2 μm

Fig. 3. AFM imaging before and after ALD cycles on SiO₂ and Pt. On SiO₂, no change was observed, while the Co film on Pt had a sub 2 nm RMS surface roughness.

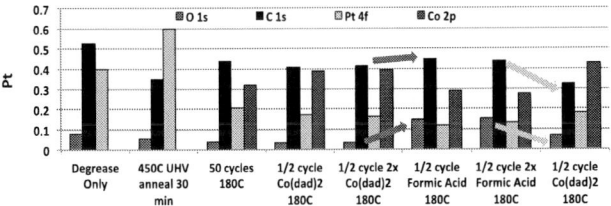

Figure 4: Saturation Study of Co(dad)₂ and HCOOH at 180°C. The self-limiting exposures were consistent with ALD. The increase in C and O after HCOOH dosing was consistent with deposition of a formate on the surface. The decrease in C and O, and increase in Co, after Co(dad)₂ dosing indicated a ligand exchange mechanism for the reaction.

the Co 2p peak that indicated the HCOOH leaving a higher binding energy component consistent with a formate on the surface that is then removed upon exposure to Co(dad)₂.

B. Selective Co from TBA

Deposition with HCOOH was attempted on Cu substrates (Fig. 6a); however, the substrate Cu signal never decreased to zero consistent with etching by HCOOH so an alkyl amine co-reactant (TBA) was also studied. For Co(dad)₂ + TBA ALD at 180°C, reduced Co metal films were deposited on Cu and Pt substrates with hyper-selectivity against SiO₂. Films as thick as 30 nm were grown on the conductors (Fig. 6b) without etching the substrates. On SiO₂, only 4% CoOₓ was deposited after an initial 50 ALD cycles. After an additional 250 ALD cycles, there was still only 4% CoOₓ consistent with saturation and hyper-selectivity (Fig. 6c). AFM imaging from ALD with TBA confirmed low surface roughness ALD on Pt and Cu, while only small (<5 nm) CoOx particles were present on SiO₂.

IV. CONCLUSION

Hyper-selective Co metal deposition was produced from Co(dad)₂ and both co-reactants (HCOOH and TBA). Utilizing HCOOH, no deposition was seen on SiO₂ consistent with infinite deposition, however HCOOH was observed to etch Cu. By switching to TBA, no Cu etching was observed and similar metallic Co films were deposited with only 4% CoOₓ on SiO₂ independent of the number of Co ALD cycles. The self-limiting deposition on SiO₂ is a novel mechanism of selectivity

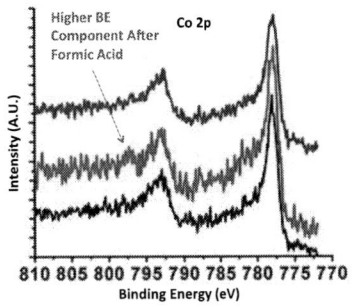

Fig. 5. Co 2p raw XPS Peaks. After formic acid dosing a higher BE component consistent with a formate deposited on the Co surface. The formate is removed after Co(dad)₂ dosing.

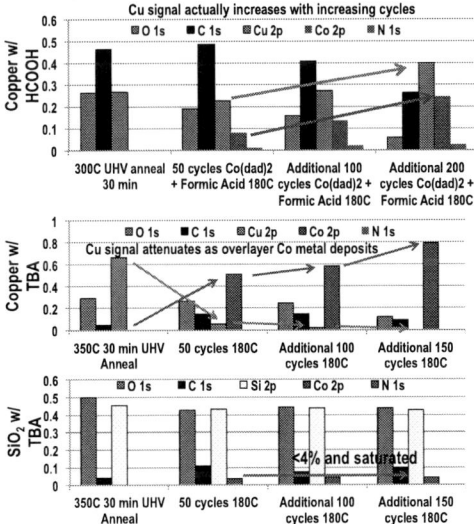

Fig. 6. Co ALD with HCOOH vs TBA on Cu vs SiO₂. (a) No attenuation of the substrate Cu signal with HCOOH was consistent with etching of Cu/CuOₓ. (b) When ALD was performed with TBA, the Cu fully buried consistent with no etching. (c) 4% CoOₓ was observed after 50 cycles of Co(dad)₂ + TBA. No additional CoOₓ was observed after 250 additional ALD cycles consistent with hyper-selectivity.

through the formation of an oxidic particulate, which results in hyper-selectivity. An extension of this chemistry for a novel Ru metal ALD with a metal-organic Ru precursor has been performed and has achieved selecitivy on metals vs. SiO₂.

ACKNOWLEDGMENT

This work was supported in part by ASCENT, one of six centers in JUMP, a Semiconductor Research Corporation (SRC) program sponsored by DARPA. The work was also supported in part by Applied Materials.

REFERENCES

[1] Yang, C-C., et al. "Characterization of copper electromigration dependence on selective chemical vapor deposited cobalt capping layer thickness." *IEEE Electron Device Letters* 32.4 (2011): 560-562.

[2] Gall, Daniel. "Electron mean free path in elemental metals." *Journal of Applied Physics* 119.8 (2016): 085101.

Novel Fine-Grain Back-Bias Assist Techniques for 14nm FDSOI Top-Tier SRAMs integrated in 3D-Monolithic

D. Bosch[1,3], F. Andrieu[1], L. Ciampolini[1,2], A. Makosiej[1], O. Weber[1,2], X. Garros[1], J. Lacord[1], J. Cluzel[1],
E. Esmanhotto[1], M. Rios[1], S. Lang[1], B. Giraud[1], R. Berthelon[2], G. Cibrario[1], L. Brunet[1], P. Batude[1],
C. Fenouillet-Béranger[1], D. Lattard[1], J. P. Colinge[1], F. Balestra[3], M. Vinet[1]

[1] CEA-LETI, Univ. Grenoble Alpes, 17 rue des Martyrs, 38054 France ; email : daphnee.bosch@cea.fr ; [2] STMicroelectronics,
850 rue Jean Monnet, F38926 Crolles ; [3]Univ. Grenoble Alpes, CNRS, Grenoble INP, IMEP-LAHC, F-38000 France

ABSTRACT

For the first time, we propose a 3D-monolithic SRAM architecture with a local back-plane for top-tier transistors enabling local back-bias assist techniques without area penalty as well as the capability to route two additional row-wise signals on individual back-planes. Experimental data are extracted from a 14nm planar Fully-Depleted-Silicon-on-Insulator (FDSOI) $0.078\mu m^2$ SRAM in order to properly model 3D top-tier cells. Simulations show this technique yields a 7% bitline capacitance reduction, a 12%/16% read/write access time improvement at V_{DD}=0.8V and a reduction of minimum operating voltage V_{min} by 60mV at 6σ w.r.t. planar SRAMs.

INTRODUCTION

Static-Random-Access-Memory (SRAM) optimization under area and performance constraints suffers from conflicting best-case conditions for read/write/retention, especially for low voltage operation. While assist techniques have been commonly adopted to address this issue [1], FDSOI offers a new degree of freedom owing to the use of *static* back-biasing as demonstrated in 28-22nm node SRAMs [2,3,4]. In planar FDSOI, *dynamic* back-bias assist suffers from a large well capacitance penalty. The bias range is also limited due to partially shared wells in the SRAM matrix between neighboring rows or columns. 3D-monolithic CoolCube™ technology [5], which consists in stacking transistors on different tiers, does not suffer from these limitations (provided local back planes are dielectrically isolated from one another). This integration was recently demonstrated to efficiently enable dynamic back-bias in standard cells [6]. This work is focused on SRAM back-bias assist techniques using the unique features of 3D-monolithic top-tier 14nm FDSOI technology.

MEASUREMENT OF PLANAR 14NM FDSOI SRAMs

14nm planar CMOS devices were fabricated featuring 6nm-thin channels, 20nm minimum gate length, SiGeB/SiP in-situ doped source/drain, 90nm Contacted Poly Pitch, 64nm Metal Pitch and $0.078\mu m^2$ SRAM minimum area [7]. High density (HD, $0.078\mu m^2$) and high current (HC, $0.098\mu m^2$) bitcell device dimensions are summarized in Fig.1. All the HC/HD transistors, *i.e.* the Pull-Up (PU) pMOS as well as the Pass-Gate (PG) and Pull-Down (PD) nMOS are built on silicon channel, with a single metal gate and single p-doped well (Figs 1-2). Excellent experimental static performance is obtained (nominal conditions are V_{DD}=0.8V, V_{well}=0) (Figs 3-5).

PLANAR SRAM CELL SENSITIVITY TO BACK BIAS

The sensitivity of SRAM *vs.* back bias (V_{well}) was also characterized experimentally. Since the well is shared between all devices in the analyzed bitcell, V_{well}<0 strengthens the Pull-Up PMOS (PU) and weakens the Pass-Gate (PG) and Pull-Down (PD) NMOS transistors, helping the PU to maintain BLTI=1 and thus BLFI=0 during the read operation (see SNM in Fig.6). Conversely, using V_{well}>0 improves the PG/PU strength ratio (and in turn the WNM) and increases the write current (governed by PG drive). As a consequence, back biasing can be used to assist both write (V_{well}>0) and read operation (V_{well}<0). In regular FDSOI, this cannot be practically achieved because a single well is common to the whole array (no column selection). Performing dynamic back-bias would, therefore, require switching a huge capacitance between read and write operation, leading to unacceptable increase of access energy and delay. 3D-monolithic integration with local back planes offers thus much greater opportunities for assist techniques than planar FDSOI.

PERFORMANCE ASSESSMENT OF 3D TOP-TIER SRAMs

A FDSOI SPICE model and a design kit were built using electrical parameters characteristics of the CoolCube™ low-temperature process [5]. The 14nm 3D-monolithic design environment includes four intermediate metal lines iML and a back plane, which follow the same design rules as a back-end metal layer (Figs 7-8).

A detailed study of the influence of independent back-bias for PU/PG/PD shows that the threshold voltage of the PU must be lowered (V_{Bpu}<0) for all figures of merit (FoM), which cannot easily be achieved using a gate-first FDSOI process with Si channel (Fig.9). This can be performed in 3D by using a PU-dedicated back plane with a constant bias (V_{Bpu}=-0.8V) applied in all operation modes. Additionally, PD (or PG) threshold voltage can be dynamically modulated according to SRAM operation to improve margins and currents. Three promising assist modes (with different V_{Bpg}, V_{Bpu}, V_{Bpd}) are selected for the write (A1) and read stability (A2) as well as for the read time (A3) assist. Using this versatile assist yields +23% WNM, +28% I_{write} with A1, +4% SNM with A2 and +28% I_{read} with A3 at V_{DD}=0.8V and V_{well}=$\pm$ V_{DD}/GND *vs.* the reference configuration with a single back-plane biased at 0V (Fig.9). Furthermore the gains are more pronounced at low supply voltage V_{DD} (Fig.10), leading to a 60mV V_{min} reduction with A2 (Fig.11). The corresponding layout (common for A1-A2-A3) has been designed, connecting two groups of local (to-the-bitcell) back planes for PD and PG through internal vias (Fig.8) without area penalty. Actually back plane lines parallel to BLs distribute a static PU bias. Moreover the two dynamic signals are routed by iML3 in the WL direction within the SRAM height (whereas wells are typically in the BL direction in planar technologies). Thus, back biasing allows boosting a selected row in top-tier without disturbing other rows.

In order to evaluate the capacitance gain provided by a local back-plane compared with a continuous one (or a single well in planar), back-end parasitics have been extracted using TCAD and included in the SPICE netlist. A 7% BL capacitance reduction and a 12/16% read/write time improvement is achieved (w.r.t. reference cell at V_{DD}=0.8V) (Figs 12-13). The demonstrated assist technique can be combined with WL underdrive, negative BL or other standard assist techniques [8] for further performance and stability improvement (Fig.14).

CONCLUSION

The presence of local back planes in 3D-monolithic technology provides an extra knob to optimize the static and dynamic bitcell performance/area of top-tier FDSOI SRAMs. This specific feature is related to the integration of row-based, local back-planes addressed by internal vias and intermediate metal lines. The optimization of back bias assist techniques in such 3D architecture allows us to reduce the read /write access time by 12/16% and V_{min} by 60mV in a 14nm HD SRAM.

978-1-7281-0943-5/19 $31.00 © 2019 IEEE

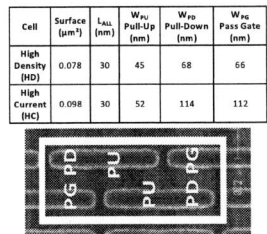

Cell	Surface (µm²)	L_{ALL} (nm)	W_{PU} Pull-Up (nm)	W_{PD} Pull-Down (nm)	W_{PG} Pass Gate (nm)
High Density (HD)	0.078	30	45	68	66
High Current (HC)	0.098	30	52	114	112

Fig.1: 14nm FDSOI 6T-SRAM. **Top:** key dimensions. **Bottom:** 14nm planar HD SRAM SEM observed at the gate level.

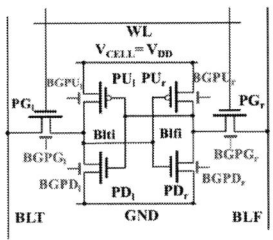

Fig.2: SRAM schematic. The additionnal terminals provided by CoolCube™ technology are highlited in red.

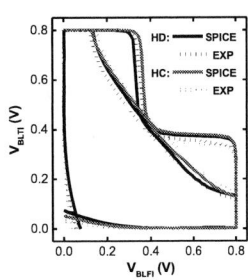

Fig.3: Exp. SNM butterfly curve and WNM at V_{DD}=0.8 V $vs.$ SPICE.

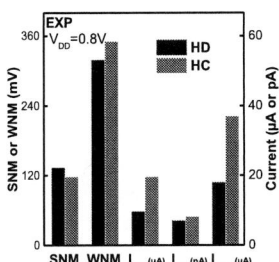

Fig.4: Exp. comparison between HD, HC cell for typical Figures-of-Merit (FoM).

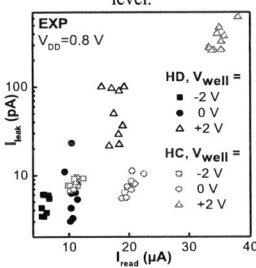

Fig.5: Exp. cell current $vs.$ cell leakage for different p-well biasing (V_{well}).

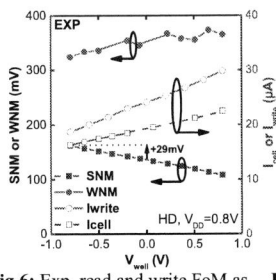

Fig.6: Exp. read and write FoM as a function of V_{well}.

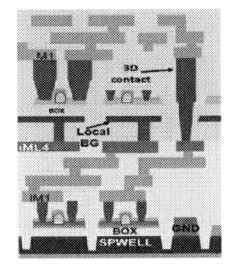

Fig.7: Schematic stack of CoolCube™ 14nm Design Kit with intermediate vias between the back plane and the upper intermediate metal line.

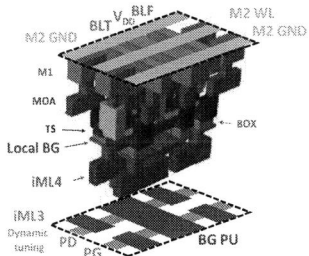

Fig.8: SRAM 3D layout view with underneath backbias connections routed in the word line direction.

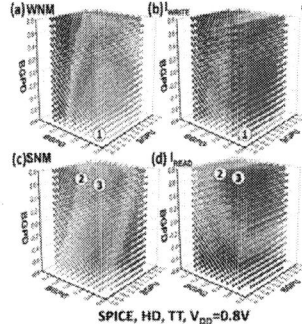

Fig.9: Sensibility (%) of (a) WNM, (b) I_{WRITE}, (c) SNM and (d) I_{READ} on independent back-biasing (on PD,PG,PU) (nominal configuration is at V_{BG}=GND) (SPICE). Three assist modes are highlighted: A1, A2 and A3.

BG bias (V)		REF	A1	A2	A3
WRITE	PG	0	0.8	0	0.8
	PU	0			0.8
	PD	0	-0.8	-0.8	0.8
READ	PG	0	0	0	0.8
	PU	0		-0.8	
	PD	0	-0.8	0.8	0.8

	FOM	REF	A1	A2	A3
WRITE	WNM (mV)	336	431	338	320
	I_{WRITE} (µA)	30.5	33	32.7	38
READ	SNM (mV)	146	171	164	124
	I_{READ} (µA)	17.4	15	18.6	22

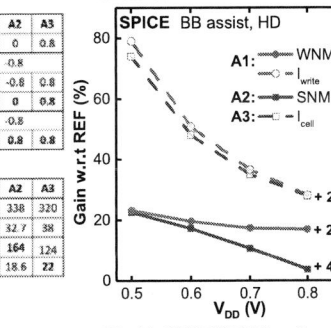

Fig.10: WNM/SNM/I_{write}/I_{read} improvement w.r.t. REF $vs.$ V_{DD} (SPICE).

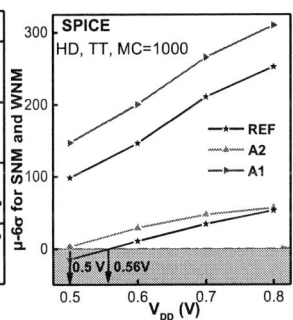

Fig.11: SNM/WNM (at µ-6σ) as a function of V_{DD}. V_{min} is lowered by 60mV with back-biasing (SPICE).

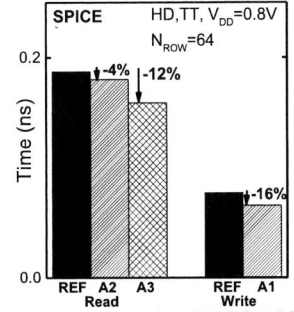

Fig.12: Bitline capacitance computation for a single bitcell with different back plane configurations (TCAD).

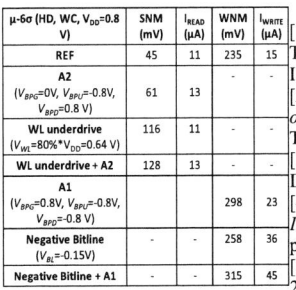

Fig.13: Read/write time (SPICE). A3 is particularly interesting to boost cell reading time.

µ-6σ (HD, WC, V_{DD}=0.8 V)	SNM (mV)	I_{READ} (µA)	WNM (mV)	I_{WRITE} (µA)
REF	45	11	235	15
A2 (V_{BPG}=0V, V_{BPU}=-0.8V, V_{BPD}=0.8 V)	61	13	-	-
WL underdrive (V_{WL}=80%*V_{DD}=0.64 V)	116	11	-	-
WL underdrive + A2	128	13	-	-
A1 (V_{BPG}=0.8V, V_{BPU}=-0.8V, V_{BPD}=0.8 V)	-	-	298	23
Negative Bitline (V_{BL}=-0.15V)	-	-	258	36
Negative Bitline + A1	-	-	315	45

Fig.14: Gain summary of different read/write assists.

REFERENCES:

[1] B. Zimmer et al., IEEE Transactions on Circuits and Systems II, 2012, pp. 853–857

[2] V. Joshi *et al.*, *2017 Symposium on VLSI Technology*, 2017, pp. T222-T223

[3] O. Thomas et al., Tech. Dig. IEDM, 2014, pp. 3.4.1–3.4.4.

[4] N. Sugii et al, *IEEE 2011 International SOI Conference*, 2011, pp. 1-19

[5] P. Batude *et al.*, Tech. Dig. IEDM, 2017, pp. 3.1.1-3.1.4.

[6] F. Andrieu et al., Tech. Dig. IEDM, 2017, p. 20.3.1–20.3.4.

[7] O. Weber et al., 2014 Symposium on VLSI Technology (VLSI-Technology): 2014, pp. 1-2.

[8] J. Wang, *et al.*, Proceedings of 13th ISLPED, 2008, p. 129.

ACKNOWLEDGMENTS: This work was supported by French Public Authorities through LabEx Minos ANR-10-LABX-55-01.

Novel Vertically-Stacked Tensily-Strained Ge$_{0.85}$Si$_{0.15}$ GAA n-Channels on a Si Channel with SS=76mV/dec, DIBL=36mV/V, and I$_{on}$/I$_{off}$=1.2E7

Yu-Shiang Huang[1], Fang-Liang Lu[1], Hung-Yu Ye[1], Ya-Jui Tsou[1], Yi-Chun Liu[1], Chien-Te Tu[1], and C. W. Liu[1,2,*]

[1]Graduate Institute of Electronics Engineering, National Taiwan University, Taipei, Taiwan,
[2]National Nano Device Laboratories, Hsinchu, Taiwan,
*E-mail: cliu@ntu.edu.tw

Abstract- The tensily strained high Ge content (85%) GeSi multi-channels stacked on the Si channel with channel size optimization have good subthreshold behaviors (SS, DIBL, and I$_{off}$), and meanwhile maintain high I$_{on}$. By narrowing GeSi channel, the bandgap of the GeSi can be increased by quantum confinement and the short channel effect can be suppressed. This also causes the more positive V$_t$ for GeSi than Si. As a result, SS=76mV/dec, DIBL=36 mV/V can be achieved due to the benefit of superior subthreshold behaviors of the Si channel underneath. The I$_{on}$/I$_{off}$ is improves to be 1.2E7. The high I$_{on}$=44μA per stack at V$_{OV}$=V$_{DS}$=1V, is contributed from both the GeSi and Si channels, and it is further enhanced to 46 μA by the external strain. The relatively low flicker noise indicates that the novel structure can mitigate the reliability issue of GeSi channels by the highly reliable Si channel underneath.

INTRODUCTION

The stacked GAA structure is a good candidate to increase I$_{on}$ for a fixed footprint and has superior electrostatic performance than FinFETs to achieve low power, high performance, and area scaling in the advanced technology nodes [1-3]. Recently, the state-of-the-art stacked Si channel GAAFETs [4] and stacked Ge channel GAAFETs are reported [5-6]. However, the Ge-based channels have poorer electrostatic control than the Si device due to its material nature of larger dielectric constant, especially at the scaled gate length. In this work, GeSi channels are used to enhance the I$_{on}$, while the Si channel is used to maintain the good subthreshold behavior.

MATERIALS AND DEVICE FABRICATION

The stacked GeSi/Ge epi layers are grown on Ge buffer on 200 mm SOI with the 70 nm-thick Si and p-type doping of 1E15cm^{-3} by RTCVD using SiH$_4$ and GeH$_4$ as the precursors (Fig. 1). All Ge layers are *in-situ* heavily P doped ([P]~3E20 cm^{-3}) to reduce the S/D parasitic resistance, while the GeSi channel layer are un-intentionally doped to suppress the impurity scattering. Two GeSi layers with thickness of 25 nm are sandwiched between Ge sacrificial layers (20 nm). The GeSi layers are fully strained (0.6%) with [Ge] = 85%, based on RSM and XRD, to improve the mobility. The key process steps for the devices with stacked SiGe channels on Si channels are shown in Fig. 2 (a). After the CVD epitaxy, the fins were then formed by e-beam lithography and anisotropic etching, followed by the channel release process of ultrasonic assist H$_2$O$_2$ wet etching [7-8]. Due to the microbridge structure of the floating channels (Fig. 2 (b)), the biaxial tensile strain from lattice mismatch between SiGe and Ge was transformed into the uniaxial strain. The strain enhancement to 1.9% (3.17x) is estimated by ANSYS simulation, to further improve the mobility. After the gate stack formation, the PMA is performed to improve the gate stack quality reduce S/D resistance. The CET is ~1.3nm. The optimized selective laser annealing is performed to further increase the activation (~2E20cm^{-3}) [9-10] at S/D using Pt as the mask. Finally, the NiGe contact is formed for reducing contact resistivity.

RESULTS AND DISCUSSION

For the device with wide GeSi channels on Si channel with channel length=60nm (Fig. 3), the GeSi channels turn on first due to the oxide charges, n-type doping, and poor short channel control (Fig.5 (a)). Large D$_{it}$ and poor short channel control of the GeSi channels result in

SS=128mV/dec and poor DIBL=121mV/V. Current mainly flows through the GeSi channels with the add-on Si channel to reach I$_{on}$=46μA. The △E$_c$ ~0 between GeSi and Si allows the current to flow along both GeSi channels and underneath Si channel. Large I$_{off}$ and low I$_{on}$/I$_{off}$ ~3E3 is observed due to low bandgap and poor short channel control for GeSi channels. The device is also expected to have poor reliability.

The V$_t$ difference between GeSi and Si and the bandgap can be tuned by the channel size with optimizing the etching recipe. By narrowing the GeSi channel size to 5nm (Fig. 4). The bandgap can be increased by quantum confinement and the short channel effect can be suppressed (Fig. 5(b)). As a result, more positive V$_t$ for GeSi than Si is observed. The Si channel turn on first with good SS=76mV/dec due to low D$_{it}$ and good short channel control. The DIBL=36mV/V is also achieved. I$_{off}$ is also dramatically reduced and the I$_{on}$/I$_{off}$ are improved to be 1.2E7 thanks to the increase of the bandgap and the good short channel control for the narrow GeSi channels. Both Ge and Si contribute the ON current. As a result, the high I$_{on}$ = 44 μA can also be maintained. The Si channel contributes 71% I$_{on}$ (Fig. 4 (b)), measured by the device with only Si channel. The TCAD simulation (Fig. 5 (c) and (d)) based on the same structure as the Fig 5 (a) and (b) is used to identify the individual contribution of the Ge channels and the underneath Si channel. The simulations have the similar behavior to the experimental data and verify the statements in Fig. 5 (a) and (b). The I$_{on}$ of the stacked narrow Ge-based channels with Si is further improved to 46μA by the external strain while the SS remains intact (Fig. 6). The devices have competitive performance in group IV nGAAFETs(Fig. 7). The stacked narrow GeSi channels with Si channel has lower flicker noise than wide GeSi channels (Fig.8). The low trapping/de-trapping due to the large tunneling barrier of SiO$_2$ layer of the Si channel, as compared to GeO$_x$ barrier of the Ge based channels, improves the noise performance, indicating enhanced reliability.

SUMMARY

By the channel size optimization, the tensily strained Ge$_{0.85}$Si$_{0.15}$ multi-channels stacked on the Si channel have good SS, DIBL, I$_{off}$, and low noise, and maintain high I$_{on}$. The bottom Si can also be a fin or nanosheets with good subthreshold behavior and reliability for real applications.

ACKNOLEDGEMENT

This work is supported by the MOST (107-2218-E-002-044-).

REFERENCE

[1] D. Yakimets *et al.*, IEDM, p. 501, 2017. [2] S. Barraud *et al.*, IEDM, p. 677, 2017. [3] Y. M. Lee *et al.*, IEDM, p. 681, 2017. [4] H. Mertens *et al.*, IEDM, p. 828, 2017. [5] J. Mitard *et al.*, VLSI, p. T83, 2018. [6] E. Capogreco *et al.*, VLSI, p. T193, 2018. [7] Y.-S. Huang *et al.*, IEDM, pp. 832-835, 2017. [8] Y.-S. Huang *et al.*, EDL, 39, 9, 2018. [9] I-H. Wong *et al.*, IEDM, pp.842-845, 2016. [10] F.-L. Lu *et al.*, TED, 65, 7, 2018. [11] H. Mertens *et al.*, VLSI, p.158, 2016. [12] H. Mertens *et al.*,IEDM, p. 524, 2016. [13] W.W. Fang *et al.*, EDL, 28, 3, 2007 . [14] I. Lauer *et al.*, VLSI, T140, 2015. [15] I-H. Wong *et al.*, IEDM, p.239, 2014. [16] S.-H. Hsu *et al.*, IEDM, p. 525, 2012. [17] Y.-J. Lee *et al.*, IEDM, p. 382, 2015. [18] H. Wu *et al.*, IEDM, p. 16, 2015.

978-1-7281-0943-5/19 $31.00 © 2019 IEEE

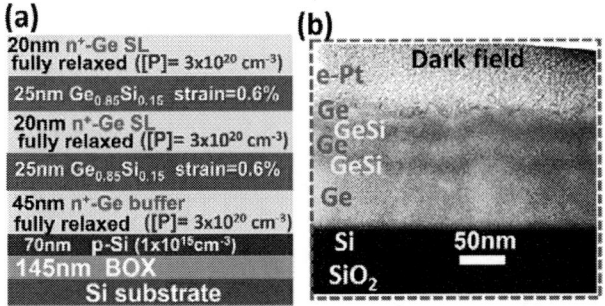

Fig.1 (a) Schematic view and (b) TEM images of epi layers. Heavily doped ([P]~3E20 cm^{-3}) for all Ge layers is used to lower S/D parasitic resistance. The GeSi channels are all unintentionally doped to suppress the impurity scattering. The 0.6% tensile strain in Ge$_{0.85}$Si$_{0.15}$ can improve the mobility.

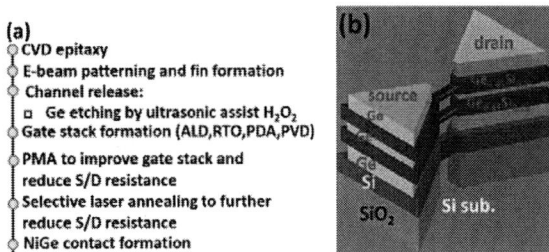

Fig.2 (a) The fabrication process flow and (b) schematic view for the structure after the channel release process. Due to the microbridge structure of the floating channels, the strain can be improved to 1.9% (3.17x), estimated by ANSYS simulation, for the improved mobility.

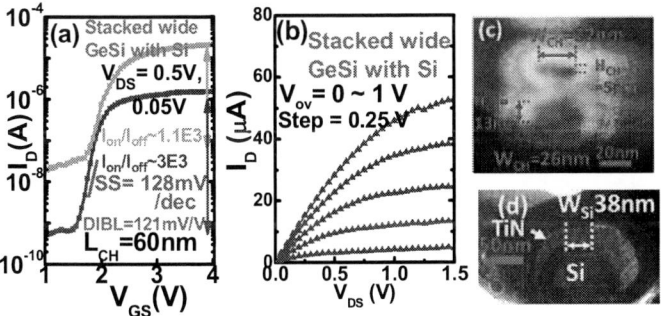

Fig.3 (a) I$_D$-V$_{GS}$, (b) I$_D$-V$_{DS}$, (c) TEM image for stacked GeSi channels, and (d) TEM image for Si channels of stacked wide GeSi channels with Si channel. Poor subthreshold behaviors are observed.

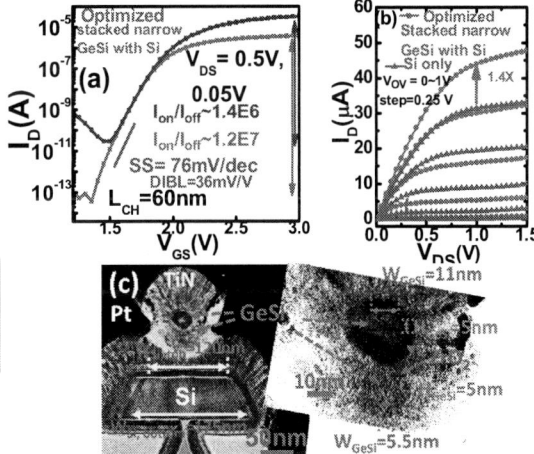

Fig.4 (a) I$_D$-V$_{GS}$, (b) I$_D$V$_D$, and (c) TEM image of optimized stacked GeSi channels with Si channel. By V$_t$ tunning using the channel size optimization, the good SS, DIBL, I$_{off}$, and high I$_{on}$ are achieved.

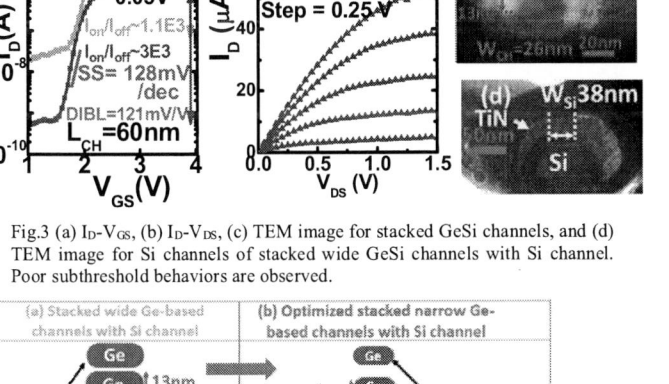

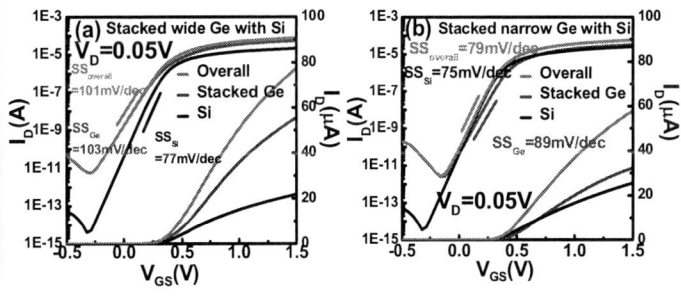

Fig.5 (a)(b) summary of the device performance with different channel dimensions. (c)(d) The TCAD simulations of the same structures as the Fig 5 (a) and (b) to identify the individual contributions of the Ge channels and the underneath Si channel. The simulations have the similar behavior to the experimental data and verify the statements in Fig 5 (a) and (b). Note that pure Ge is used in the channel to simplify the simulation.

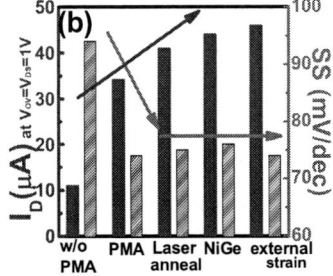

Fig.6 I$_{on}$/SS vs. process condition of the device in Fig. 4. The I$_{on}$ can be further improved to 46µA by the external strain while the SS remains intact.

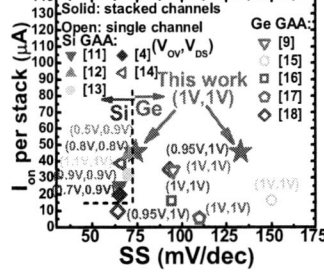

Fig.7 Benchmark of I$_{on}$ per stack v.s. SS for in group IV nGAAFETs. The devices in this work have competitive performance.

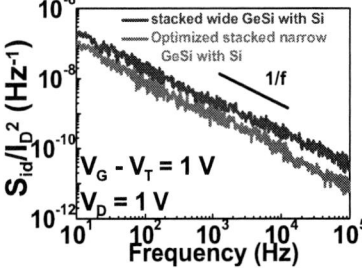

Fig.8 Noise spectra. Stacked narrow GeSi channels with Si channel has lower noise level due to the larger tunneling barrier of SiO$_2$ on Si with lower trapping-detrapping, indicating good reliability.

978-1-7281-0943-5/19 $31.00 © 2019 IEEE   113

The First GeSn Gate-All-Around Nanowire P-FET on the GeSnOI Substrate with Channel Length of 20 nm and Subthreshold Swing of 74 mV/decade

Yuye Kang,[1] Kaizhen Han,[1] Eugene Y.-J. Kong,[1] Dian Lei,[1] Shengqiang Xu,[1] Ying Wu,[1]
Yi-Chiau Huang,[2] and Xiao Gong.[1,*]

[1] Department of Electrical and Computer Engineering, National University of Singapore (NUS), Singapore.
[2] Applied Materials Inc. Sunnyvale, California, United States.
*Phone: +65 6516-7871, Fax: 65 6516-1689, E-mail: elegong@nus.edu.sg

ABSTRACT

We realized the first germanium-tin (GeSn) gate-all-around (GAA) p-channel field-effect-transistors (p-FETs) on a 200 mm GeSn-on-insulator (GeSnOI) substrate, achieving subthreshold swing (SS) of 74 mV/decade for a device with channel length (L_{CH}) of 60 nm. The GAA structure provided excellent control of short channel effects and also enabled the realization of a transistor with L_{CH} of 20 nm with decent electrical characteristics.

I. INTRODUCTION

In the past few years, innovations in transistor technologies such as strained silicon, high-k gate dielectrics, metal gate, and multi-gate FETs have led to ever-increasing transistor performance to push the Moore's law to the end-of-road-map. GeSn, due to its higher hole mobility than Si, SiGe, and Ge, is a promising channel material for future p-FETs [1-9]. For the GeSn-based p-FETs, significant effort has been made to address the challenges in gate stack and source/drain (S/D) formation. In addition, for extremely scaled FETs in future, multi-gate transistors are required to suppress short channel effects (SCEs). GeSn p-FinFETs with small SS and high transconductance have been demonstrated recently [7, 9]. Stacked GeSn GAA p-FETs on the Si substrate with Ge as the buffer layer were also realized [10].

In this work, for the first time, we demonstrate GeSn nanowire p-FETs on a GeSnOI substrate which was formed using direct wafer bonding and layer transfer technique with 200 mm wafer size. The 3D schematic of fabricated GeSn GAAFETs is shown in Fig. 1. The combination of inductively-coupled plasma (ICP) dry etch for fin formation and wet etch to remove the underneath SiO_2 to release the nanowire was employed to realize GeSn nanowires with the smallest width of 15 nm. Due to the excellent control of the SCEs by a GAA structure and suppression of sub-surface leakage current using the GeSnOI substrate, GeSn GAA pFETs with L_{CH} down to 20 nm were realized with decent electrical characteristics.

II. FORMATION OF GESNOI SUBSTRATE

The GeSnOI substrate was formed by direct wafer bonding and layer transfer technique [11], which is shown in Fig. 2. The as-grown wafer before bonding comprises of ~100 nm-thick CVD grown GeSn and a ~1 µm-thick Ge buffer layer on a Si substrate. The GeSn layer was fully compressively strained to the Ge buffer layer as confirmed using XRD.

After forming the GeSnOI substrate, the GeSn layer was thinned down to ~30 nm. The GeSn surface has a small root-mean-square (RMS) roughness of 0.566 nm (Fig. 3), as measured using AFM. The GeSn layer has very good crystalline quality before and after the GeSnOI substrate formation. The transmission electron microscopy (TEM) image of the GeSnOI substrate is shown in Fig. 4(a). High resolution TEM images in Fig. 4(b) and Fig. 4(c) show a sharp GeSn/SiO2 interface and excellent crystalline quality of GeSn film, respectively. Raman spectroscopy (Fig. 5) shows that the GeSn peak positions before and after GeSnOI formation are identical, indicating that the strain and Sn composition remain unchanged after GeSnOI formation. The GeSn peaks in XRD curves (Fig. 6) indicate the good crystalline quality of the donor wafer and the formed GeSnOI substrate.

III. FABRICATION OF GESN NANOWIRE P-FET

The key process steps for fabricating GeSn nanowire p-FETs are listed in Fig. 7. After the GeSnOI substrate was formed, the channel region was defined using hydrogen silsesquioxane (HSQ) by electron beam lithography (EBL). TEM image in Fig. 8(a) shows the device with smallest L_{CH} of 20 nm. The GeSn nanowire was surrounded by high-k and gate metal. Boron implantation was performed with a dose of 1×10^{15} cm^{-2} and the energy of 10 keV. This was followed by dopant activation at 400 °C for 1 minute. S/D and nanowire regions were defined with HSQ by EBL and dry etched by Cl-based plasma. After DHF (1:100) dipping for 2 minutes, GeSn nanowires were released. This was followed by high-k gate dielectrics deposition by atomic layer deposition (ALD). Metal gate comprising of Ti/Ni was deposited by e-beam evaporation and was lifted off. Finally, metal S/D contacts were formed. The tilted-view SEM images in Fig. 8(b) and (c) show released GeSn nanowires and a completed device with 5 nanowires in parallel, respectively.

IV. ELECTRICAL CHARACTERIZATION AND BENCHMARK

The fabricated GeSn nanowire p-FETs have L_{CH} from 20 nm to 300 nm and W_{CH} from 15 to 20 nm. Fig. 9 shows the transfer characteristics of the device with a L_{CH} of 20 nm and W_{CH} of 15 nm. SS of 110 mV/decade is obtained. This channel length is the smallest for Ge-based p-FETs. For a device with L_{CH} of 60 nm, the lowest SS achieved is 74 mV/decade, as shown in Fig. 10. This indicates good control of short channel effects. The extrinsic transconductance ($G_{m,ext}$) of this device is shown in Fig. 11. To improve $G_{m,ext}$, further optimization of nanowire release process is required. The Peak $G_{m,ext}$ of 244 µS/µm is extracted at a high V_{DS} of -0.5 V. The S/D resistance (R_{SD}) of ~ 1.7 kΩ is extracted from the transfer characteristics. Normalized drain conductance G_D of 74 µS/µm is extracted from output characteristics. By taking into the consideration of the effect of R_{SD} and G_D, peak intrinsic transconductance ($G_{m,int}$) is extracted to be 299 µS/µm. It should be noted that drive current and transconductance are normalized by the total perimeter of the nanowires. Fig. 12 shows the benchmark of SS as a function of Sn compositions for GeSn p-FETs. The red star represents the GeSn nanowire p-FETs realized in this work. This device has the lowest SS among all the reported GeSn nanowire p-FETs [4, 10].

IV. CONCLUSION

GeSn GAA p-FETs with L_{CH} down to 20 nm were realized. For L_{CH} of 60 nm, the smallest SS of 74 mV/decade was achieved. GeSn high mobility channel, together with the good control of the short channel effects by employing a GAA architecture and capability of suppressing the sub-surface leakage current by using GeSnOI substrate holds promise for future high performance and low power logic applications.

Acknowledgement. The authors acknowledge support from R-263-000-C58-133 and R-263-000-B43-733.

References:
[1] G. Han et al., IEDM 2011, p. 402.
[2] P. Guo et al., JAP 114, 044510 (2013).
[3] S. Gupta et al., IEDM 2011, p. 398.
[4] X. Gong et al., VLSI 2013, p. 34.
[5] X. Gong et al., EDL 34, 339 (2013).
[6] D. Lei et al., JAP 119, 024502 (2016).
[7] D. Lei et al., VLSI 2017, p. 198.
[8] Y.-S. Huang et al., IEDM 2016, p. 822.
[9] D. Lei et al., VLSI 2018, p. 197.
[10] Y.-S. Huang et al., IEDM 2017, p. 832.
[11] D. Lei et al., APL 109, 022106 (2016).
[12] L. Wang et al., SSE 83, 66 (2013).
[13] M. Liu et al., VLSI 2014, p. 100.
[14] C. Zhan et al., VLSI-TSA 2013, p. 82.

978-1-7281-0943-5/19 $31.00 © 2019 IEEE

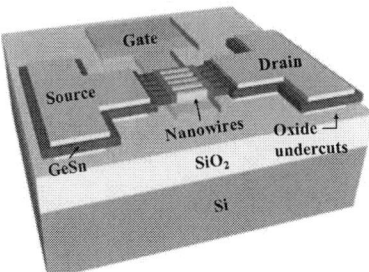

Fig. 1. 3D schematic showing the GeSn GAAFETs that are realized in this work. The GeSn nanowires are suspended and wrapped around by the high-k and gate metal.

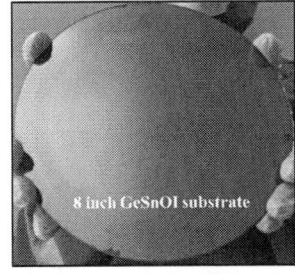

Fig. 2. The photograph of an 8 inch GeSnOI substrate formed by direct wafer bonding and layer transfer technique.

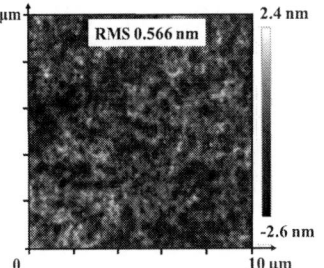

Fig. 3. AFM scan of the GeSnOI surface shows a smooth GeSn surface with small RMS of 0.566 nm.

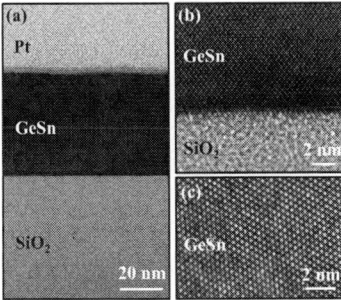

Fig. 4. (a) TEM image of the GeSnOI substrate. High resolution TEM images show (b) a sharp interface between GeSn and SiO_2 and (c) excellent crystalline quality of GeSn.

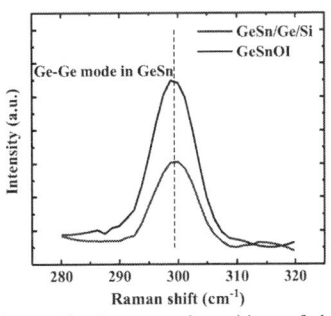

Fig. 5. The Raman peak positions of the GeSn/Ge/Si sample and of the GeSnOI substrate are the same, indicating no change in the strain and Sn composition after the GeSnOI formation.

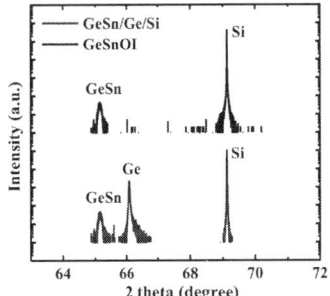

Fig. 6. The GeSn peaks in XRD curves show the good crystalline quality of GeSn film before and after bonding process.

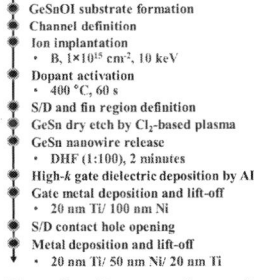

- GeSnOI substrate formation
- Channel definition
- Ion implantation
  - B, 1×10^{15} cm^{-2}, 10 keV
- Dopant activation
  - 400 °C, 60 s
- S/D and fin region definition
- GeSn dry etch by Cl_2-based plasma
- GeSn nanowire release
  - DHF (1:100), 2 minutes
- High-k gate dielectric deposition by ALD
- Gate metal deposition and lift-off
  - 20 nm Ti/ 100 nm Ni
- S/D contact hole opening
- Metal deposition and lift-off
  - 20 nm Ti/ 50 nm Ni/ 20 nm Ti

Fig. 7. Process flow for fabricating GeSn nanowire p-FETs on the GeSnOI substrate. GeSn nanowires were released by wet etch using DHF.

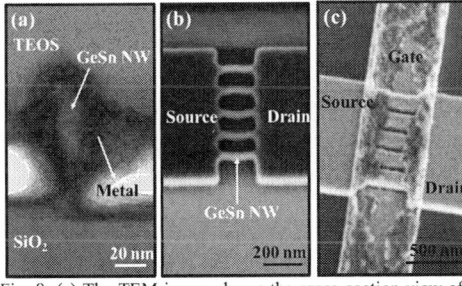

Fig. 8. (a) The TEM image shows the cross-section view of a device with L_{CH} of 20 nm and W_{CH} of 14.8 nm. The tilted-view SEM images show (b) released GeSn nanowires and (c) a complete device with 5 GeSn nanowires.

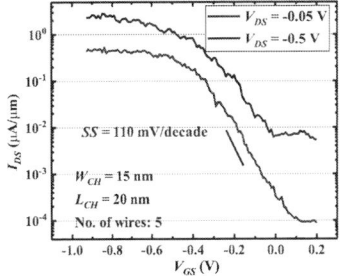

Fig. 9. Transfer characteristics of the device with minimum L_{CH} = 20 nm showing SS of 110 mV/decade.

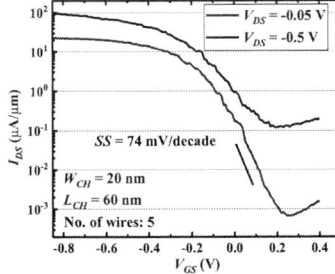

Fig. 10. I_{DS}-V_{GS} of GeSn nanowire p-FETs with L_{CH} = 60 nm having the smallest SS value of 74 mV/decade for GeSn nanowire FETs.

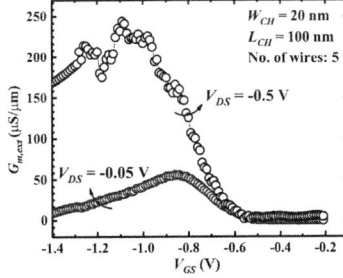

Fig. 11. $G_{m,ext}$ vs. V_{GS} for a device with L_{CH} of 100 nm shows the peak $G_{m,ext}$ of 244 μS/μm at V_{DS} of -0.5 V.

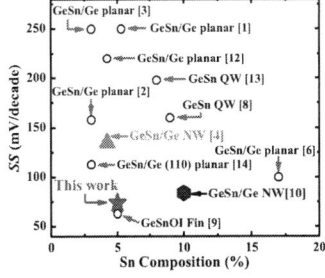

Fig. 12. The benchmark of SS vs. Sn compositions shows the smallest SS achieved in this work among all the reported GeSn nanowire p-FETs.

978-1-7281-0943-5/19 $31.00 © 2019 IEEE

Comparison of Vertically Double Stacked Poly-Si Nanosheet Junctionless Field Effect Transistors with Gate-all-around and Multi-gate Structure

Meng-Ju Tsai, Kang-Hui Peng, Yu-Ru Lin, and Yung-Chun Wu*

Department of Engineering and System Science, National Tsing Hua University, Hsinchu, Taiwan
*E-mail: ycwu@ess.nthu.edu.tw

ABSTRACT

This work demonstrates the double-stacked nanosheet (NS) p-channel vertically junctionless field-effect transistors (VJ-FET) with gate all around (GAA) structure. The stacked NS device shows superior electrical properties, including high Ion/Ioff ratio ($>10^8$), subthreshold slope (SS) =100 mV/dec, and normally off at Vg=0V. More, this work also discusses three different gate structure in electrical properties. The 3D TCAD simulation is applied for analysis of physical characteristics of the proposed devices.

INTRODUCTION

Junctionless-FET with high and uniform doping in channel and source/drain regions have much attraction [1]. Also, it can avoid complex source and drain doping engineering, has a low thermal budget for flexible high-k metal gate adoption and it is easily integrated for 3D IC and NAND [2], [3]. Recently, the stacked nanosheet FET can enhance Ion and device's performance, and still keep the same layout footprint. Therefore, it highly promises that it can extend FET to sub-5 nm nodes [4], [5].

FABRICATION

Initially, a 400 nm SiO2 layer was deposited on 6-inch bulk silicon wafers. Then, an amorphous silicon (α-Si) layer was deposited. After solid-phase recrystallized (SPC) process, the first poly-Si layer was implanted with 35-keV boron difluoride (BF2) at 2×10^{14} cm^{-2}, followed by annealing. A 30 nm-thick SiO2 layer deposited on first poly-Si layer. Next, second active layer was formed with SPC process. Then same implanting with first active layer and furnace annealing. The channel of nanosheets were defined by electron beam lithography (EBL) and reactive ion etching (RIE). Subsequently, the inter layer and the buried oxide were etched by dilute HF for Omega-gate and pure HF for GAA. A 8nm-thick thermal oxide was deposited by LPCVD as the gate oxide layer, and a 200 nm-thick in-situ doped N$^+$ poly-silicon was patterned by EBL and RIE as a gate electrode. Then, oxide layer was deposited as passivation and Al-Si-Cu metallization was performed. Finally, all devices were sintered at 400°C for 30 minutes for repairing of dangling bond.

RESULTS AND DISCUSSION

The double-stacked NS VJ-FET devices process flow are illustrated in Fig. 1, and Fig. 2(a-c) shows the TEM image of tri-gate, omega-gate, and GAA with critical dimensions. Fig. 3 presents the plots of transfer I_D–V_G characteristics of the double stacked NS VJ-FET with Tri-gate and Omega-gate at V_D= - 1V. The I_{ON}/I_{OFF} is 1.5×10^7 for the stacked NS VJ-FET with Omega-gate and is 7.6×10^6 for the stacked NS VJ-FET with Tri-gate. The sub-threshold swing (SS) of Omega-gate and Tri-gate is 149 mV/dec. and 187 mV/dec., respectively. The threshold voltage (V_{TH}), which extracted by constant-current at $I_D = 10^{-8}$A, are -0.24V and -0.16 V for Omega-gate and Tri-gate. Obviously, the stacked NS VJ-FET with Omega-gate has

great gate controllability to improve SS and enhance current drivability. Fig. 4 statistical scheme shows DIBL and SS with Omega-gate and Tri-gate at L_G=300, 600, 800, and 1000 nm. Because of the less electric field at channel edge to suppress barrier induced, (a) V_{TH} of p-type transistor gradually changes to positive value from L_G= 1000 nm to 300 nm in both Omega-gate and Tri-gate. The SS increases from 149 to 251 mV/dec. for Omega-gate and increases from 187 to 305 mV/dec. for Tri-gate, while the gate length is shortened from 1000 to 300 nm. Also, the SS increases continuously with temperature in both the stacked NS VJ-FET. Fig. 5(a) plots statistical SS value of the stacked NS VJ-FET at various temperatures from 50 ℃ to 200 ℃ and at L_G= 1000 nm, V_D= - 1V. Besides the V_{TH} value is changed from negative to positive in p-type stacked NS VJ-FET with Omega-gate and Tri-gate in high temperature as shown in Fig. 5(b). Fig. 6 shows experimental I_D-V_G characteristics of double-stacked GAA NS VJ-FET. The I_{ON}/I_{OFF} is 2×10^8, SS is 100 mV/dec., and V_{TH} is extracted by constant-current at $I_D = 10^{-8}$A, is -1.3 V. Fig. 7(a) shows the statistical SS values of GAA NS VJ-FET of 20 devices with different L_G. This proposed device has very low I_{off}, and steep SS~ 100 V/dec, owing to extremely robust GAA control of NS channels. For more understanding the performance double stacked NS VJ-FET, 3D TCAD simulation [6] was applied. Figure 8(a) illustrates simulated NS VJ-FET structure as the same dimension of the experimental device. Double-stacked NS VJ-FET simulation results of hole concentration are shown in (b) on-state, and (c) off-state. The results confirm the truth of body current for both I_{ON} and I_{OFF} as well as the GAA shows the strongest gate controllability. In addition, the GAA reveals the lowest hole density, which is capable of being fully depleted with junctionless device in off state and the GAA with NS structure ensures that low leakage current (I_{off}) at Vgs = 0V. Comparison of the I_D-V_G curves of double stacked NS VJ-FET with L_G=1000nm and V_D= -1V are shown as Fig. 8(d). Same as the experiment, the GAA of stacked NS VJ-FET shows the highest I_{ON}, lowest I_{OFF} and lowest SS.

CONCLUSION

This stacked NS VJ-FET is fabricated through a simple process and showing high electrical performance. Therefore, this proposed device is high potential for the monolithic multilayer 3D stacked integrated circuit applications.

ACKNOWLEDGEMENTS

This work was supported in part by the Ministry of Science and Technology, Taiwan, under Contract MOST 107-2221-E-007-059-, and in part by the Taiwan National Nano Device Laboratories.

REFERENCE

[1] J. P. Colinge et al., Nature Nanotech., vol. 5, pp. 225, 2010.
[2] S.-Y. Kim et al., VLSI, p.115, 2014.
[3] H .-T. Lue, et al., VLSI, p.152, 2013.
[4] L. Chen, et al., EDL, p. 1256-1258, 2017.
[5] N. Loubet, et al., VLSI, p. 1 T230-T231, 2017.
[6] Sentaurus TCAD Ver. 2015, Synopsys, CA, USA.

978-1-7281-0943-5/19 $31.00 © 2019 IEEE

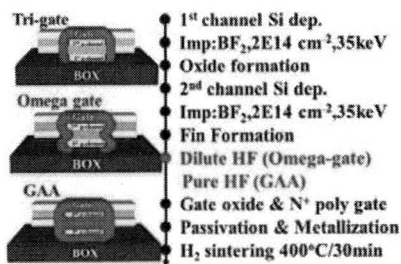

- 1st channel Si dep.
- Imp:BF_2,2E14 cm^2,35keV
- Oxide formation
- 2nd channel Si dep.
- Imp:BF_2,2E14 cm^2,35keV
- Fin Formation
- Dilute HF (Omega-gate)
- Pure HF (GAA)
- Gate oxide & N^+ poly gate
- Passivation & Metallization
- H_2 sintering 400°C/30min

Fig. 1. Process flows of double-stacked NS VJ-FET.

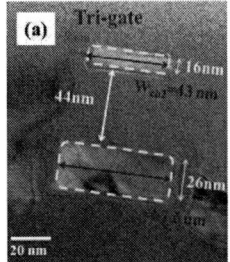
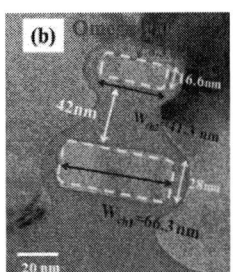
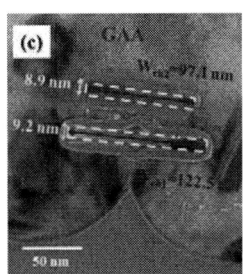

Fig. 2. TEM image of (a) tri-gate, (b) omega-gate, and (c) GAA with double-stacked structure NS VJ-FET.

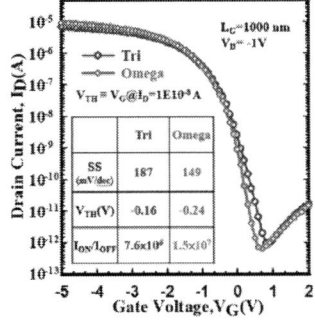

Fig. 3. Experimental I_D-V_G characteristics of Tri-gate and omega-gate NS VJ-FET.

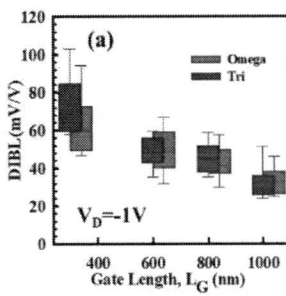
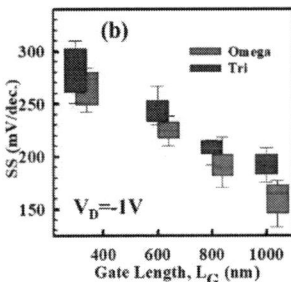

Fig. 4. Statistical (a) DIBL, and (b) SS of Tri-gate and omega-gate NS VJ-FET of 20 devices with different L_G.

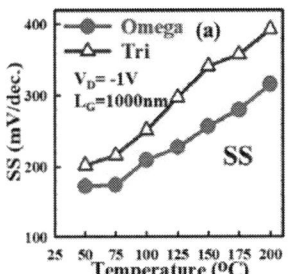

Fig. 5. Statistical (a) SS of both structures with different temperature.

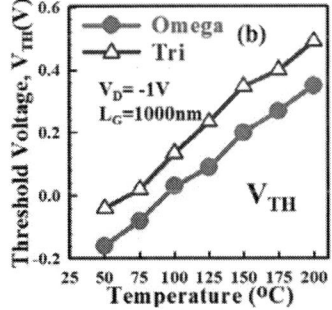

Fig. 5. Statistical (b) V_{TH} of both structures with different temperature.

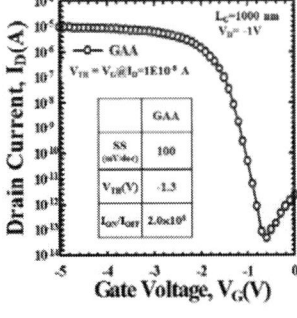

Fig. 6. Experimental I_D-V_G characteristics of GAA NS VJ-FET.

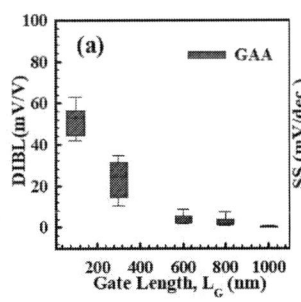
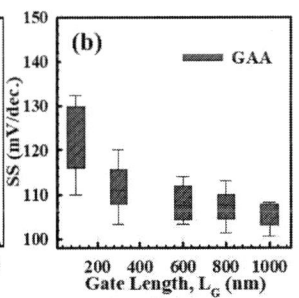

Fig. 7. Statistical (a) DIBL, and (b) SS of GAA NS VJ-FET of 20 devices with different L_G.

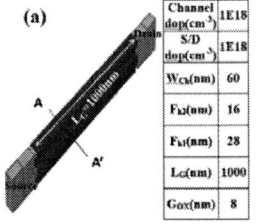
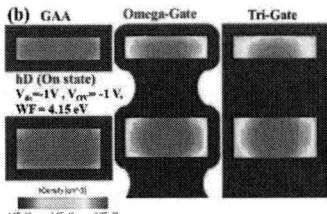
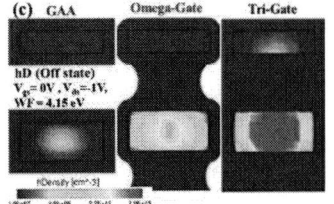
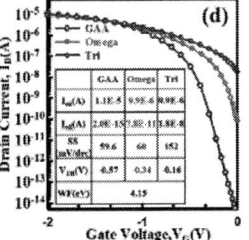

Fig. 8. (a) double stacked NS VJ-FET structure for 3D TCAD simulation. Device hole concentration in (b) on-state, and (c) off-state of GAA, omega-gate, and tri-gate. (d) simulated I_D-V_G for above structures.

978-1-7281-0943-5/19 $31.00 © 2019 IEEE

Virtual Source based I-V Model for Cryogenic CMOS Devices

Hazem Elgabra[*], Brandon Buonacorsi[*], Christopher Chen[**], Jeff Watt[**], Jonathan Baugh[*], Lan Wei[*]

[*]The University of Waterloo, Waterloo, Canada [**]Intel Programmable Solutions Group, San Jose, USA

I. INTRODUCTION

Cryogenic CMOS has drawn increasing attention in the recent years as a potential candidate to control quantum bit (qubit) devices for scalable quantum computing systems. In such systems, the control electronics is expected to operate in the proximity of the qubits at extremely low temperature (e.g. ≤ 4 K). While MOSFET devices from commercial technologies would still function as transistors at cryogenic temperatures, their behaviors deviate from what are expected by the current models developed and characterized for higher temperatures. In this paper, we extend the virtual-source model [1] to include important phenomenon at cryogenic temperatures, such as increase of subthreshold slope. The model shows good agreement with measured data from foundry 65-nm process devices.

II. VIRTUAL SOURCE MODEL

The MIT virtual source (VS) model is a simple model for short-channel MOSFETs that self-consistently describes the current and charge behavior of the transistors [1]. The model requires only a few fitting parameters, most of which have a physical significance. The model has been proven to work well for bulk and silicon-on-insulator (SOI) MOSFET [1]-[3], III-V transistors [4] and has also been extended to graphene [5]-[6], carbon-nanotube [7], and Gallium-nitride transistors [8]. In [9], the basic VS model has been further upgraded to account for drain-dependent gate charge and carrier density dependent velocity with a few more fitting parameters. These effects are non-trivial in the devices with low carrier effective mass and carrier degeneracy (e.g. III-V devices). However, since the bulk transistors measured in this work are not expected to show significant impacts from these effects, we choose to use the original VS model as the starting point for simplicity and fewer fitting parameters.

In a nutshell, the VS model uses the charge-sheet approximation to describe the current as the product of the charge density $\left(Q_{ix_o}\right)$ and local carrier velocity $\left(v_{x_o}\right)$ at the top of the barrier, or "virtual source" as described in eq. (1) in Fig. 1. The figure describes the basic concept of the VS model with its key equations. In bulk devices where the channel carriers are not degenerated, v_{x_o} is nearly independent of V_{DS} and V_{GS}. Q_{ix_o} is approximated by eq. (2), where the fermi transition function $\left(F_f\right)$, given by eq. (3), provides smooth transition between subthreshold and above threshold regions. To describe the current in the triode region, the saturation function (F_s), given by eq. (4), is employed to smoothly transition between 0 (at intrinsic drain-to-source voltage, $V'_{DS} = 0$) and 1 (at $V'_{DS} \geq V_{DSAT}$) using transitioning factor (β), where the saturation voltage V_{DSAT} is described by eq. (5).

III. MODELING AND VERIFICATION

There are several recent attempts to extend conventional MOSFET compact models (e.g. PSP-based [10], EKV-based [11] to cryogenic temperatures). The work presented in this paper is an improved VS model with expanded temperature range to include cryogenic temperatures. The model can be easily used to simulate a wide variety of MOSFET technologies which makes it a great tool for performance projection and technology benchmarking/selection for applications like quantum computing. To demonstrate the cryogenic effects and their simulation, foundry 65-nm NMOS and PMOS devices were measured at 300 K and 4 K. The devices' width, channel length, and gate capacitance are 4.8 μm, 65 nm, and 1.67×10^{-6} F/cm², respectively. The measurement setup introduces probe resistance of around 300

Ωμm (NMOS)/700 Ωμm (PMOS) added to the source and drain series resistance (R_S).

The main addition to the existing VS model is addressing the unexpected change in the subthreshold swing (SS) shown in Fig. 2. Predicted from the classic thermionic emission model, SS proportional to temperature (T), should see a decrease of 75x from 300 K to 4 K. However, the measurement shows a mere ~5 times decrease (Fig. 3) in SS from 300 K to 4 K. This is because the temperature dependence of $SS(T) = \ln(10) nk_BT/q$ is not solely due to T, but also due to the subthreshold factor n. This was previously noticed by [10] and [12], where it is explained by the effect of interface states charges (N_{it}), which starts affecting n when temperature drops below 77 K, adding qN_{it}/C_{ox} to the well-known expression $n = 1 + C_d/C_{ox}$ as shown in eq. (8). In the VS model an extra term of $n_dV'_{DS}$ is also added to n to account for punch-through (eq. (6)). The key contribution in this work is analyzing and modeling this phenomenon to bring the VS model closer into accurately simulating the cryogenic effects on MOSFET devices. (See Figs. 1 and 3 for equations and parameter definitions).

Closer inspection of the SS at 4 K reveals that the slope is not as constant as seen in the 300 K case, which signifies possible N_{it} dependence on bias voltages due to quantum capacitance. This can be explained as the bias potentials cause the interfacial states to become occupied or freed. Measured n values are plotted against V_{GS}, V_{DS}, and T as demonstrated in Fig. 3, highlighting the dependence on bias voltages and conditional dependence on temperature. To the first order, N_{it} increase exponentially with V_{GS} and the ripple-shape dependence of N_{it} on V_{GS} is attributed to quantum capacitance. However, in our model, we find that an empirical exponential function of N_{it} fits the subthreshold behavior well enough and it avoids introducing iterative calculation to account for quantum capacitance. With the assumption that N_{it} is completely responsible for this behavior, this phenomenon is modeled according to eqs. (7) and (8) in Fig. 3. Two additional fitting parameters, $N_{it_{inv}}$ (interfacial charge density at onset of inversion) and η (V_{GS} dependence fitting parameter) are added. In strong inversion, N_{it} is capped at $N_{it_{inv}}$, as the traps would be filled or shielded in this region.

Despite this addition, the low temperature model (e.g. ~4 K) still exhibits numerical singularity and discontinuity due to the very low ϕ_T values, especially when it is present in the denominator of an exponential function like F_f. Since, n greatly increases at low T, the empirical fitting parameter α is changed to $n\alpha$, without affecting high temperature simulations where $n \approx 1$. Finally, the modified VS model is used to model 65-nm CMOS devices showing good agreement with measured data, without compromising fittings at high temperature. This is shown in Figs. 4 and 5 where the model parameters are listed in Table 1 (the probe resistance is not included in R_S, and V_{T0} is the threshold voltage at $V_{DS} = V_{BS} = 0$).

CONCLUSION

A model for the effect of interfacial capacitances in the subthreshold regime at cryogenic temperatures is developed and embedded into the simple VS model. The functionality of the model is verified against sample data from foundry 65-nm devices measured at 300 K and 4 K with good agreement. This improvement is an important step towards realizing cryogenic circuit simulations and design.

References: [1] A. Khakifirooz et al., *IEEE TED*, vol. 56, no. 8, pp. 1674-1680, 2009. [2] L. Wei et al., *IEEE TED*, vol. 59, no. 5, pp. 1263–1271, 2012. [3] A. Majumdar et al., *IEEE TED*, vol. 61, no. 2, pp. 351–358, Feb 2014. [4] D. Kim et al., *IEEE IEDM*, Dec 2009. [5] S. Rakheja et al., *IEEE IEDM*, 2013. [6] S. Rakheja et al., *IEEE TNANO*, 2014. [7] C. -. Lee et al., *IEEE TED*, vol. 62, no. 9, pp. 3061-3069, Sept. 2015. [8] U. Radhakrishna et al., *IEEE IEDM*, San Francisco, CA, 2014.

[9] S. Rakheja et al., *IEEE TED*, vol. 62, no. 9, pp. 2786-2793, Sept. 2015. [10] R. M. Incandela et al., *IEEE J-EDS*, vol. 6, pp. 996-1006, 2018. [11] A. Beckers et al., *EUROSOI-ULIS*, Granada, 2018, pp. 1-4. [12] I. Hafez et al., *JAP*, vol. 67, no. 4, pp. 1950-1952, 1990.

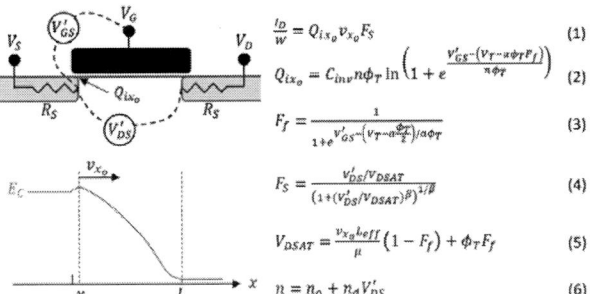

$$\frac{I_D}{W} = Q_{ix_o} v_{x_o} F_S \tag{1}$$

$$Q_{ix_o} = C_{inv} n\phi_T \ln\left(1 + e^{\frac{v'_{GS}-(v_T-\alpha\phi_T F_f)}{n\phi_T}}\right) \tag{2}$$

$$F_f = \frac{1}{1+e^{\frac{v'_{GS}-(v_T-\alpha\frac{\phi_t}{2})/\alpha\phi_T}}} \tag{3}$$

$$F_S = \frac{v'_{DS}/V_{DSAT}}{\left(1+(v'_{DS}/V_{DSAT})^\beta\right)^{1/\beta}} \tag{4}$$

$$V_{DSAT} = \frac{v_{x_o} L_{eff}}{\mu}(1-F_f) + \phi_T F_f \tag{5}$$

$$n = n_o + n_d V'_{DS} \tag{6}$$

Fig. 1. Visualization of the VS model where the VS velocity v_{x_o} and VS charge density Q_{ix_o} are defined at x_o. W is transistor width. C_{inv} is the effective gate-to-channel capacitance in strong inversion, n is the subthreshold factor, and ϕ_T is equal to $k_B T/q$, where k_B is Boltzmann constant, T is temperature in Kelvin, and q is the single electron charge. V'_{GS} and V'_{DS} are the internal gate-to-source and drain-to-source voltages, respectively, excluding the voltage drop across the source/drain series resistances R_S. V_T is the threshold voltage after accounting for body effect and drain-induced-barrier-lowering (DIBL). The term $\alpha\phi_t F_f$ models the shift of V_T between strong and weak inversion the empirical factor α, C_d is the depletion capacitance and V_{DSAT} is the saturation voltage which varies between weak ($V_{DSAT} \approx \phi_t$) and strong inversion ($V_{DSAT} \approx v_{x_o} L_{eff}/\mu$), where L_{eff} is the effective channel length and μ is the low field mobility. Finally, n_o is the portion of n due to depletion capacitance while n_d is the punchthrough factor.

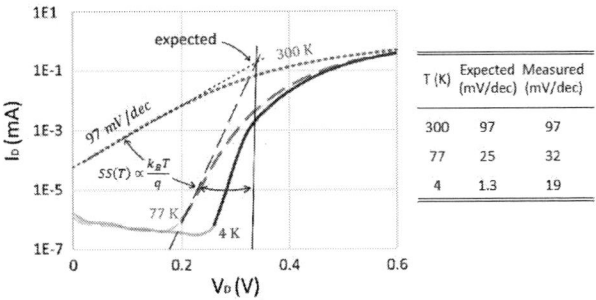

Fig. 2. Measured transfer characteristics (thick lines) of a PMOS at $V_{DS} = 1\ V$ and 300 K (dotted), 77 K (dashed), and 4 K (solid). Expected SS (thin lines) with purely thermionic emission assumption (i.e. $SS \propto T$) are drawn based on the 300 K measurements. The table shows of measured SS at current levels 10^{-10} to 10^{-8} for 4 K and 77 K and 10^{-8} to 10^{-6} at 300 K alongside expected SS.

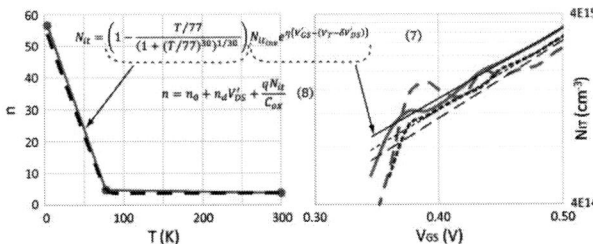

$$N_{it} = \left(1 - \frac{T/77}{(1+(T/77)^{30})^{1/38}}\right) N_{it_{inv}} e^{\eta(v'_{GS}-(v_T-\delta V'_{DS}))} \tag{7}$$

$$n = n_o + n_d V'_{DS} + \frac{q N_{it}}{C_{ox}} \tag{8}$$

Fig. 3. (a) Measured subthreshold factor n (solid with symbol) from Fig. 2, and its modeled temperature dependence (dashed). (b) Measured non-ideality factor n at $V_{DS} = 50$ mV (dashed), 0.8 V (dotted), and 1 V (solid) of NMOS at 4 K (thick lines), and their exponential approximation (thin lines). C_{ox} is the gate capacitance and δ is the DIBL coefficient. $N_{it_{inv}}$ and η are the effective interface traps at onset of inversion and a fitting parameter, respectively, obtained by fitting the model to experimental data.

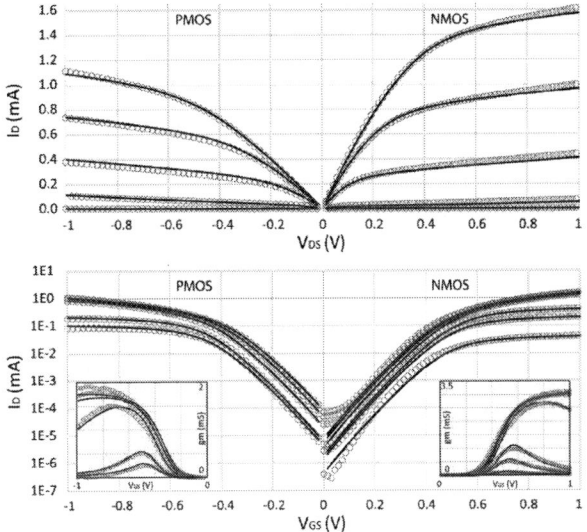

Fig. 4. Measured (symbols) and simulated (line) $I_D V_{DS}$ (top), $I_D V_{GS}$ (bottom), and g_m-V_{GS} (bottom insets) characteristics of a 65 nm NMOS and PMOS at 300 K. The voltage bias for $I_D V_{DS}$ and $I_D V_{GS}$ are $|V_{GS}|$ = (0.2, 0.4, 0.6, 0.8, and 1) and $|V_{DS}|$ = (0.01, 0.05, 0.1, 0.5, 0.8, 1), respectively.

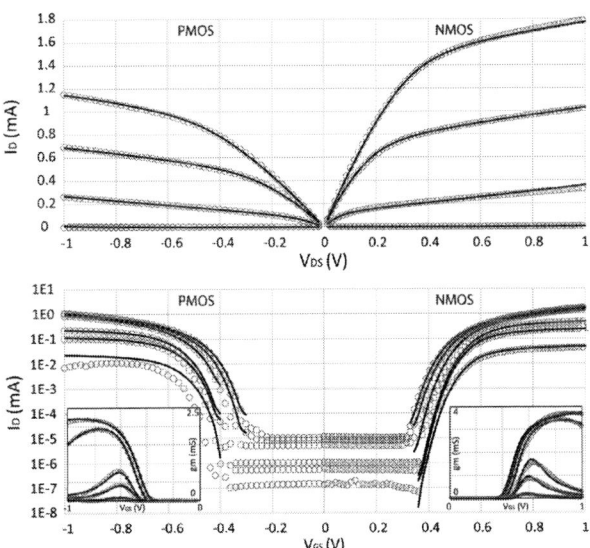

Fig. 5. Measured (symbols) and simulated (line) $I_D V_{DS}$ (top), $I_D V_{GS}$ (bottom), and g_m-V_{GS} (bottom insets) characteristics of a 65 nm NMOS and PMOS at 4 K. The voltage bias for $I_D V_{DS}$ and $I_D V_{GS}$ are $|V_{GS}|$ = (0.2, 0.4, 0.6, 0.8, and 1) and $|V_{DS}|$ = (0.01, 0.05, 0.1, 0.5, 0.8, 1), respectively.

TABLE I
MODEL PARAMETERS FOR THE MEASURED FOUNDRY 65-NM MOSFETS AT 300 K AND 4K.

Type	T (K)	R_S ($\Omega\mu$m)	μ (cm²/Vs)	V_{t0} (mV)	DIBL (mV/V)	n_o	v_{xo} (1e7 cm/s)	n_d	$N_{it_{inv}}$ (1e6 cm⁻³)	η
NMOS	300	153	180	586	59	1.5	0.68	0.00	2.38	2.43
	4	84	268	626	87	1.5	0.81	0.00	2.38	2.43
PMOS	300	250	84	536	107	1.5	0.46	0.03	3.39	2.59
	4	120	122	673	106	1.5	0.73	0.40	3.39	2.59

978-1-7281-0943-5/19 $31.00 © 2019 IEEE

Core-shell TFET Developments and TFET Limitations

M. Passlack, P. Ramvall, T. Vasen, A. Afzalian, C. Thelander*,
K.A. Dick*, L.-E. Wernersson*, G. Doornbos, and M. Holland

TSMC Corporate Research, Belgium
*Lund University, Sweden
E-mail: matthias_passlack@tsmc.com

ABSTRACT

Tunneling field-effect transistors (TFET) based on a vertical gate-all-around (VGAA) nanowire (NW) architecture with a core-shell (CS) structure have been explored for future CMOS applications. Performance predictions based on a tight-binding mode-space NEGF technique include a drive current I_{on} of 6.7 μA (NW diameter d = 10.2 nm) at V_{dd} = 0.3 V under low power (LP) conditions (I_{off} = 1 pA) for an InAs/GaSb CS TFET. This compares to Si nMOSFET I_{on} = 2.3 μA at V_{dd} = 0.55 V (d = 6 nm). On the experimental side, scaling of vertical CS NWs resulted in smallest dimensions of d_c = 17 nm (GaSb core) and t_{sh} = 3 nm (InAs shell) for a total diameter of 23 nm. VGAA CS nFETs demonstrated drive current of up to 40 μA (V_d = 0.3 V) and subthreshold swing SS = 40 mV/dec (V_d = 10 mV) for NW diameters between 35 - 50 nm. Although key TFET properties such as current drive and subthermal SS have been demonstrated using a VGAA CS architecture for the first time, experimental results still lag predictions. An intrinsic relationship between band-to band-tunneling (BTBT) and D_{it} related trap assisted tunneling (TAT) was found which imposes challenging D_{it} requirements, in particular for LP I_{off} specifications. Complexity of fabrication and a material system foreign to CMOS manufacturing further impact prospects of TFET technology.

INTRODUCTION

TFETs have been considered as a candidate for future CMOS technologies due to their fundamental potential of subthermal SS [1]. High drive current I_{on} requires an extremely short BTBT distance and small effective tunneling mass. Both requirements favor broken or staggered band alignment in a III-V material system such as GaSb/InAs. In the past, axial TFET architectures were widely explored [2] while reports on VGAA CS architecture TFETs have been sparse because of more complex manufacturing requirements.

However, CS architectures have been speculated and recently predicted [3] to exhibit substantially enhanced electrical properties compared to their axial counterparts due to more efficient gate control of the tunnel junction and electrostatics independent of NW (core) diameter (refer to Fig. 1 which illustrates the structure and the electron flow in a VGAA CS architecture). These attributes warranted further exploration.

EXPERIMENT AND MODEL

CS GaSb/InAs NWs were grown by metal organic chemical vapor deposition (MOCVD) using the vapor liquid solid method from Au particle catalysts on an InAs substrate. Fig. 2a shows a high-resolution TEM image overlaid with HAADF/EDX data of a CS InAs/GaSb NW with the smallest features size obtained (d_c = 17 nm, t_{sh} = 3 nm). The SEM image in Fig. 2b shows a group of 4 NWs with d_c = 35 nm and t_{sh} = 3 nm grown in predefined locations which are used for VGAA TFET manufacturing. Key fabrication steps include (i) ALD of a high-k dielectric layer of AlO$_x$ of 5 - 8 nm, (ii) sputter deposition of 35 nm TiN, (iii) resist planarization and O$_2$ thinning to reveal the top of the NW, (iv) wet etching of TiN (Fig. 3a shows a SEM image after this step), (v) formation of TiN gate pads, (vi) wet etching of the high-k layer, (vii) digital wet etching of the InAs shell, (viii) ALD of 5 nm AlO$_x$ to protect the revealed GaSb segment, (ix) resist planarization and patterning to open vias to the gate and substrate, (x) resist hard baking and thinning to 50 nm above the TiN gate edge (top spacer between the gate and top contact), (xi) etching of the protective AlO$_x$ layer, (xii) top contact deposition of 10 nm Ni, 5 nm Ti, and 200 nm Au and pattering of top contact pads. Fig. 3b shows a fabricated VGAA NW GaSb/InAs CS TFET with GaSb core diameter of 53 nm and InAs shell of 4.5 nm.

Fig. 4 shows I_d-V_d characteristics of a CS nTFET with GaSb core diameter of 35 nm, InAs shell thickness of approximately 4 nm, and a gate length L_g of 35 nm. Typical FET operation is observed with

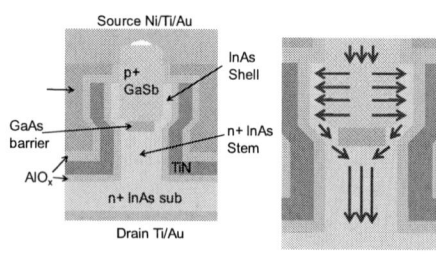

FIGURE 1. (left) Schematics of core-shell TFET architecture. (right) Illustration of electron flow from source to drain in a VGAA CS architecture.

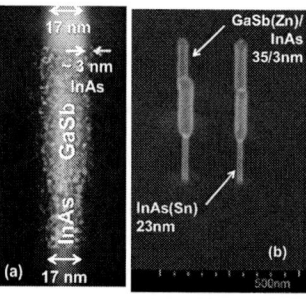

FIGURE 2. (a) HRTEM image overlaid with HAADF/EDX data of smallest grown CS NW. (b) SEM image of a group of CS nanowires grown in predefined locations for device fabrication.

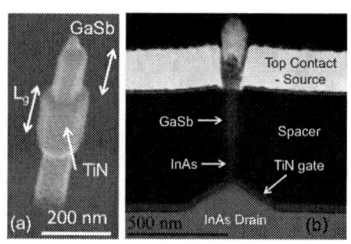

FIGURE 3. (a) CS NW after etching of TiN on top portion of NW. (b) CS NW after completion of device fabrication.

978-1-7281-0943-5/19 $31.00 © 2019 IEEE

minimum superlinear onset. Fig. 5 shows I_d-V_g characteristics in subthreshold regime of a similar device. The steepest measured subthreshold swing is approximately 40 mV/dec. Hysteresis is also shown and is approximately 10 mV where S = 40 mV/dec. Distinct steps are observed which may be attributed to oxide defects. The minimum average swing over one decade in current is degraded from 42 to 56 mV/dec due to the presence of a step. Fig. 6 shows I_d-V_g (linear scale) for V_d = 0.5, 0.3, 0.15, and 0.05 V for a device with the highest measured current drive up to 40 µA/NW and 60 µA/wire at significant gate overdrive for V_d = 0.3 and 0.5 V, respectively. The device has a GaSb core diameter of approximately 50 nm. Further

details can be found in [4].

The presented data establish the potential of this technology to obtain high current drive and subthermal swing *in principle*. However, the ability of high current drive is associated with substantially increased SS in the experimental data, an observation which points to an intrinsic relationship between effectiveness of BTBT on one hand and D_{it} related TAT on the other hand. Fig. 7 demonstrates fitting of a measured CS TFET I_d-V_g below threshold using the simulator Sentaurus Device [5]. An interface state density $D_{it} = 1.4 \times 10^{13}$ cm^{-2}eV^{-1} and a capture cross section $\sigma = 10^{-14}$ cm^2 are used in the simulations. The intrinsic BTBT - D_{it}/TAT relationship and implications for a practical TFET technology will be further elucidated at the conference.

Finally, the performance potential of the TFET CS architecture is outlined. I-V curves are predicted based on a tight-binding mode-space NEGF technique. Table I summarizes simulated CS data, compares the data to axial TFET data in an identical material system, Si FET data, and future IRDS targets [6].

	VGAA NW		InAs/GaSb VGAA nTFET	
	Si nFET	IRDS	Core-shell	Axial
d (nm)	6.0	6.0	10.2	5.5
V_{dd} (V)	0.55	0.55	0.30	0.30
I_{off} (pA)	0.6	0.6	1.0	0.6
I_{on} (µA)	2.3	3.8	6.7	1.0

TABLE I. Drive currents predicted using a tight-binding mode-space NEGF technique for CS and axial InAs/GaSb VGAA nTFETs and Si VGAA NW nMOSFETs compared to IRDS year 2033 targets.

REFERENCES

[1] U.E. Avci et al., *Energy efficiency comparison of nanowire heterojunction TFET and Si MOSFET at L_g = 13 nm, including P-TFET and variation considerations.* IEEE International Electron Devices Meeting, p. 33.4.1-33 (2013).

[2] X. Zhao et al., *Sub-Thermal Subthreshold Characteristics in Top-Down InGaAs/InAs Heterojunction Vertical Nanowire Tunnel FETs.* IEEE Elec. Dev. Lett., vol. 38, p. 855 (2017).

[3] A. Afzalian et al., *A High Performance InAs/GaSb Core-Shell Nanowire Line Tunneling TFET: An Atomistic Mode Space NEGF Study.* IEEE Journal Elec. Dev. Soc., 2018.

[4] T. Vasen et al., *Vertical Gate-All-Around Nanowire GaSb-InAs Core-Shell n-Type Tunnel FETs.* Scientific Reports, 2019.

[5] *Sentaurus Device*, version m-2016.12.

[6] https://irds.ieee.org/roadmap-2017

ACKNOWLEDGEMENT

The authors acknowledge support from the technical staff at Lund University Nanolab, and IMEC, Leuven, Belgium. We thank T.-M. Shen, Y.-C. Yeo, C.H. Diaz, J. Wu, Y.C. Sun (all TSMC) and L. Samuelson (Lund University) for support.

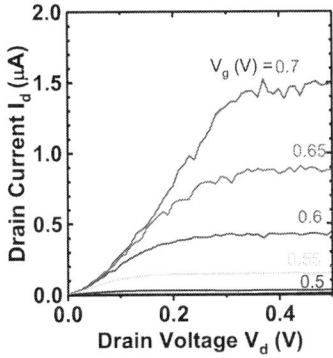

FIGURE 4. I_d-V_d of a CS nTFET with GaSb core diameter of 35 nm, InAs shell thickness of approximately 4 nm, and a gate length L_g of 35 nm.

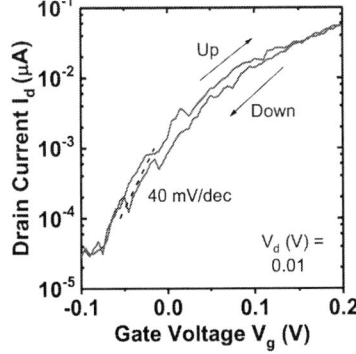

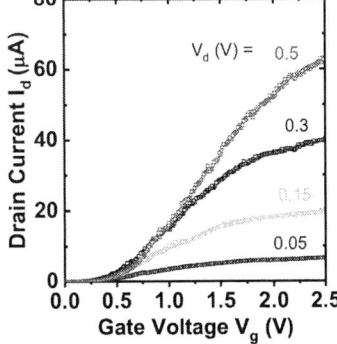

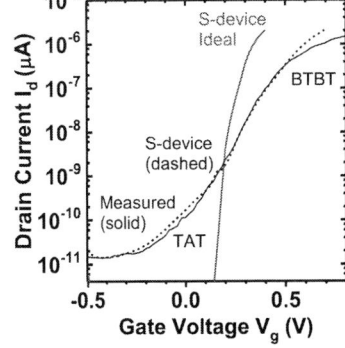

FIGURE 5. I_d-V_g at V_d = 10 mV for the CS nTFET with the steepest measured subthreshold swing of approximately 40 mV/dec.

FIGURE 6. I_d-V_g for V_d = 0.5, 0.3, 0.15, and 0.05 V for a CS nTFET with the highest measured current drive up to 40 µA/NW at V_d = 0.3 V. The high required gate overdrive is indicative of high D_{it}.

FIGURE 7. Measured (solid line), fitted model (dashed line), and ideal (D_{it} = 0) I_d-V_g of CS nTFET. The stretch-out of the experimental curve in the subthreshold region is due to TAT originating from D_{it}.

978-1-7281-0943-5/19 $31.00 © 2019 IEEE

3D Integration

Piyush Gupta

Qualcomm

Three trends are playing out actively:

1) Silicon Process Node Adoption is slowing down and requiring higher investment and even less vendors are investing.

2) Number of wireless end markets are moving beyond mobile (Industrial, wearables, AI, gaming, consumer IOT) and each market requires it's on optimal feature set (no more one size fits all SOC approach).

3) End Device complexity and form factor constraints continue to increase requiring higher and higher OEM investment to bring product to increase (3G —> 4G —> 5G). Moreover, increasing functionality is impeding on battery and/or device size.

We could try and solve each of these challenges separately. Instead there is synergy with them through a technology path of package and silicon 2.5/3D integration where we can tackle all three together. The path is through heterogenous silicon integration through various package level integrations (RDL transitions, silicon to silicon, etc...). Above enables new product creation through silicon partition upgrades to meet market requirements. We will explore these in detail.

Advanced Stacking Technologies for Heterogeneous Device Integration

Ionut RADU

SOITEC

Vertical (3D) integration creates new paths for functional diversification of advanced VLSI. Several applications can fully take the advantage of using the monolithic 3D stacking technology. This talk will present an overview of advanced wafer-level stacking of single crystal films enabling 3D monolithic integration of electronic devices. The monolithic stacking technology based on low temperature Smart Cut[TM] leads to the front end integration of large variety of devices with nanometer alignment capability; therefore it provides more degree of freedom for the designers and integration for high density and better performance.

High Density W2W and D2W DBI® Hybrid Bonding for Stacked Applications

Sitaram Arkalgud

Xperi Corporation
3025, Orchard Parkway, San Jose, CA 95134 USA
E-mail: sitaram.arkalgud@xperi.com

ABSTRACT

Chip and wafer stacking have made major inroads into several semiconductor segments, including CMOS image sensors, MEMS and HPC. As lithographic scaling slows down and functionality requirements increase, stacking becomes a necessity and continues to grow into DRAM, RF and other markets. Wafer or die bonding is an essential part of stacking. While several bonding technologies are used in HVM, microbumps are today's bonded interconnect of choice. However, they face issues with scalability and reliability, together with thermo-mechanical issues introduced by organic underfill and polyimide layers, which only worsen as dies become larger. Xperi's room temperature Direct Bond Interconnect (DBI®) hybrid bonding process [1,2] eliminates microbumps and associated organic layers, resulting in high reliability, low cost and high-density interconnects with improved thermo-mechanical performance. This talk will focus on our DBI hybrid bonding technology and describe the process flow, electrical and reliability results.

INTRODUCTION

Stacking of wafers or die typically requires bonding, thinning, and TSVs for front to back interconnects. Bonding is a critical component, and MEMS is a perfect example where several different bonding technologies are in High Volume Manufacturing today, including fusion, anodic, glass frit, adhesive, eutectic and metal thermocompression bonding [3]. Microbump technology is the current bond interconnect of choice, but its scalability below 40μm pitch is questionable. Hybrid bonding, with its reliance on damascene metallization, results in a cost-effective bonding process with superior thermal, electrical [4] and reliability performance, which can scale below 1μm interconnect pitch [5].

DIRECT BOND INTERCONNECT (DBI®)

The concept of DBI was pioneered by Ziptronix in the early 1990s [1,2]. Table 1 shows an example of a wafer to wafer DBI process, where two metallized CMOS wafers each receive a mirrored, Cu damascene DBI pad layer. The wafer surfaces are activated by a plasma process, following which the two wafers are aligned in a wafer aligner and bonded at room temperature in atmosphere. Subsequently, the wafers are batch annealed at a low temperature (150°C and above) where the dielectric surfaces convert to a strong covalent bond and the differential CTE between the dielectric and the Cu causes the mirrored Cu pads to expand into each other and bond without the need for external pressure, in contrast to metal thermocompression bonding.

While wafer to wafer DBI is in high volume manufacturing, stacked DRAM (HBM and 3DS) and 2.5D are attractive candidate applications for die to wafer DBI stacking. HBM is targeting 12 and 16 high stacks, from its current 4 and 8 high stacks, and DBI offers the elimination of microbumps and underfill, leading to improved electrical, thermal and reliability performance, while also reducing cost and maintaining package height. HPC with 2.5D would benefit from utilizing DBI due to the pitch scalability (higher bandwidth between the processor and memory dies) and improved SI/PI, thermal and reliability performance.

Xperi has also developed a die to wafer DBI process [5] for the above applications, and anywhere else where dies or wafers are dissimilar, or where yield is a major concern. Table 2 provides an overview of the process flow. As with the wafer to wafer flow, the wafers are processed through the damascene DBI pad formation step, following which the wafers are diced. The dies are cleaned and activated on the tape frame, picked by a die bonder and bonded to a wafer or a die. The die stack on the wafer is batch annealed in a similar fashion to the wafer to wafer flow, and then molded and processed through final assembly and test.

Figure 1 shows the cross section of a DBI bond, and Fig 2 shows a 15 high stacked die image from our work at Xperi. We routinely average around 92% yield on daisy chains with 31,356 DBI links in a Class 1000 laboratory cleanroom (much higher yield expected in HVM). Parts passed automotive reliability testing with temperature cycling test (TCT) and high temperature storage (HTS) with no issues. The conditions are shown in Table 3. Figure 3 shows the cross section of a DBI bond after 2000 temperature cycles.

CONCLUSION

In summary, this talk will discuss the importance of bonding in 3D stacking and the cost, scalability, electrical, thermal and reliability benefits of DBI hybrid bonding for CMOS 3D integrated circuits.

REFERENCES

[1] P.Enquist, "High density direct bond interconnect (DBI) technology for three dimensional integrated circuit applications" MRS Proceedings, 970, 0970-Y01-04, 2006.

[2] P. Enquist, G. Fountain, C. Petteway, A. Hollingsworth, "Low cost of ownership scalable copper Direct Bond Interconnect 3D IC technology for three-dimensional integrated circuit applications", 2009 IEEE International Conference on 3D System Integration.

[3] S. Arkalgud, "Application of direct bonding in MEMS", MEMS Engineer Forum, Tokyo, Japan, April 25, 2018.

[4] A. Agarwal, S. Huang, G. Gao, L. Wang, J. DeLaCruz and L. Mirkarimi, "Thermal and electrical performance of direct bond interconnect technology for 2.5D and 3D integrated circuits", IEEE 67th Electronic Components and Technology Conference, pp 989-998, 2017

[5] G. Gao, L. Mirkarimi, G. Fountain, L. Wang, C. Uzoh, T. Workman, G. Guevara, C. Mandalapu, B. Lee and R. Katkar, "Scaling package interconnects below 20mm pitch with hybrid bonding", 2018 IEEE 68th Electronic Components and Technology Conference, pp 314-322, 2018.

Process Step
Incoming CMOS Wafers
Cu DBI pad formation
Plasma Activation
Wafer align and bond
Low temperature anneal
Post processing

Table 1: Wafer to Wafer DBI process flow

Process Step
Incoming CMOS Wafers with DBI pads
Dicing
Clean die on frame
Die pick, align and bond
(Repeat for multilevel stack)
Low temperature anneal

Table 2: Die to Wafer DBI process flow

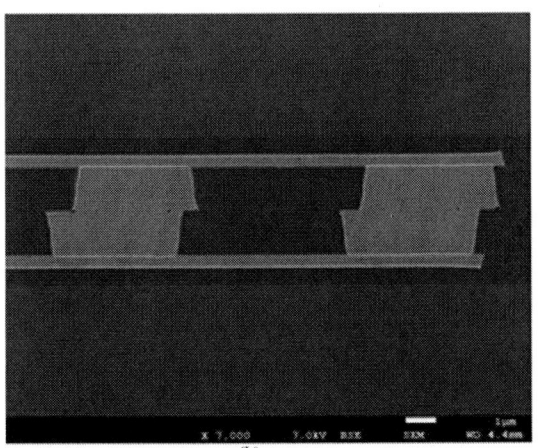

Figure 1: SEM cross section of an "as-bonded" DBI daisy chain with 5μm circular pads on 10μm pitch.

Figure 2: Optical images of 15 dies stacked on a wafer to demonstrate the ease of stacking with DBI.

Test	Standard	Test Condition	Part type	Sample Size	Status of Test Cycle / hrs	Result
Temperature Cycling	JESD22-A104D Condition M	-40°C to 150°C, 1000 cycle	5um pad DBI (31k chain)	45	2000 cycl	100% pass
High Temperature Storage	JESD22-103D	225°C, 1000 hours	5um pad DBI	22	2000 hrs	100% pass

Table 3: Reliability test results with automotive conditions, showing no failures at 2000 cycles (TCT) and 2000 hours (HTS).

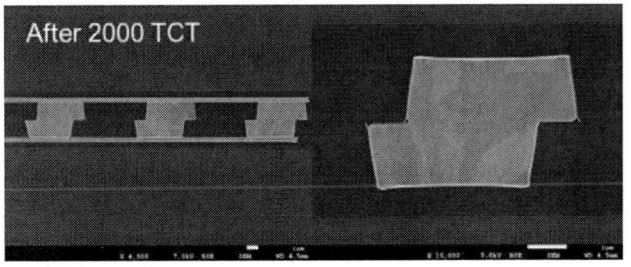

(a)

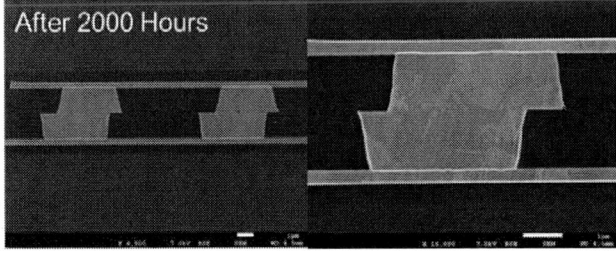

(b)

Figure 3: SEM cross sections of DBI daisy chains, showing no change compared to an "as-bonded" DBI daisy chain, after reliability testing under automotive conditions (a) 2000 temperature cycles and (b) 2000 hours of high temperature storage.

Silicon Photonics as a Post Moore Technology

Koji Yamada

National Institute of Advanced Industrial Science and Technology, Japan
E-mail: yamada.koji@aist.go.jp

ABSTRACT

Silicon photonics is a high-density photonic integration technology based on silicon CMOS fabrication technology. Thanks to the silicon's superior material characteristics both in photonics and electronics, this technology can provide us compact, functional and energy-efficient photonic-electronic convergent circuits. Moreover, superior performance in reliability and volume production of silicon CMOS-based fabrication technology can make photonic circuits significantly cost-effective. Hence, silicon photonics, as a post Moore photonics technology, can provide us immediate solutions to the explosive growth of network traffic and data processing volume in advanced information-oriented society, which must consume an infinite number of photonic and electronic devices and circuits.

INTRODUCTION

Advanced information-oriented societies such as named Industrie 4.0 and Society 5.0, are based on communication network systems and advanced computing systems, such as artificial intelligence. In order to widely develop these advanced information-oriented societies, we must deeply consider environmental and economic issues.

In communication networks, we are now facing an explosive traffic increase. Within ten years, energy consumption and capital/operating expenditures in network systems will increase tenfold or more; however, from environmental and economic viewpoints, such a large and rapid growth is not sustainable. Moreover, data transmission capacity per fiber has now reached its physical limit, or nonlinear Shannon limit[1]. In other words, we are now entering the Post Moore era in data communication technology.

In computing systems, data processing capacity in data centers is explosively increasing with a growth rate of around 100% a year. In order to support such an explosive growth in computing, we need a data interconnection technology with an ultra-large bandwidth, ultra-low power consumption, low latency and low cost. Moreover, in advanced information-oriented society, we need infinite number of intelligent edge computing systems with ultra-low power consumption, ultra-low cost and low latency. However, we are in the Post Moore era.

One of the breakthrough technologies, or Post Moore technology, for dealing with explosion both in communication and computing systems is high-density photonic integration with compact photonic devices. The smaller the device is, the less the power required, which directly reduces the power consumption of data transmission modules. Moreover, integrated photonic functions with redundancy enable flexible and energy-efficient operation of ultra-high volume low-latency network and interconnection[2]. Furthermore, compact photonic circuits can be integrated with electronics circuits, and can improve total performance of information systems. Optically-assisted computing with integrated photonic circuits might provide low latency computing systems. For example, optical path-gate logic architecture is proposed[3]. Photonic neutral networks (PNN) based on integrated photonic circuits also attracts wide attentions as advanced computing systems[4].

As described above, the high-density photonic integration can be a breakthrough technology in advanced information systems, and silicon photonics is one of the most promising photonic platforms for the high-density photonic integration. In silicon photonics, the ultra-fine and highly-reliable fabrication technologies that have matured in the silicon electronics industry are applied to the fabrication of compact photonic devices.

In this article, silicon photonics technology is reviewed as a context of Post Moore technology, or More Moore, More than Moore and novel-principle devices/architecture, or replacement of electronics with photonics.

FEATURES OF SILICON PHOTONICS

Si photonics has the following superior features as an integrated photonic technology. Optically, Si is transparent and shows a large refractive index in the 1200−1700-nm telecommunications wavelength band. Generally, silicon optical waveguides consists of silicon core with an index of ~3.5 and silica (SiO_2) cladding with an index of ~1.45. Since the index contrast at the Si/SiO_2 interface is very large, the optical filed can be tightly confined in the Si core. Therefore, very small waveguide core and very tight bending section can be constructed and make photonic devices very compact. Electrically, Si is a semiconductor, and controls carrier density by applying some electronic device structures. Since carriers in semiconductor can change the refractive index and absorption coefficient of the material, Si provides various active functions, such as optical modulation. Germanium (Ge) devices on Si can also provide photodetectors (PDs) and light sources. Moreover, electronic circuits can be integrated with Si photonic circuits, which would provide intelligence and flexibility into photonics. Physically, Si devices and substrate are robust, and are tough for various types of heterogeneous integration, which provides further functionalities. Of course, the Si platform is very reliable, and has superior capabilities for cost-effective mass production.

POST MOORE APPLICATIONS OF SILICON PHOTONICS

More Moore (geometrical integration): Thanks to the very strong optical confinement ability of silicon photonics technology, photonic circuits can be miniaturized considerably, and data transmission capacity per unit chip area can be increased. A typical example is shown in [5] for an integrated WDM receiver chip, which consists of a multi-channel wavelength filters, PD array, and electrodes for signal output. By using silicon photonics, an arrayed-waveguide-grating type wavelength filter, which is a standard multi-channel wavelength filter, can be miniaturized to 1-mm square. Moreover, Ge PDs and through-silicon via (TSV) electronic wiring can be scattered over the whole the chip. Thus, the required area of a 100-ch WDM receiver chip will be reduced to 1 cm^2, which is 1/100 of the area of chips based on conventional technology. Assuming PD operation at 25 Gbps, total capacity will reach 2.5 $Tbps/cm^2$. In parallel transmission systems, where no wavelength filter is required, total bandwidth can be increased up to 30 $Tbps/cm^2$[6]

978-1-7281-0943-5/19 $31.00 © 2019 IEEE

More than Moore (Functionality integration): Integration with electronics is the most impactful post-Moore technology for photonic circuits. In particular, integration with modulation drivers, transimpedance amplifiers for PDs, and various control circuits on a silicon photonics chip can significantly reduce the size of photonic-electronic integrated modules. Moreover, since electronic circuits can be placed very close to photonic devices, high-frequency performance can be significantly improved[7].

Since silicon photonics is based on silicon electronics technology and the silicon platform is reliable and robust, electronic circuits can be integrated by both monolithic and hybrid approaches, as shown in Fig. 1. The monolithic approach is very attractive from the viewpoint of ultra-high-volume production, and some short-range data transmission modules have already been commercialized[8]. However, we must consider that device performance might be degraded because of narrow margins in the fabrication process for photonics-electronics convergence. For example, in monolithic photonics-electronics convergence, the dark current of germanium PDs is likely to increase. Moreover, typical CMOS electronics technology cannot provide the high-speed electronics required for high-bit-rate optical data transmission.

For improving device performance both in photonic and electronic circuits, the hybrid approach is attractive. Since both types of circuits can be fabricated by their respective optimized fabrication process, high-performance photonics-electronics convergence can be achieved. Hybrid integration is performed by using various wafer-bonding/die-bonding techniques with TSV and micro-solder bump technologies[9]. Since hybrid integration technology is a kind of 3D integration technology, it can also contribute to geometrical integration in the More Moore approach.

Replacement of electronics with photonics: This approach can be seen in the dynamic optical path network (DOPN) using large-scale (32x32) Mach-Zehndar interferometer (MZI) type optical switch matrixes.[2,10] Conventional switching systems for network are made of electronic devices, where numerous small packets are independently routed to a number of unspecified terminals. Thus, power consumption for the switching increases proportionally to the data traffic. Moreover, complicated and frequent packet routing seriously increases the latency. In the DOPN, dedicated optical transmission channels are temporarily constructed by an optical switching matrix. Since packet switching is not utilized in the DOPN, the power consumption for bulky data transmission can be greatly reduced. The latency is also greatly reduced because it is determined only by the transit time through optical circuits. Recently, a DPON field trial using a silicon-photonics-based optical switch matrix has been carried out[2]. Of course, DOPN also contributes much to the power reduction and bandwidth increase in intra-datacenter network.

Optically-assisted computing is also promising approach for advanced computing systems. This approach can be seen in optical path gate logic circuits[3]. Figure 2 shows the schematic diagram the optical path-gate logic circuit for 4-bit adder. The circuit is a kind of cascaded 2x2 MZI switches. An MZI can work as an XOR, and adding carry path from the 1st MZI to the 2nd one, we can construct a unit adder. In this circuit, once setting condition of MZIs electrically, computing is performed only by transmitting light through the optical circuits. The latency per gate can be a few picoseconds, which is significant smaller than those of state of art CMOS gate.

SUMMARY

Silicon photonics is a high-density photonic integration technology based on silicon CMOS fabrication technology. The main features of the technology are compactness, functionality integration, photonics electronics convergence and low-cost volume production.

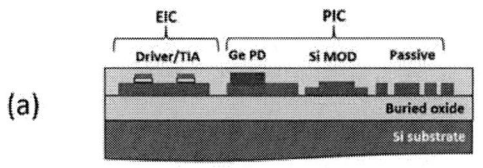

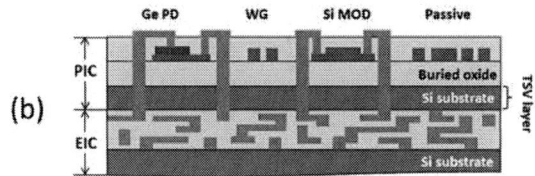

FIGURE 1. PHOTONICS-ELECTRONICS INTEGRATION WITH SI PHOTONICS, (A) MONOLITHIC AND (B) TSV-BASED HYBRID INTEGRATION.

Simultaneous realization of these features means a paradigm shift in

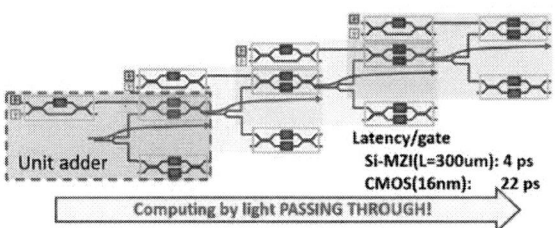

FIGURE 2. SCHEMATIC DIAGRAM OF A 4-BIT ADDER CIRCUIT BASED ON OPTICAL PATH-GATE LOGIC.

photonic technology, and silicon photonics can be a breakthrough in the development of advanced information technology. The development approaches here are similar to those in the Post Moore electronics. As it turns out, paradigm-shift technologies are created through these common approaches and total system performance will continue to improve following Moore's law.

REFERENCES

[1] D.J. Richardson, Science **330**, p.327, 2010.

[2] J. Kurumida et al., the 40th European Conference on Optical Communication (ECOC 2014), Cannes, PDP 1.3, Sept. 2014.

[3] A. Shinya et al., NTT Technical Review 16, pp.33-38, 2018.

[4] Y. Shen *et al.*, Nature Photonics 11, p.441, 2017.

[5] K. Yamada, PIC Magazine, 1st December 2016, https://picmagazine.net/article/101212/Silicon_Photonics_For_A_Post-Moore_Era/feature, Angel Business Communications publication.

[6] Y. Urino et al., the 39th European Conference on Optical Communication (ECOC 2013), Mo.4.B.2F, London, Sept. 2013.

[7] M. Usui et al., Proc. ICEP-IAAC 2015, Kyoto, pp.660-665, 2015.

[8] T. Pinguet et al., IEEE the 9th International Conference on Group IV Photonics (GFP 2012), ThC1, San Diego, Aug. 2012.

[9] J-M. Fedeli et al., IEEE J. Selected topics in Quantum Electronics 20, p. 8201909, 2014.

[10] K. Suzuki et al., the 41th Optical Networking and Communication Conference and Exhibition (OFC 2018), San Diego, Th4B.5, March 2018.

VLSI Researches for Machine Learning and Neuromorphic Computing

Atsuya Okazaki

IBM Research - Tokyo

To accelerate performance per power in demanding machine learning applications in data centers, recently graphics processing units (GPU), field-programmable gate array (FPGA) and applicationspecific integrated circuit (ASIC) are broadly utilized by equipping massively parallel digital multiply-accumulators, where multiply-accumulation is an arithmetic bottleneck in software workloads using neural networks. Some of intelligent edge devices also become equipping ASIC or IP cores including such massively parallel multiply-accumulators optimized for specific machine learning applications at low-power consumption. This presentation introduces our current research activities in analog-based machine learning accelerators and analog-based spiking neural network processors aiming at high performance per power in machine learning applications by taking advantage of high-speed and low-power analog multiply-accumulation arithmetic over densely-packed synaptic non-volatile memory (NVM) resistive device arrays.

Machine Learning Solutions for Process Control in Semiconductor Manufacturing

Eugen Foca

Carl Zeiss SMT, PCS Group

Reliable and robust process control solutions are pivotal in the era of advanced technology nodes where the stochastic effects dominate the pattern characteristics. The key ingredients of the modern process control solutions are huge amount of data and smart data analytics algorithms. The state-of-the-art machine learning methods allow screening large amounts of data for statistically relevant, but so far unknown and unexpected correlations, hence providing new options for process optimizations.

In this talk the authors focus on combining the power of novel imaging technologies with smart algorithms for data analytics. The advantages of machine learning for defect detection are shown. Additionally, it is shown how measurements artefacts can be efficiently sorted out or data can be reconstructed by applying different modes of artificial intelligence. In the case of 3D tomography, it will be shown how an efficient object detector and context-based segmentation help to navigate through the data and identify anomalies in the images for further in detail analysis. The talk will introduce a novel approach to identify relevant CD measurement strategies by segregating samples from normal and defect classes.

Deploying New Nodes Faster with Machine Learning for IC Design and Manufacturing

Chris Schuermyer

Synopsys, USA
E-mail: chris.schuermyer@synopsys.com

Over the past few years there has been renaissance of machine learning being applied to problems as varied as predicting wind turbine failure to automating drug discovery. Semiconductor design and manufacturing is no exception: new publications on semiconductor machine learning nearly tripled between 2017 and 2018. This work provides a roadmap for machine learning that acts as a guide for understanding the current advancements in our field as well as a path toward our future.

INTRODUCTION AND BACKGROUND

To elaborate on how Machine Learning (ML) will evolve our work in semiconductors, we must first give it a strict definition. ML should be thought of as 'The Five Levels of Machine Learning', similar to how there are five levels of autonomous cars, each with increasing complexity. The motivation for thinking of ML in these terms is so we can foresee the future of ML in our industry by having a roadmap and populate it with examples from other industries that may be further along. Every ML application needs be thought of in the context of how it improves human action.

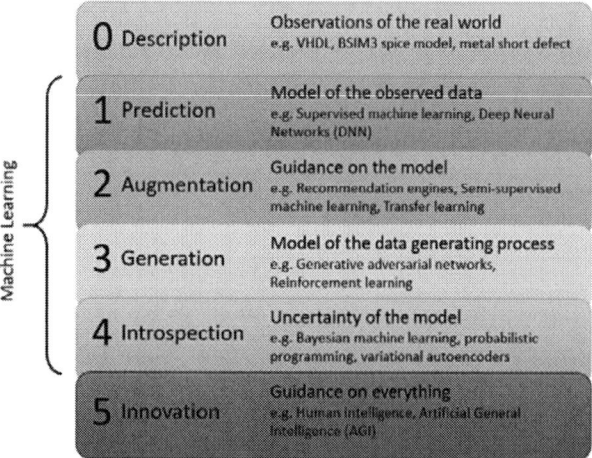

At present there are many people making dire predictions of how Artificial Intelligence will displace human workers and eliminate jobs and industries. This thinking focuses on the wrong problem. The emergence of ML should be thought of as a model for how we structure our problems, so we can be our most innovative selves. The power of AI is not replicating human behavior, it's the ability to augment humans by automating routine tasks. One of the earliest applications of AI was in the field of Optical Character Recognition (OCR). Prior to modern OCR, a lawyer might have had to read through many documents to identify information relevant to his case. OCR made documents searchable so that a lawyer can spend the same amount of time, but on high-value tasks like working with clients and strategizing.

Engineers in IC design and manufacturing are presented with a similar opportunity. Designers can be guided through a given design flow by leveraging machine learning rather than a guess-and-check approach that involves running expensive simulations. Process engineers can be assisted by technology that automatically generates hypotheses about causal mechanisms for yield loss. AI should free up time so that instead of doing detailed manual analysis engineers can instead focus on what they're best at: improving and innovating.

The power of machine learning is not about applying the latest-and-greatest algorithm or hardware. The power is a disciplined approach to problem solving. For any problem within the engineering discipline we must first answer a series of questions:

- What are the inputs?
- What are the constraints?
- How do we measure success?
- What will be done with the results?

Regardless of whether we are using linear regression, XGBoost, or implementing a deep neural network (DNN) we must answer these questions. Engineers tend to focus on one of these questions at a time; applying ML requires that we answer them all right when we're defining the problem.

APPLICATIONS OF MACHINE LEARNING IN DESIGN AND MANUFACTURING

What follow are several examples from the various levels of machine learning being applied to IC design and manufacturing.

Prediction: Deep Learning and Device Simulation

In addition to guiding the design process via things like recommendation engines, machine learning can also be used to scale up existing capabilities. The field of deep-learning has been shown to do a good job of identifying complex relationships in high-dimensional data like identifying objects in images and performing natural language understanding. Due to the vast amount of data available in semiconductor design and manufacturing, applications for deep learning are getting increased attention.

A recent example of applying deep-learning was given in [1] as a method to reduce Self-heat (SH) simulation runtime. The problem of SH can manifest itself as a real reliability problem in silicon device and thus needs to be properly predicted. However exhaustive HSPICE simulation is not practical to simulate for billions of transistors. To resolve this, the authors use an HSPICE simulator to generate the self-heat data on a large number of cells along with the simulated value of the self-heat. Then the authors trained a deep neural network on the simulated data to create a high-dimensional, but non-physics based, model. The (left-out) test results correlated well to the simulated results, but with a 100X runtime improvement.

978-1-7281-0943-5/19 $31.00 © 2019 IEEE

The improvement in the design process doesn't come from the speedup, it comes from what the speedup enables. The designers were never going to be able to simulate all the cells in the design. By using the deep neural network prediction, they were able to sort-order the data such that they only exhaustively simulated the worst 1% of the cells. Thus, the total runtime is the same: 100X speedup, exhaustively simulate 1% of the cells), however the end-result is much better because the simulated cells are not chosen at random: they are the cells that are most likely to have an issue with self-heating according to the ML prediction. Again, ML doesn't replace the designer, it augments the designer by making their time more productive.

Augmentation: Recommendation Engine for Physical Design

One ubiquitous application of machine learning in our lives is that of the recommendation engine. Whether it's Netflix recommending a new show, a product recommendation on shopee.tw, or articles in a Facebook news feed, they're all using machine learning recommendation engines to provide guidance. This recommendation isn't eliminating anyone's *job*, it's merely sorting the information, in order to help people make decisions faster.

This concept is now being extended to providing recommendations to chip designers. In [2], the authors present a supervised machine learning framework that learns the effect of various placement flows and provides recommendations given a set of design goals. The training data consists of previous place-and-route (P&R) runs, and the predictions are which settings can provide the best result.

From a physical design engineer's perspective, the main takeaway should be the importance of collection, archiving, and access to a corpus of circuits, P&R settings, and results. This kind of extensive database combined with the power of machine learning enables a guided path for the design engineers and reduces the design cycle-time. By automating routine tasks like 'what is the effect of increasing the number of I/O connections', the designers can spend more time on the questions like 'why is this flow generating the optimal results and can I do even better'?

Generation: OPC with Generative Adversarial Networks

The next step in ML complexity is developing generative models of the underlying process that produces observed data. Currently the Generative Adversarial Network is one of the most popular techniques being applied in image generation, speech synthesis, and natural language generation.

An example of generative modeling for Optical Proximity Correction (OPC) is demonstrated in [3]. The advantage demonstrated is the ability to fit a model based on estimates from Inverse Lithography Technique (ILT). This application of machine learning moves beyond fitting a model to the data and into the realm of fitting the model to the data-generating process.

Introspection: Prediction of PFA Success

One of the primary methods that's used to drive yield learning in modern processes is Physical Failure Analysis (PFA). With each successive technology generation PFA is becoming more challenging. The defects are increasingly subtle and require more and more sophisticated techniques such as nanoprobing, TEM imaging, thermal laser stimulation, etc. Additionally, there are an increasing number of design-induced yield limiters that don't behave like traditional defects. For example, critical path timing marginalities or ATPG overkill may be the root cause of the yield loss, but no amount of PFA will be able to find a defect.

Historically there has been a divide between the fab and the fabless company wherein data is 'thrown over the wall' and neither side has a complete view. This makes sense from an IP protection standpoint, however it has a tendency of leading to a lack of ownership. For example, what happens when an unknown timing marginality causes yield loss on a product? The fabless customer might send samples to the fab for PFA, but the fab returns the result No Defect Found (NDF). At this point the companies are at an impasse; the fabless company is relying to the fab to solve their yield problem, but the fab doesn't have access to the design data required to prove that it's not a physical defect.

The right answer is to reframe this as a machine learning problem with uncertainty quantification. The value to predict is the *probability* that PFA will uncover a defect. The predictors can consist of data about the device such as which softbin is failing, is the device failing only at certain voltage conditions, what is being reported by the scan diagnostics or memory bitmapping, what type of ATPG pattern is being applied, etc.

Again, the most important thing is to standardize on the data and the communication. The fabless company should always provide requests in the same standardized format including as much information is possible without compromising any intellectual property. Likewise, the fab should always provide PFA results in the same standardized format that includes information about the defect location, layer, size, and behavior. This enables both sides to automatically train a machine learning model for PFA and predict the probability of a successful PFA *prior* to attempting the PFA job. If the prediction for a given device has a low probability of finding a defect, say less than 80%, the engineers can avoid sending the device for PFA. This kind of thinking evolves the role of PFA from 'tell me what the defect is', to 'why can't I find a defect'. What's more, an introspective model means that things like sensitivity analysis or ranked feature importance are directly part of the model estimation.

CONCLUSION

Machine learning is an exciting area that will continue to find more and more application in semiconductor design and manufacturing. Machine learning fills the role of automating routine tasks in design and yield engineering so that we may focus more time on answering the more difficult question: why, rather than what. The first and most important impact of adopting a machine learning approach is enforcing the standardization of data collection, data access, and success criteria. To continue the pace of technological progress we've delivered over the past 50 years it demands us to invest in new and innovative ideas. We don't have time to spend on the mundane; let the machines do that work.

REFERENCES

[1] C. K. Lee, et al., "Full Chip Self-Heat Prediction Using Machine Learning," in *Synopsys User Group Silicon Valley*, 2018.

[2] G. Grewal, et al., "Automatic Flow Selection and Quality-of-Result Estimation for FPGA Placement," in *IEEE International Parallel and Distributed Processing Symposium Workshops*, 2017.

[3] H. Yang, et al., "GAN-OPC: Mask Optimization with Lithography-guided," in *Design Automation Conference*, 2018.

Designing and Modeling Analog Neural Network Training Accelerators

Sapan Agarwal[1], Robin B. Jacobs-Gedrim[2], Christopher Bennett[2], Alex Hsia[2], Michael S. Van Heukelom[2], David Hughart[2], Elliot Fuller[1], Yiyang Li[1], A. Alec Talin[1], Matthew J. Marinella[2]

[1]Sandia National Laboratories, Livermore, CA
[2]Sandia National Laboratories, Albuquerque, NM
E-mail: sagarwa@sandia.gov

ABSTRACT

Analog crossbars have the potential to reduce the energy and latency required to train a neural network by three orders of magnitude when compared to an optimized digital ASIC. The crossbar simulator, CrossSim, can be used to model device nonidealities and determine what device properties are needed to create an accurate neural network accelerator. Experimentally measured device statistics are used to simulate neural network training accuracy and compare different classes of devices including TaOx ReRAM, $Li_{1-x}Co_xO_2$ devices, and conventional floating gate SONOS memories. A technique called "Periodic Carry" can overcomes device non-idealities by using a positional number system while maintaining the benefit of parallel analog matrix operations.

INTRODUCTION

Training a neural network is dominated by data movement. Even in an optimized digital ASIC accelerator data must constantly be moved between local caches and the computational unit. Analog crossbars can eliminate most of this data movement and can potentially reduce the energy and latency required to train a neural network by three orders of magnitude when compared to an optimized digital ASIC[2]. They accelerate three key operations that are the bulk of the computation in a neural network: vector matrix multiplication (VMM), matrix vector multiplication (MVM), and outer product rank 1 updates (OPU) [3]. For each operation, the computations are performed in a single parallel step in memory. Thus, for an NxN array, the CV^2 and $I \times V$ energy scale as the array size, $O(N^2)$ [3]. This is $O(N)$ better than trying to read or write a digital memory. Each row of an NxN digital memory array must be accessed sequentially, resulting in N columns of length $O(N)$ being charged N times, requiring $O(N^3)$ energy to read a digital memory.

Unfortunately, analog devices are noisy and suffer from several non-idealities including read noise, write noise and write nonlinearity. Analog arrays suffer from parasitic voltage drops. Furthermore, analog systems tend to have limited bit precision on the inputs and outputs to a crossbar, with the fewer bits used, the faster and more energy efficient an analog system is. All of these issues will impact the final classification accuracy of a neural network. To compensate for these issues and take advantage of large gains in energy and latency enabled by analog systems, neural algorithms will need to be designed specifically to overcome the hardware limitations. This will require new co-design tools where the impact of device level properties on algorithmic performance can be assessed so that analog device development can be driven by algorithmic requirements. Consequently, we have developed a new open source simulation tool called CrossSim [4] to quantify the impact of device level properties on algorithmic performance.

REQUIRED DEVICE PROPERTIES

In [1] we use CrossSim to evaluate the impact of read noise, write noise and write nonlinearity on training a two layer neural network for MNIST (recognizing handwritten digits). Training or classifying with analog devices that have a read noise standard deviation (σ) up to 5% of the total conductance range does not significantly degrade the accuracy (~1%). Neural network training requires a smaller write noise with $\sigma < 0.4\%$ of the weight range. Nevertheless, this can still be 3X larger than a typical update as training a neural network requires small updates on the order of 0.1% of the weight range. This will vary slightly depending on the dataset and the neural network architecture. Even a slightly asymmetric write nonlinearity substantially degrades classification accuracy, as shown in Fig 1. To compute in a large energy efficient crossbar, resistive memories must also have a high on-state resistance. Given that scaled wires at a 10nm half pitch can only handle 10 μA before electromigration occurs, reading 100 devices in parallel already sets a limit of 100 nA per-device current draw. Assuming a 1V read, this suggests a minimum on-state resistance of 10 MΩ. Parasitic voltage drops also become an issue for higher currents or larger arrays.

EVALUATING EXPERIMENTAL DEVICES

To understand how different types of analog devices perform in a training accelerator, we compare three different analog devices: a TaOx ReRAM [5], a battery inspired $Li_{1-x}Co_xO_2$ device[6], and a conventional floating gate SONOS (Silicon-oxygen-nitrogen-oxygen-silicon) memory[7]. The respective structures are illustrated in Fig 2. The analog write noise statistics and write nonlinearity are measured and directly used by CrossSim to simulate the accuracy of a neural network training accelerator built on those devices as illustrated in Fig 3. Both $Li_{1-x}Co_xO_2$ and SONOS can train to high accuracy, while the TaOx device is limited to an accuracy of ~80%.

PERIODIC CARRY

In order to compensate for the remaining device non-idealities a technique called periodic carry [8] can be used. Multiple devices can be used to represent a weight with a positional number system, such as base 2 or base 10, exponentially increases the number of levels with the number of devices. However, this is not compatible with a parallel write as carries need to be performed between digits. This can be overcome by allowing devices to store extra levels and periodically (every 100-1000 updates) reading the device and performing any necessary carries. This allows noisy, nonlinear TaOx devices that previously trained to 80% accuracy on MNIST, to achieve 97% accuracy, only 1% away from the ideal numeric accuracy of 98%. In addition, both the SONOS and $Li_{1-x}Co_xO_2$ devices can achieve ideal accuracy using periodic carry.

ARCHITECTURE COMPARISON

To understand the potential advantages of accelerators built on different devices, we compare kernel level energy, latency and area for 4 accelerator architectures[2, 7]: digital SRAM, digital ReRAM, analog ReRAM and analog SONOS. The energy and latency advantages strongly depend on the bit precision of the accelerator.

Analog ReRAM has the greatest possible advantages over a digital SRAM based ASIC of 11X in area, 430X in energy and 34X in latency. ReRAM based memories are also starting to be integrated in commercial foundries but are too noisy and nonlinear to train to high accuracies. Single device per weight accelerators are limited in accuracy to around 80%, necessitating the use of techniques like periodic carry to help compensate for poor device properties. As a nearer term option, SONOS devices are currently available in commercial foundries but typically require long μs to ms write pulses and high voltages around 10V to program. Nevertheless, SONOS

978-1-7281-0943-5/19 $31.00 © 2019 IEEE

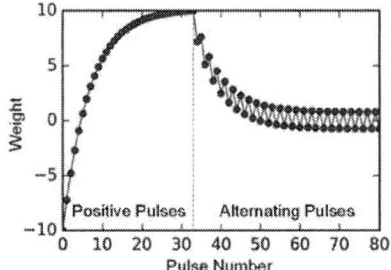

Fig. 1: (from [4]) Applying identical alternating positive and negative pulses causes the weight to decay towards a center value when it should remain constant. When the weight is near the maximum, a positive pulse does not change the weight much, but a negative pulse significantly decreases it. The opposite holds for weights near the minimum weight.

still has area, energy and latency advantages of 4X, 120X and 2X respectively over a digital ASIC. $Li_{1-x}Co_xO_2$ devices also have the potential to be highly accurate and efficient but additional research is needed in fundamental device physics and process integration before their full potential is realized.

ACKNOWLEDGEMENTS

This work was supported by the Department of the Defense, Defense Threat Reduction Agency, under Grant HDTRA1-17-1-0038, the Department of Energy (DOE) Advanced Manufacturing Office and the Laboratory Directed Research and Development program at Sandia National Laboratories, a multimission laboratory managed and operated by National Technology and Engineering Solutions of Sandia, LLC., a wholly owned subsidiary of Honeywell International, Inc., for the U.S. DOE's National Nuclear Security Administration under contract DE-NA-0003525. This paper describes objective technical results and analysis. Any subjective opinions do not necessarily represent the views of the U.S. DOE or the US Govt.

REFERENCES

[1] S. Agarwal *et al.*, "Resistive memory device requirements for a neural algorithm accelerator," in *2016 International Joint Conference on Neural Networks (IJCNN)*, 2016, pp. 929-938.
[2] M. J. Marinella *et al.*, "Multiscale Co-Design Analysis of Energy, Latency, Area, and Accuracy of a ReRAM Analog Neural Training Accelerator," *IEEE Journal on Emerging and Selected Topics in Circuits and Systems*, 2018.
[3] S. Agarwal *et al.*, "Energy Scaling Advantages of Resistive Memory Crossbar Based Computation and its Application to Sparse Coding," *Frontiers in Neuroscience*, vol. 9, p. 484, 2016, Art. no. 484.
[4] S. Agarwal *et al.* (2017). *CrossSim*. Available: http://cross-sim.sandia.gov
[5] R. B. Jacobs-Gedrim *et al.*, "Impact of Linearity and Write Noise of Analog Resistive Memory Devices in a Neural Algorithm Accelerator," presented at the IEEE International Conference on Rebooting Computing (ICRC) Washington, DC, November 2017.
[6] E. J. Fuller *et al.*, "Li-Ion Synaptic Transistor for Low Power Analog Computing," *Advanced Materials*, vol. 29, no. 4, p. 1604310, 2017.

[7] S. Agarwal *et al.*, "Using Floating Gate Memory to Train Ideal Accuracy Neural Networks," *IEEE Journal of Exploratory Solid-State Computational Devices and Circuits*, 2019.
[8] S. Agarwal *et al.*, "Achieving ideal accuracies in analog neuromorphic computing using periodic carry," in *VLSI Technology, 2017 Symposium on*, 2017, pp. T174-T175: IEEE.

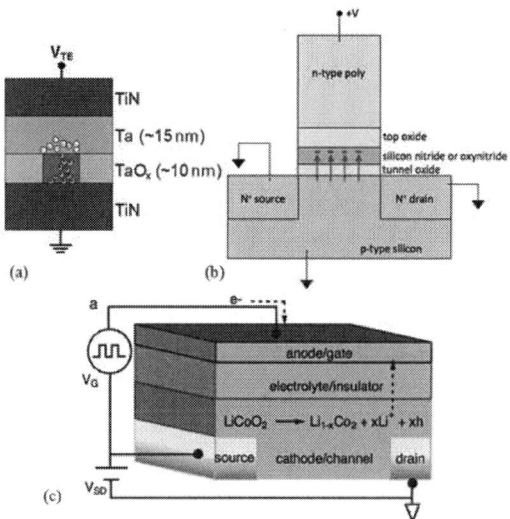

Fig 2: we compare three different analog devices: (a) TaOx ReRAM [5], (b) conventional floating gate SONOS (Silicon-oxygen-nitrogen-oygen-silicon) memory[7] and (c) battery inspired $Li_{1-x}CoO_2$ devices[6].

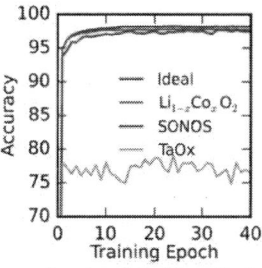

Fig 3: Neural accelerators based on SONOS or $Li_{1-x}Co_xO_2$ devices can reach near ideal accuracies while accelerators based on TaOx can reach around 80% accuracy on MNIST.

TABLE I
AREA COMPARISONS

	8 bit	4 bit	2 bit
Digital SRAM (μm^2)	836,000	814,000	800,000
Digital ReRAM (μm^2)	137,000	114,000	101,000
Analog ReRAM (μm^2)	75,000	46,000	41,000
Analog SONOS (μm^2)	195,000	166,000	161,000

TABLE II
ENERGY AND LATENCY COMPARISONS

	VMM			MVM			OPU			Total		
	8 bit	4 bit	2 bit	8 bit	4 bit	2 bit	8 bit	4 bit	2 bit	8 bit	4 bit	2 bit
Energy – Digital SRAM (nJ)	2850	2237	1848	4855	4241	3852	4300	3673	3274	12,000	10,150	8974
Energy – Digital ReRAM (nJ)	2139	1502	1098	2139	1502	1098	3246	2572	2143	7525	5577	4339
Energy – Analog ReRAM (nJ)	12.8	1.00	0.44	12.8	1.00	0.44	2.2	1.00	0.46	27.9	2.66	1.35
Energy – Analog SONOS (nJ)	14.4	2.25	1.5	14.4	2.25	1.5	71.5	30.9	10.6	100	35.4	13.6
Latency – Digital SRAM (μs)	4	4	4	32	32	32	8	8	8	44	44	44
Latency – Digital ReRAM (μs)	176	176	176	176	176	176	340	340	340	692	692	692
Latency – Analog ReRAM (μs)	0.384	0.024	0.011	0.384	0.024	0.011	0.512	0.032	0.032	1.28	0.080	0.054
Latency – Analog SONOS (μs)	0.402	0.032	0.014	0.402	0.032	0.014	20	20	20	20.80	20.06	20.02

Full Memory Encryption With Magnetoelectric In-Memory Computing

Albert Lee, Kang -L. Wang

University of California, Los Angeles, USA
E-mail: alee0618@ucla.edu

ABSTRACT

We propose an in-memory computing architecture based on the magneto-electric random access memory (MeRAM). The unique precessional magnetism of MeRAM is utilized to carry out XOR encryption of the device state with a key, allowing for encryption without readout of the device state, thus saving a significant energy and delay over computing-in-memory architectures. Furthermore, parallel encryption of memory blocks provides additional orders of improvement in encryption time. We simulate the proposed encryption scheme using a 28nm CMOS process and a macrospin LLG MTJ model, then evaluate the energy and latency of full-memory encryption on a 512x512 array. The proposed scheme achieves up to 7.4x and 1024x in energy and latency over computing in memory architectures, respectively.

INTRODUCTION

With the growth of internet-based systems, data security is becoming increasingly critical. The widespread distribution of information across IOT and mobile devices has given attackers more opportunities to gain physical access to nodes through methods such as theft, loss, or second-hand/disposed devices [1]. Those with intention can easily extract sensitive personal information such as passwords and financial data from the device memory, even potentially gain access to data stored on the cloud.

The adoption of new, embedded nonvolatile memory technologies to retain data without supply further aggravates the security issue [2]. While these memories enable zero-leakage, instantaneous deep-sleep/wakeup systems, retaining information during sleep and shutdown provides vulnerability where the system states are available long after operation. Without a method to protect the memory, stored information can be easily compromised.

An important method to mitigate this vulnerability is full disk encryption: before power off or standby, the data must be stored in a way that is incomprehensible (encrypted) [3]. In this case, even when the attacker gains physical access to the device, it will be significantly more difficult to recover any useful information, allowing the original owner time to take countermeasures against the attack (e.g. change passwords, disable credit cards, etc).

However, full-disk encryption incurs processing of the entire memory. This operation is extremely costly in terms of energy and latency, making it impractical for mobile devices with limited power budget. In traditional architectures, the data is read from the memory, encrypted by the CPU, then stored back into the memory; resulting in large overhead from I/O transmission. Recently, computing-in-memory (CIM) structures [4] carry out this computation within the memory macro to eliminate the I/O transmission. Nevertheless, it still requires sequential readout of all the contents in the array.

In-memory computing is a memory structure in which the computing is performed within the memory cell itself. In this case, no readout process is required during the encryption, providing a substantial boost in latency and energy efficiency. Furthermore, as each cell is capable of operating simultaneously, the computation process can be parallelized. A comparison of the three architectures is shown in Fig.1.

MeRAM IN-MEMORY XOR ENCRYPTION

The MeRAM utilizes the voltage-controlled magnetic anisotropy (VCMA) effect [5] to switch the magnetic orientation of an magnetic tunneling junction (MTJ). When an electric field is placed across a MTJ, charge accumulated on the tunneling barrier can modulate the interfacial perpendicular anisotropy (PMA). A positive voltage reduces the PMA and energy barrier (Eb) between the two stable states, resulting in precessional motion of the free layer magnetization and therefore oscillation of the device state. When the voltage is removed, the PMA and Eb is restored, stabilizing the magnetization. In most applications, the voltage pulse applied is timed to half the precesional period to switch the device state. An illustration of this process is shown in Fig.2

The XOR operator is commonly used in encryption, either by itself or as a component of more complex ciphers. During encryption, a stream of data is processed by applying bitwise XOR with a key [1], [6]. During decryption, the original data is recovered by applying bitwise XOR of the encrypted data with the same key. An illustration of this process is shown in Fig.3.

To achieve in-memory encryption, the state of the memory device after encryption needs to be the XOR result of the original memory state with the external key. This operation should not involve knowledge of the device state. This is achieved by utilizing the unique precessional magnetization of the MeRAM device. Here, we assign the AP and P state to logic 1 and 0, respectively. When the received key is 1, a pulse of half the precessional period is applied on the MTJ, causing the next state to be the inverted value of the current state (i.e. if the original device state is logic 1/0; the encrypted state becomes 0/1). On the other hand, when the key received is 0, we do not apply a pulse, and the encrypted state is identical to the original state. The truth table of the XOR in-memory encryption of a bit of data is shown in Table I.

SIMULATION AND ANALYSIS

We demonstrate the encryption of a 4x4 memory array in a 28nm technology process with the LLG MeRAM model described in [7]. Parameters are extracted from devices with precessional half period of ~1.2ns and switching voltage of 0.7V. The structure of the memory macro is identical to a standard 1-transistor 1-MTJ memory array. The BL/ SL of each column is first set to 1V/0V if the corresponding bit in the key is a 1; and 0V/0V if the corresponding bit in the key is a 0. Then, a pulse is applied to all WLs to encrypt (Fig.4). The simulation for encrypting [0001, 0110, 1010, 1001] with the key [0101] is shown in Fig.5. As can be seen, the entire array can be successfully encoded and decoded within a single 2ns cycle.

To evaluate the performance of the proposed in-memory encryption, we build the critical path of a 512x512 (256kb) array and compare the performance with a computing-in-memory macro. Since encryption algorithms may use different keys for each memory block, we evaluate the energy and delay over the granularity in which different keys are applied (e.g. a different key for a block size of every 1, 2, 4…256 WLs). For small block size of 1 WL, the proposed method consumes 14.3 fJ/bit and a latency of 512 cycles, while that of computing in memory consumes 33.8fJ and 1024 cycles, showing an improvement of 2.4x and 2x. For large block size of 512 WLs, the

978-1-7281-0943-5/19 $31.00 © 2019 IEEE

proposed method consumes 4.6fJ/bit and 1 cycle, while the cost of computing in memory remains the same. The improvement is increased to 7.4x and 1024x, respectively. As the array size becomes larger, the amount of improvement also increases.

CONCLUSION

Memory encryption is indispensable to protect nonvolatile memories against security attacks, however comes at a large cost. In-memory computing not only removes readout latency and energy, but also provides the capability of parallel processing, demonstrating orders of magnitude improvement over state-of-the-art memory architectures.

REFERENCES

[1] V. Young, *ACM SIGARCH Comput. Archit. News,* vol. 43, no. 1, pp. 33–44, 2015. [2] S. Swami, in *Proceedings of the Design Automation Conference (DAC)*, pp. 1–6, 2016. [3] D. Kaplan, *White Paper*, 2016. [4] S.Li, in *Proceedings of the Design Automation Conference*, p. 173, 2016. [5] W.-G. Wang, *Nat. Mater.*, vol. 11, no. 1, pp. 64–68, 2011. [6] C. Yan, in *ACM SIGARCH Computer Architecture News*, vol. 34, no. 2, pp. 179–190, 2006. [7] H. Lee, in *IEEE Trans. Magn*, vol. 54, no. 4, 2018.

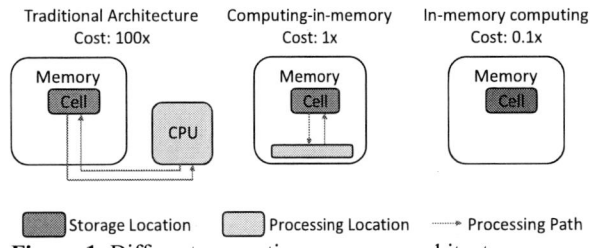

Figure 1: Different computing - memory architectures.

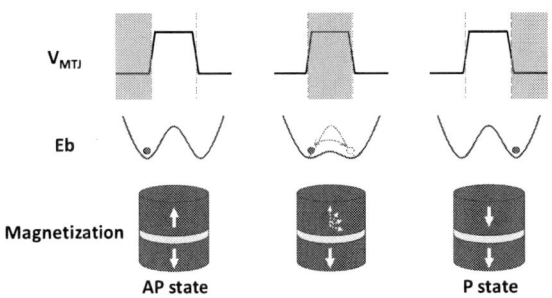

Figure 2: Switching of the MeRAM device. With a voltage applied, the MTJ oscillates between the two states.

Memory State	Key	Apply Pulse	Memory Next State
0	0	No	0
0	1	Yes	1
1	0	No	1
1	1	Yes	0

Table I. Truth table and operation conditions of the MeRAM in-memory XOR operation

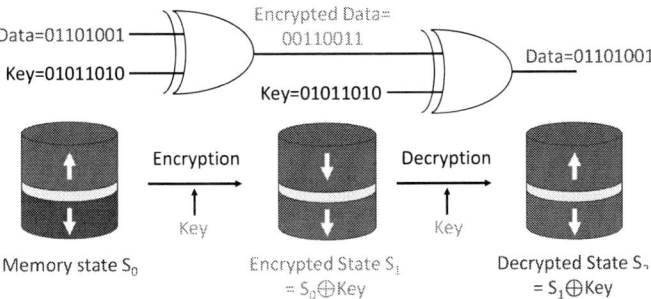

Figure 3: Encryption and decryption using the MeRAM device. The encrypted state of the device is the XOR of the original state with a Key. The original state can be recovered by another XOR of the encrypted state with the same key.

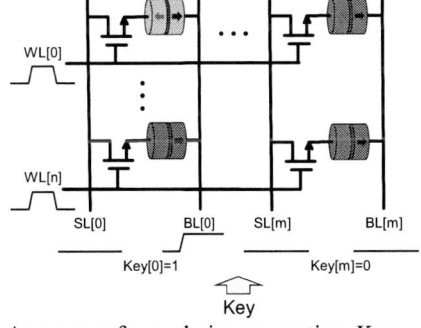

Fig.4 Array waveforms during encryption. Keys are applied to the SL/BL, then WLs to encrypt are enabled.

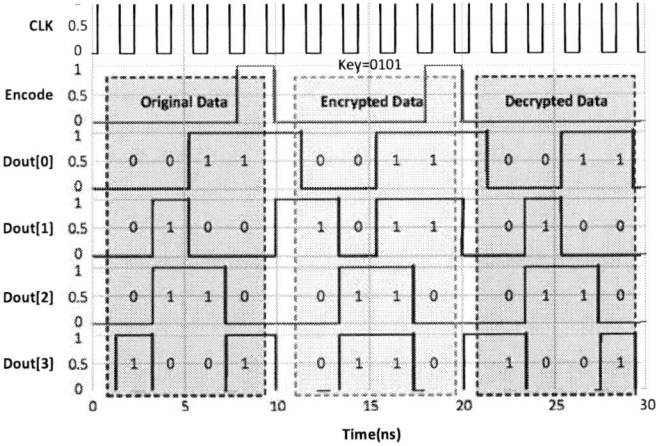

Fig.5 Simulation of a 4x4 full-array encryption and decryption. Each row of the data in the array is read out consecutively, followed by a 2-ns full-array encryption (enabled by the signal "Encode") and readout of the encrypted data. The data is then decoded and verified.

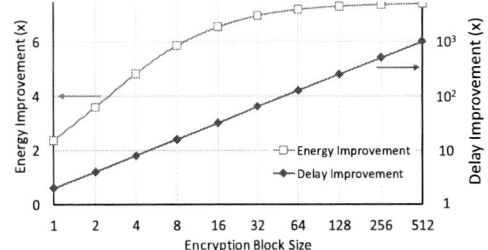

Fig.6 Energy and delay improvement of the proposed scheme compared to CIM methods

Integrated Photonics of Transistor Laser, Detector and Active Load for All Optical NOR Gate

Ardy Winoto, Junyi Qiu, Dufei Wu, Yu-Ting Peng, and Milton Feng

Micro and Nanotechnology Laboratory, Department of Electrical and Computer Engineering
University of Illinois at Urbana-Champaign, Urbana, Illinois, U.S.A., 61801 (mfeng@illinois.edu)

Abstract: The transistor laser is a promising candidate for high-speed integrated optoelectronics for its capability to operate under both base current modulation and collector tunneling modulation simultaneously, which fundamentally enables the monolithic integration of electrical and optical signal processing on the device level. This work demonstrates one of such applications by building an optical-to-optical NOR gate with the transistor laser in an integrated form.

I. Introduction

The transistor laser (TL) invented by Milton Feng and Nick Holonyak, Jr. [1,2] inherits the benefits from heterojunction bipolar transistor (HBT) with quantum-wells inserted in the base for faster recombination process and coherent optical output. Different from the typical diode laser, the TL has a tilted charge profile such that instead of waiting for carriers to "pile-up" for recombination, the carriers will be swept by the electric field towards the collector. Therefore, the recombination time has been reduced from nanosecond range as in the typical diode laser to picosecond range in the TL, resulting a resonance-free frequency response and high modulation bandwidth [3]. Apart from the fast recombination, the TL's four-port functionality is also a unique trait as opposed to the typical two-port operation of diode lasers. The four-port operation of the TL is realized by the intra-cavity photon-assisted tunneling (ICPAT) in the collector junction acting as a built-in voltage-controlled optical absorber as well as a source for carrier re-supply in the base and collector current output, which makes the direct voltage modulation possible in addition to the common current-injection modulation from the base [3].

Compared with current transistor IC technology such as CMOS process, the TL not only can be modulated without any external modulators or drivers, reducing the complexity of the circuitry, but also can be fabricated monolithically for both transmitter and receiver on the same substrate, which is essential for integration. The receiver can be implemented by using an additional intrinsic layer or the base-collector junction of the TL. An example of the wafer-level integration for both transmitter and receiver (transceiver) is shown in Fig. 1 [5]. Aside from the typical application in optical data transmission, the TL has been proposed for an optical NOR gate logic

[4]. In this work, we report the fabrication and the characterization of an all-optical NOR gate based on TL.

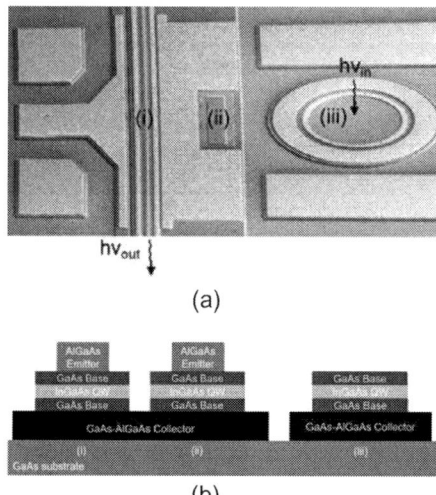

(a)

(b)

Fig. 1. Top view (a) and cross-section diagram (b) of a transistor laser integrated circuit containing a transistor laser with an optical cavity and cleaved facets (i), an HBT without an optical cavity (ii), and a vertical p-i-n photodiode (iii).

II. Theory of Operation

The circuitry of an all-optical TL-based NOR gate is shown in Fig 2. TL2 is a regular TL similar as the one in previous work [5]. TL1 is an HBT with an open base configuration, acting as an active load to increase the output impedance of the left half. TL0 is a P-i-N photodetector formed with the base-collector junction of transistor laser. The functionality of an optical NOR gate is explained as follows: when there is no optical input at TL0, TL2 will be in the lasing state with a constant optical output ("Logic 1") at an appropriate base current bias level [4]; the inverted topology of "Logic 0" is realized when there is one or multiple optical inputs detected at TL0, such that the increased collector-emitter voltage at TL2 shifts the operation point of the TL2 along the load line, causing the laser light output to decrease. The operation of the NOR gate is simulated with Keysight ADS and shown in Fig. 3. Theoretically, the optical NOR gate design can function more than two inputs since each additional input only serves to increase the collector-emitter voltage at TL2 and therefore the optical output at TL2 diminishes [5]. The threshold of logic '0' should be set at the output level of a single active input.

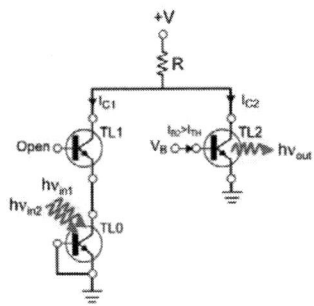

Fig. 2. Circuit diagram of a transistor based optical NOR gate.

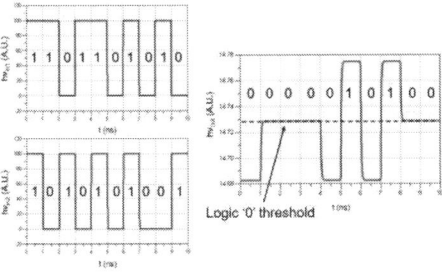

Fig. 3. Simulated logic diagram for a two-input optical NOR gate at 1 Gb/s.

III. Fabrication Process

The TL epitaxial structure is grown by MOCVD on a semi-insulating GaAs substrate. The emitter is formed with the n-type AlGaAs and the base of the TL is formed with p-type GaAs. There is an InGaAs quantum-well inside the base to promote recombination process. The device layout and the lateral cross-section view are shown in Fig. 4. The metal interconnect is placed after planarization with benzocyclobutane (BCB) and the double via process formed by dry-etching. After the top metal has been established, the wafer is thinned down to 150 μm and cleaved into stripes to form the facets of edge-emitting lasers. Each stripe is then bonded with indium to a copper substrate for testing [5].

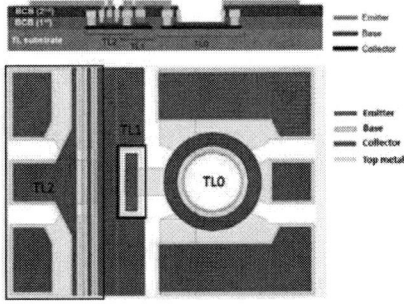

Fig. 4. Lateral cross-section and top view diagrams of the transistor laser all optical NOR gate.

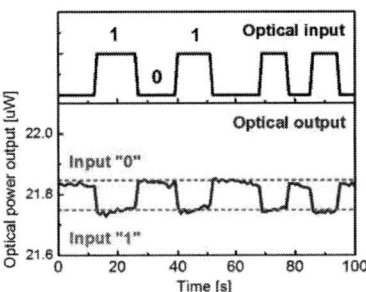

Fig. 5. Logic timing diagram for the optical NOR gate. It is noted that the optical input illustrated is a square wave of 10 mW amplitude.

IV. Characterization

The logic functionality of the TL- based optical NOR gate is performed with a base current of 50 mA as input to maximize the optical output. The voltage supply is set at 4V with the external resistor of 11.3 Ω, resulting a collector-emitter voltage of 2.3V at TL2. The optical input to the TL0 is coupled from an external modular laser with a square wave of 10 mW amplitude. The responsivity of the T0 has been characterized as 0.0346 A/W. After the signal is coupled into TL0 and switched on and off to provide a square input signal, the generated optical output at TL2 is shown in the blue line in Fig.5 [5]. It should be noted that in this case the NOR gate acts like an inverter due to the single input. The functionality of NOR gate can be realized by feeding a multi-level optical input signal or two optical inputs. The threshold of logic 0 is 21.75 μW and that of logic 1 is 21.85 μW.

Acknowledgement

This project is also supported by the National Science Foundation under grant 1640196, and the Nanoelectronics Research Corporation (NERC), a wholly-owned subsidiary of the Semiconductor Research Corporation (SRC), through Electronic-Photonic Integration Using the Transistor Laser for Energy-Efficient Computing, a Nanoelectronics Research Initiative (SRC-NRI) under Research Task ID 2697.001-Transistor Laser Based Circuits and Devices.

References

[1] M. Feng and N. Holonyak, Jr. "The Metamorphosis of the transistor into a laser", Optics & Photonics News (OPN), Optical Society of America page 44-49 (OSA), (2011)

[2] M. Feng, N. Holonyak, Jr., H.W. Then, C. H. Wu, G. Walter, "Tunnel junction transistor laser," Applied Physics Letters, 94, 0411181 (29th Jan, 2009).

[3] H. W. Then, M. Feng and N. Holonyak, Jr., "The Transistor Laser: Theory and Experiment," Proceedings of the IEEE, Special Issues on 50th Anniversary of the LED vol.10, no 101, p 2271-2298, Oct. (2013) (Invited Paper)

[4] M. Feng, A. Winoto, J. Qiu, Y.-T. Peng and N. Holonyak, Jr., "All optical NOR gate via Tunnel-Junction Transistor Lasers for High Speed Optical Logic Processors", VLSI-TSA, 2018

[5] A. Winoto, J. Qiu, D.Wu, M. Feng , "Transistor Laser Integrated
Photonics for Optical Logics", IEEE Nanotechnology , 2018

A Novel RRAM Based Watermark Technique Utilizing the Impact of Forming Conditions on Reset Distribution

Yachuan Pang[1], Huaqiang Wu[1]*, Bin Gao[1], Bohan Lin[1], and He Qian[1]

[1]Institute of Microelectronics, Tsinghua University, Beijing, China, 100084

E-mail: wuhq@tsinghua.edu.cn

ABSTRACT

Watermark is a very important technique for information security. In this work, a RRAM-based electronic watermark technique leveraging the impact of Forming conditions on Reset distribution is experimentally demonstrated. The Forming process is indispensable for almost all RRAM device. The Forming conditions can affect the shape and size of the conductance filament, and further affects the resistance distribution at Reset. The tested data shows that the stronger Forming condition can get smaller mean value of resistance, and the use of RRAM as memory is not affected. In this design, two different Forming conditions are used to write watermark data '0' and '1'. A resistance in parallel technique is applied to enhance reliability. The experimental results show that the bit error rate (BER) reduced from ~30% to ~7% after using the optimized method. And no reliability degradation occurs after 200 normal Reset/Set cycles.

INTRODUCTION

Hardware security is challenged because of its own flaws with an increasing number of attack methods (e.g. IP piracy, side-channel attacks, counterfeiting, etc.). Watermark is a very important technique for information security (e.g. digital watermark for electronic information security). Watermark can provide unique identifier and certification support for each product. Resistance random access memory (RRAM) has some unique characteristics (e.g. cycle to cycle variation and device to device variation) [1], making it adapt for hardware security applications, such as physical unclonable function (PUF) [2-4] and true random number generator (TRNG) [5]. But none of these applications can be used as both security circuit and memory. In this work, we proposed a novel RRAM based electronic watermark technique to meet this challenge.

The test results show that the different Forming conditions will affect the resistance distribution after single-pulse Reset, and the larger Forming condition can get smaller mean value of resistance. In this case, one stronger condition (condition 1) is defined as watermark '1', and another weaker condition is defined as watermark '0'. We can read-out the watermark data when needed, and the RRAM can be used as normal memory in other time. This proposed method is implemented in a 1Kb RRAM array, and a resistance in parallel technique is applied to enhance reliability.

RRAM WATERMARK DESIGN

Fig. 1 shows the fabricated 1Kb 1-transistor-1-resistor (1T1R) array (include a 128 row×8 columns RRAM array and a WL decoder) and a schematic of 1T1R structure. The RRAM structure uses a TiN/TaOx/HfOx/TiN stack, and the RRAM cell is integrated between M4 to top metal. Typical I-V of the RRAM cell is shown in Fig. 2. To implement different resistance distribution with the single-pulse Reset scheme, two typical Forming conditions (condition 1: BL (5V)/ WL (1.2-2.2V, Step 0.05V), and condition 2: BL (3.5V)/ WL (1.2-2.2V, Step 0.01V)) are tested. The resistance distribution after Forming is shown in Fig. 3. We can find that the condition 2 has tight distribution. Fig. 4 shows the normal Reset/Set (multi-pulse, with verify) resistance distribution respectively, these two conditions show no significant difference. As for single-pulse Reset, a 1.8V Reset pulse (pulse width: 50 ns) is applied to all cells. Corresponding resistance distribution is shown in Fig. 5. The test results show that the Condition 2 gets larger resistance mean value than condition 1 (as shown in the inset figure of Fig. 5). Endurance is a very importance factor for watermark application. There are two different endurance tests: one is used to evaluate the BER with the number of read-out times (Set all cells to LRS with verify before next single-pulse Reset process), and the other is used to assess the impact of the normal Reset/Set on the watermark read-out BER (a certain amount of normal Reset/Set cycles are applied before next single-pulse Reset). Fig. 6 shows the resistance changes with the single-pulse Reset. It shows that the relation of the two resistance values can be maintained for more than 50 cycles. Fig. 7 shows the resistance change after single-pulse Reset with the normal Reset/Set cycles. The normal Reset/Set has no significant effect to the relation of the two resistance values.

To implement watermark function in the RRAM array, we propose a RRAM based watermark circuit. It is shown in Fig. 8. It includes memory part (up) and watermark read-out part (down), and two major functions are controlled by switch S (S open: RRAM array used as memory, S close: read-out watermark)

RRAM WATERMARK OPTIMIZATION

Fig. 6 and Fig. 7 show that the resistance distribution at two conditions have larger overlap, which will cause large BER (close to 30%) in watermark read-out mode. To address this challenge, we propose a novel resistance in parallel technique to decrease the overlap. Fig. 9 shows resistance distribution changes with the increasing of the number of resistance in parallel (normalization). We can find that the resistance overlap gradually decreases, and gets smaller overlap in 8-R in parallel. And the BER shows the same trend. Fig. 10 shows the BER reduced from ~30% to ~7% after using the optimized method. Fig. 11 shows the tested data mapping changes at different process. Left figure shows the target data mapping (Forming each cell using condition 1/condition 2 at initialization according to target data). The 1st read-out shows only two wrong points in the watermark data mapping after initialization. The 2nd read-out is the watermark data after one time normal Reset/Set, and the last figure is the watermark data after 200 times normal Reset/Set cycles. We find that the watermark has good reliability after 200 normal Reset/Set cycles.

CONCLUSION

A novel RRAM based watermark technique is proposed and evaluated in a 1Kb RRAM array. The reliability of read-out is optimized using 8-R in parallel technique. The optimized watermark can achieve BER close to 7% after 200 normal Reset/Set cycles.

ACKNOWLEDGEMENTS

This work is supported in part by the NSFC (61674092), MOST of China (2016YFA0201803), Beijing Innovation Center for Future Chip (ICFC), Beijing Municipal Science and Technology Project (Z181100003218001).

REFERENCES

[1] I. G. Baek, et al., IEDM, 587-590, 2004. [2] Y. Pang, et al., IEDM, 12.2.1-12.2.4, 2017. [3] Y. Pang, et al., IEEE EDL, vol. 38, no. 2, pp. 168-171, 2017. [4] R. Liu, et al., IEEE EDL, vol. 36, no. 12, pp. 1380-1383, 2015. [5] H. Jiang, et al., Nature Communications, vol. 8, no. 1, pp. 882, 2017.

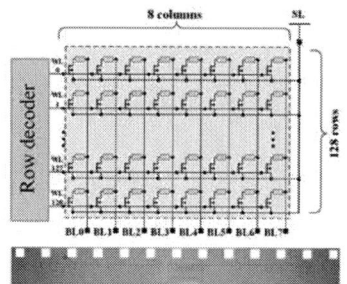

Fig. 1 Photo image of the fabricated 1kb 1T1R RRAM array with array details (including WL decoder and RRAM array), and a schematic of the 1T1R cell structure.

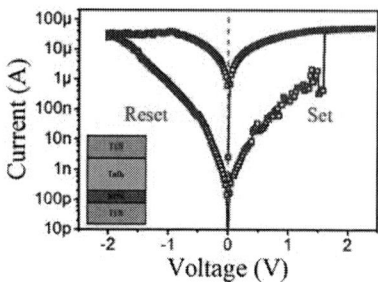

Fig. 2 Typical measured I-V curves of 1T1R RRAM cells with a RRAM stack inserted.

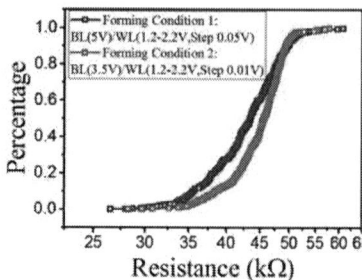

Fig. 3 Resistance distribution after Forming process with two different conditions (1: BL(5V)/ WL(1.2-2.2V, Step 0.05V), 2: BL(3.5V)/ WL(1.2-2.2V, Step 0.01V)).

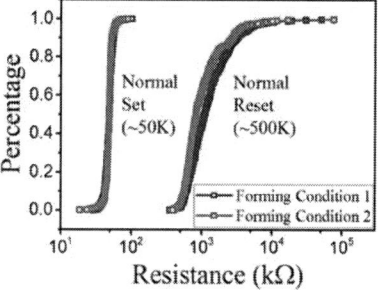

Fig. 4 Resistance distribution in normal Set/Reset state (multi-pulse, with verify), corresponding to two Forming conditions (red: condition 1, blue: condition 2).

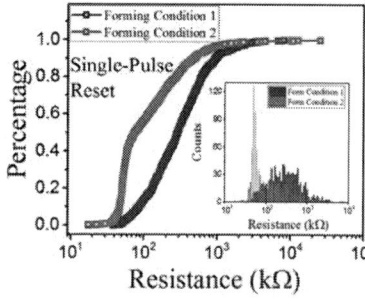

Fig. 5 Resistance distribution in single-pulse Reset state (without verify), and Condition 2 gets larger resistance mean value than condition 1.

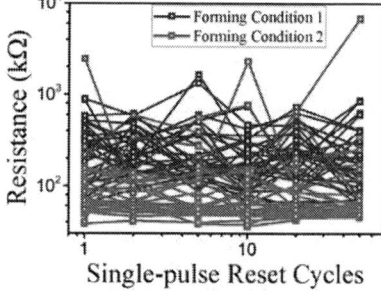

Fig. 6 Resistance changes with the single-pulse Reset cycles (Set all cells to LRS before next single-pulse Reset process), Condition 2 keeps larger mean value during 50 cycles.

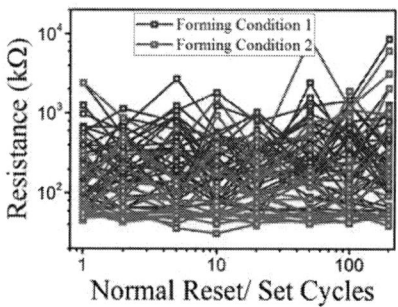

Fig. 7 Resistance change after single-pulse Reset with the normal Reset/Set cycles (a certain amount of normal Reset/Set cycles are applied before next single-pulse Reset).

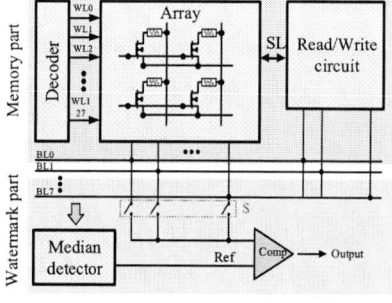

Fig. 8 Implementation of RRAM based watermark circuit. Two major functions are controlled by switch S (S open: RRAM array used as memory, S close: read-out watermark).

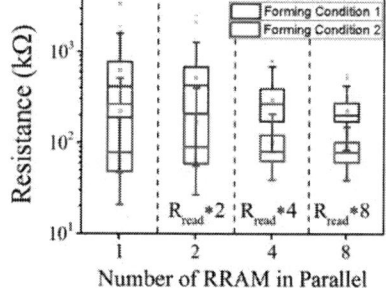

Fig. 9 Resistance distribution changes with the increasing of the number of resistance in parallel (normalization), and 8-R in parallel method shows smallest overlap.

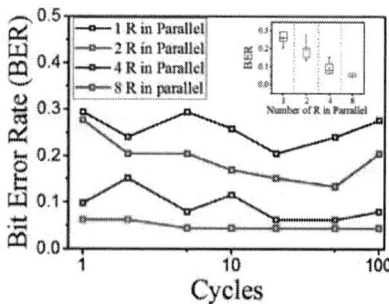

Fig. 10 BER changes with the increasing of the number of resistance in parallel, and 8-R in parallel method shows best reliability.

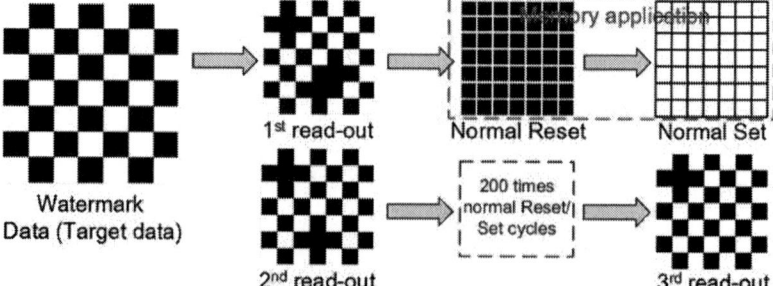

Fig. 11 Tested data mapping changes in different process. Left figure shows the target data mapping. The 1st read-out shows the data after initialization. The 2nd read-out is the data after one time normal Reset/Set. The last figure is the data after 200 times normal Reset/Set cycles.

978-1-7281-0943-5/19 $31.00 © 2019 IEEE

AUTHOR INDEX

Afzalian, A. 120
Agarwal, Sapan 132
Ahles, Christopher 41, 67
Amita 100
Andrieu, F. 110
Ansari, Md. Hasan Raza 53
Arakawa, Yasuhiko 21
Arkalgud, Sitaram 124
Balestra, F. 110
Batude, P. 110
Baugh, Jonathan 118
Bennett, Christopher 132
Berthelon, R. 110
Besnard, Guillaume 102
Bharti, Deepak 55
Bi, Chong 82
Bosch, D. 110
Breeden, Mike 108
Brunet, L. 110
Buonacorsi, Brandon 118
Chai, Yang 71, 73
Chan, Mansun 43
Chang, Li Fung 2
Chang, Ting-Chang 59
Chang, Wen Hsin 106
Chang, Wen-Hao 79
Chang, Y. J. 51
Chang, Y. W. 27
Chao, T.-S. 97
Chen, Christopher 118
Chen, G. L. 51
Chen, G. Y. 31
Chen, Kuang-Chao 27
Chen, T. J. 93
Chen, T. P. 93
Chen, Yi-Ju 65
Cheng, C. C. 27
Cheng, Chuan-Cheng 5
Cheng, Haitao 5
Chenz, Osbert 93
Chidambaram, Pr. Chidi 5
Choe, Hwan Sung 91
Choi, Jong 41, 67
Chung, Steve S. 93
Ciampolini, L. 110
Cibrario, G. 110
Ciou, J.-Y. 97
Cluzel, J. 110
Colinge, J. P. 110
Collaert, Nadine 102
Dateki, Takashi 8
Davies, Mike 20
Dick, K. A. 120
Doornbos, G. 120
Dunworth, Jeremy 5

Ecarnot, Ludovic 102
Elgabra, Hazem 118
Endo, Hiroki 49
Ernst, Thomas 19
Esmanhotto, E. 110
Fang, C.-C. 97
Feng, Milton 136
Fenouillet-Béranger, C. 110
Foca, Eugen 129
Fukuzawa, Hideaki 85
Fuller, Elliot 132
Furnemont, Arnaud 15
Ganguly, Swaroop 35
Ganguly, Udayan 35, 100
Gao, Bin 139
Garg, Chirag 87
Garros, X. 110
Gaudin, Gweltaz 102
George, Steven M. 63, 99
Giraud, B. 110
Gong, Xiao 114
Gorad, Ajinkya 100
Guo, Jyh-Chyurn 104
Gupta, Piyush 122
Hamzaoglu, Fatih 16
Han, Kaizhen 114
Haq, Jesmin 85
Hatsuda, Kosuke 13
He, Jr-Hau 80
He, Renren 85
Holland, M. 120
Hong, T.-C. 97
Hoya, Katsuhiko 13
Hsia, Alex 132
Hsieh, E. R. 93
Hsieh, K. Y. 31
Hsieh, Peter 11
Hsin, Y. C. 51
Hsu, H.-S. 97
Hsu, K. C. 31
Hsueh, F.-K. 97
Hu, Chenming 91
Huang, H.-F. 97
Huang, K. C. 51
Huang, K.-P. 97
Huang, S. A. 93
Huang, Tongshuang 45
Huang, Y.-C. 97
Huang, Yi-Chiau 114
Huang, Yu-Shiang 112
Hughart, David 132
Hughes, Brian 87
Hung, Raymond 41
Hung, Yu-Ting 25
Hwang, S. W. 27

AUTHOR INDEX

Ida, Jiro 49
Irisawa, Toshifumi 106
Ishii, Hiroyuki 106
Iwata-Harms, Jodi 85
Jan, Guenole 85
Janardhanan, Shankaran 4
Jao, C.-Y. 97
Kaiser, Winfried 10
Kalpat, Sriram 5
Kanai, Tatsunori 13
Kang, Lixing 75
Kang, Yuye 114
Kao, K.-H. 97
Kao, Yun-Feng 33
Kim, Namsung 41
King, Ya-Chin 25, 33
Kong, Eugene Y.-J. 114
Kranti, Abhinav 53
Ku, S. H. 27
Kuehn, Wilhelm 61
Kulkarni, Jaydeep P. 29
Kummel, Andrew 41, 67, 108
Kuo, Y. C. 51
Kutty, R. Govindan 75
Lacord, J. 110
Lam, Vinh 85
Lang, S. 110
Lattard, D. 110
Le, Son 85
Lee, Albert 134
Lee, Chieh 25
Lee, D. Y. 31
Lee, F. M. 31
Lee, Feng-Min 18
Lee, H. H. 51
Lee, M. H. 31
Lee, Y.-J. 97
Lee, Yuan-Jen 85
Lei, Dian 114
Li, F. L. 93
Li, J.-H. 97
Li, J.-Y. 97
Li, S. H. 51
Li, Weisheng 71
Li, Xiang 82
Li, Yan 17
Li, Yiyang 132
Li, Yujun 3
Li, Zu-Cheng 104
Liao, Hsiu-Hsien 65
Liao, T.-H. 97
Lin, Bohan 139
Lin, C. S. 51
Lin, Cheng-Jun 25

Lin, Chia-Chen 37
Lin, Chrong Jung 25, 33
Lin, Jinq-Min 104
Lin, Jyi-Tsong 53
Lin, Shy-Jay 82
Lin, Wei-Liang 27
Lin, Y. Y. 31
Lin, Yu-Ru 116
Liu, C. H. 93
Liu, C. W. 112
Liu, Gang 5
Liu, Huanlong 85
Liu, Lenvis 27
Liu, Yi-Chun 112
Liu, Zheng 75
Lu, C. Y. 31
Lu, Chih-Yuan 27
Lu, Chun-Chang 27
Lu, D. D. 97
Lu, Fang-Liang 112
Lu, Po-Sheng 37
Lu, Tao-Cheng 27
Lung, H. L. 31
Luo, S.-X. 97
Luo, Y. C. 93
Ma, W. C.-Y. 97
Ma, Xueli 57
Ma, Zichao 43
Maeda, Tatsuro 106
Mahato, Ajay Kumar 55
Makosiej, A. 110
Marinella, Matthew J. 132
Maruta, Yasushi 9
Maruyama, Shigeo 69
Matsukawa, Takashi 95
McVey, Shawn 61
Migita, Shinji 95
Mori, Takayuki 49
Murdzek, Jessica A. 63
Nagashio, Kosuke 77
Navlakha, Nupur 53
Nemani, Srinivas 41, 67
Nguyen, Bich-Yen 102
Nibhanupudi, S. S. Teja 29
Ohsawa, Takashi 39, 45
Okazaki, Atsuya 128
Okumura, Yukihiko 7
Ossieur, Peter 23
Ota, Hiroyuki 95
Pang, Yachuan 139
Passlack, M. 120
Patel, Sahil 85
Peng, Kang-Hui 116
Peng, Yu-Ting 136
Phung, Timothy 87

AUTHOR INDEX

Prawoto, Clarissa 43
Qian, He 139
Qiu, Junyi 136
Radu, Ionut 102, 123
Raghuwanshi, Vivek 55
Rahaman, S. Z. 51
Ramvall, P. 120
Rettner, Charles 87
Rios, M. 110
Sahota, Kamal 5
Salahuddin, Sayeef 91
Schuermyer, Chris 130
Schwarzenbach, Walter 102
Serrano-Guisan, Santiago 85
Shahidi, Ghavam 1
Shang, Yaqi 39
Shen, Dongna 85
Sherman, Alan 24
Shi, Yi 71
Shirota, Yusuke 13
Simsek, Telem 82
Song, M. 82
Su, C.-J 97
Su, J. W. 51
Su, Pin 37, 47, 89
Su, Y. H. 51
Sundar, Vignesh 85
Sung, P.-J. 97
Sy, Wing 5
Tai, Mao Chou 59
Talin, A. Alec 132
Tan, Ava J. 91
Tang, Denny 81
Teng, Jeffrey 85
Thelander, C. 120
Thomas, Luc 85
Tiwari, Shree Prakash 55
Tong, Ru-Ying 85
Toriumi, Akira 95
Tsai, Meng-Ju 116
Tsai, Wen-Jer 27
Tsai, Wilman 82
Tsao, Yu-Ching 59
Tseng, Kuei-Yang 47
Tseng, P. H. 31
Tseng, Tseung-Yuen 27
Tsou, Ya-Jui 112
Tsuchida, Kenji 13
Tsui, Bing-Yue 65
Tu, Chien-Te 112
Uchida, Noriyuki 106
Ueda, Scott 108
Van Heukelom, Michael S. 132
Vandooren, Anne 102
Vardhan, P. Harsha 35

Varun, Ishan 55
Vasen, T. 120
Vinet, M. 110
Wang, C.-J. 97
Wang, D. Y. 51
Wang, I. J. 51
Wang, Jian-Ping 84
Wang, Joseph 5
Wang, K. C. 31
Wang, Kang -L. 134
Wang, Longyan 43
Wang, Po-Kang 85
Wang, Shan X. 82
Wang, Wenwu 57
Wang, Xiaolei 57
Wang, Xiaowei 75
Wang, Xinran 71
Wang, Y.-H. 97
Wang, Y.-S. 97
Wang, Yu-Jen 85
Watanabe, Yohji 13
Watt, Jeff 118
Weber, O. 110
Wei, J. H. 51
Wei, Lan 118
Wernersson, L.-E 120
Winoto, Ardy 136
Wolf, Steven 108
Wong, Keith 67
Wu, C. I. 51
Wu, C.-T 97
Wu, Dufei 136
Wu, Guan-Wei 27
Wu, Huaqiang 139
Wu, M. C 31
Wu, Ming C 22
Wu, W.-F. 97
Wu, Ying 114
Wu, Yung-Chun 116
Xia, Deying 61
Xiang, Jinjuan 57
Xiang, Rong 69
Xie, Yuan 12
Xu, Shengqiang 114
Yamada, Koji 126
Yang, Ming-Ta 5
Yang, S. Y. 51
Yang, See-Hun 87
Yang, Yi 85
Ye, Hung-Yu 112
Ye, Tianchun 57
Yeh, W.-K. 97
Yoon, Alex 91
You, Wei-Xiang 47, 89
Yu, Zhihao 71

AUTHOR INDEX

Zhang, Gang..71
Zhang, Lining..43
Zhang, Xintong..43
Zhao, Chao 57
Zhong, Tom 85
Zhou, Jiadong...75
Zhou, Lixing 57
Zhu, Jian...85
Zhu, Ying..71
Zhu, Zhongwei 91

IEEE
445 Hoes Lane
Piscataway, NJ 08854-4141

ISBN 978-1-7281-0943-5